Aiki Secrets

The Aiki Codex:
The Secret to Circular Aiki

William Dockery

Copyright © 2019 William B Dockery
All Rights Reserved
AikiSecrets.com
ISBN: 1-7334615-0-7
ISBN-13: 978-1-7334615-0-4
LCCN: 2019912335

No part of this document may be reproduced or transmitted in any form or by any means, electronic, mechanical, photocopying, recording, or otherwise, without prior written permission of the publisher

Note to the Reader

The information in this book is distributed without warranty and is only presented as a means of preserving a unique aspect of the heritage of the martial arts. All information and techniques are to be used at the reader's sole discretion. While every precaution has been taken in preparation of this book, neither the author nor publisher shall have any liability to any person or entity with respect to liability, loss, or damage caused or alleged to be caused directly or indirectly by the contents contained in this book or by the procedures or processes described herein. There is no guarantee that the techniques described or shown in this book will be safe or effective in any self-defense or medical situation, or otherwise. The activities, physical and otherwise, may be too strenuous or dangerous for some people. You may be injured if you apply or train in the techniques described in this book. We therefore suggest you use restraint when practicing all techniques contained within this book, wear safety gear, and practice only under the supervision of a qualified instructor. Specific self-defense responses illustrated in this book may not be justified in any particular situation in view of all the circumstances or under the applicable federal, state, or local laws.

*Dedicated to the four angels that can throw me with just a word;
three, when they say "Daddy";
the other, every time she says my name.*

Keep Going…

CONTENTS

Forewords viii
Preface xii
Acknowledgements xiii
How to read this book xvi

Basic Terms 1
    Willowed 1
    Bounce it off the Rock 2
    Drop step, walking in a pool 7
    Silk-Reeling and the Drop Step 12
    Flying the Plane 20
    The Timing of Tick-Tick, Whoosh! 22
    Link your shoulders 24
    Cheating the Arc (Cheating the Circle) 25
    Peeling Tape Off the Wall 31
    Rolling the Sleeves 32
    The Plank 33
    Rolling the Tack 36
    The Snow Shovel 37
    Crush Box, Stretch Box, the Teeter, the Back Teeter, Twisted, and Checked 39

Advanced Concepts 61

Where Uke is Weak 63
    The Secret of the 20 63
    (Three-legged) Dog Spots 71
    Secret to Aiki Joint Locking 89

Where We are Strong **153**
    Presence 153
    Tori's Hypotenuse 156
    Fork and Knife 168
    Ligamentary /Fascial Grasping 172
    Zhan Zhuang (According to Bill and Joel) 181
    The Transmission 218

Advanced Connection Builders **233**
    The Albatross 233
    The Broom and Shovel 240
    Shield Theory 260
    Rolling 284
    Orbits 322

Closing **335**

Appendix **336**

Foreword by Moe Stevens

I had always been a very good athlete with excellent kinetics awareness, which is an advantage for physical activities. Another advantage, or disadvantage (depending on one's point of view), was that I questioned everything I was taught. These traits greatly enhanced the ability to physically replicate what my teachers were doing as they practiced and taught Aikido. My problem was that I could not verbally explain to others exactly what I was doing with my body. This led to frustration for both my students and for myself.

After returning to class post a hiatus due to having started his family, Bill Dockery asked me to do something that made a profound change in my understanding and practice of Aikido. He asked me to give names to everything we did. Bill insisted emphatically that ideas, motions, and intentions had to have names (any name will do) for a concept to exist in our minds. The names helped us share our experiences.

I was aware of my lack of communication of physical movements and principles. I even named my Aikido club "Just This Aikido" because I would often say, "The motion is JUST THIS." When trying to come up with vocabulary, if I so much as paused, Bill would suggest words so that we could sharpen their meaning, saying that if Confucius desired not just words, but precision in words, then so too should we.

Bill's simple request changed the way I approached Aikido. He, along with his partners, Joel Copeland and Chris Parkerson, have made me a much better Aikidoka and, more importantly, a better teacher.

Thank you for your request that changed my Aikido life, Bill Dockery.

Moe Stevens
7th Degree Black Belt
Shihan, Tomiki Aikido of the Americas

Foreword by Chris Parkerson

In my 46 adult years of training and teaching martial arts, my most challenging and fulfilling experience has been to be a participant in "The Crucible," a working group of four yudansha (Moe Stevens, Joel Copeland, Bill Dockery and myself) that began meeting 15 years ago (twice a week) at Shihan Moe Stevens' dojo.

At this dojo, we have allowed scientific modeling to inform us about all forms of overt and subtle body mechanics that are present in any pugilistic or grappling art, especially in the realm of kinematics (two bodies in motion).

We reserve the right to respectfully call "Balderdash" on modern or classical movement, knowing well that the classical model of transmission from master to student, without an independent objective standard, is a mathematical model that most often results in diminishing returns as one generation of masters rarely progresses beyond his or her own master from generation to generation. Therefore, we have strived to study deeply, test the waza, and find the physical principles that make it work most efficiently by using modern disciplines like mechanical engineering, biomechanics, physics (in specific kinematics), sports theory, etc.

Bill Dockery is a vital part of this group. As a professional engineer, 20+ adult year martial arts practitioner and yudansha in Aikido, he brings an acute ability to perform the tasks of breaking down human movement into its many layered parts, developing working models that represent each physical principle involved, placing a simple name and explanation to the action and establishing a sequential taxonomy to the whole process.

Bill is also very passionate about our mission – to question all assumptions and dogmas using the scientific process with an objective of either supporting or improving upon martial movement as it is currently being taught.

Those who have participated as guests at Crucible sessions will wholly affirm that we will question, stress test, break down, reverse engineer and fine tune actions presented to us from most any art that comes our way without fear of traditional boundaries that are often set by a classical teacher-student relationship. It is Bill who often carries the heavier load by bringing new insights and creative modelling into the Crucible when some of us begin to rely too much on our past insights, practices and achievements. But that is, indeed, part of the mission and practice of The Crucible. A student's insight and skill should surpass the master. This is the only algorithm that will maintain depth in the arts over a series of generations.

We expect this book to be of significant value to both conceptual learners and kinesthetic learners. The reader will be able to adopt a vocabulary and sequential learning models to help one's students understand what you want them to do. Natural athletes may not need such models but those who teach must find ways to communicate with all styles of learning and all levels of talent.

This book is Bill's second publication. What I find most rewarding is that the first book, Aiki Secrets: Six Precepts and the Dynamic COB, published in 2011, was so acute in its modeling and so thoroughly stress tested in content that our group believes it needs no revision, though all of us have continued to grow in martial wisdom, insight and skills over the last 8 years since its publication. I am confident that, due to Bill's discipline in condensing and writing, that this current publication may become the "best in class" and possibly the Rosetta Stone for many students to study martial movement whether it be a so-called internal or external; soft or hard art.

Two caveats must be addressed. First, this book focuses upon circular, (2-dimensional) movement for the most part. To have included spiral motions (3-dimensional movement) would have made the book twice as long. So I hope the third volume in Bill's trilogy will come when he has the time to compile and edit such data. Second, these books take time for Bill to write, edit and publish and therefore they do not represent our ever growing quiver of discoveries, reverse engineering experiments, mechanical modeling ideas and teaching curriculum content – especially in the more subtle movements we have worked with in our studies of Aikijutsu, Aikido, Hsing-I, Bagua, Taiji, Silat and Kenpo. Yet, at the same time, all four of us remain perpetual students willing to have our own limitations, assumptions and dogmas exposed using that same scientific process.

As for me, I certainly hope that Bill continues to write and publish so that The Crucible's work is recorded and presented publicly for others to evaluate.

Chris Parkerson

Foreword by Joel Copeland

I began my martial journey over thirty years ago and have explored many arts including Kenpo, Kungfu, Silat, and Aikido.

During this time, I met Bill, who had a similar background as I and the same growing frustrations with the martial arts: the inability to explain the "how" and the "why".

Together we were determined to become accomplished Martial Arts students.

We stopped accepting hollow explanations, began questioning and, as best as we could, started constructing concise definitions that we could use to improve the questions and gain skill. The unintended result is that it became a method to rapidly teach others the skills that took us so long to develop.

The task was made easier, in that, Bill excels at breaking down complex theory into written word, scientific modeling, and real-world representation.

So, if you are looking for something deeper than the common "Step here, do this" training, you came to the right place!

This book uses a principle-based approach that was developed by observing, participating, questioning, digging for the truth in the answers, and continually testing as we built models that can be taught to students and masters alike.

These models strive to explain mechanics, physics of the body, and the dynamic relationship between martial combatants.

Our sincere intention is to offer new ways to look at martial abilities, so that others might also gain insights, be inspired, and take a step forward toward additional understanding.

Ultimately, we hope that you improve upon these concepts, increase your skillset, and enhance your teaching abilities.

Afterall, the measure of complete mastery is not just performance, it includes the ability to share what you do with others so that they can perform as well.

Joel Copeland: Student Eternal

Preface

We are back and we have gone much deeper into the precepts. In fact, 'Six Precepts...' was only the first few years of the investigation. This one covers the six (or seven) that followed the writing of the first book.

Beginners beware! There is something for practitioners at all levels of experience in Aiki(do/Aikijutsu), but intermediate students are our intended audience. It will be assumed that the reader has a foundation in the six precepts (especially the COB [Center of Balance]) and understands the 'Secret to Aiki'. You will not get completely lost because you are not fully versed in the first book, but we are going to use many of the previously introduced concepts and we are going to refer to them with just a word. In short, expect some frustration and a lack of depth of understanding without having first read 'Six Precepts' at least once. Note: Originally, I started calling out every time we came across words you will find in 'Six Precepts', but found it tedious to read and removed a lot of the 'go read 'Six Precepts' highlighting, I instead pointed out when there was fundamental benefit of looking at the crossovers, so, if you come to a word or concept you do not understand, be sure to check 'Six Precepts'. I am sure it will cost me a star in a book review (inside Mokuren joke), but so be it!

For those amply prepared, we are going to expand on the first book, but have no fear, this book is not a rehash. We start with some new takes on old concepts, but will quickly wade into the deep end and introduce some incredible new concepts; e.g., orbits, the hypotenuse, transmission, rolling, Aiki joint locking, and more...

As you read, please keep in mind my sincere intention that this book helps you practice and/or teach Aiki(do) to others, but please read with an eye for critique, post your experiences, and expand on these concepts. I have deeply appreciated the feedback received thus far. Thank you in advance for your constructive and creative feedback!

I must add; even with the great amount that I have learned by decoding both of my Sensei's movements, I am still nowhere as adept in this practice as Moe and Chris. I am constantly humbled by their expertise and strive with all I have to reach such heights. I am forever indebted to them both.

Ready? Because I am! I cannot wait to tell you about a great many things but we should always start with remembering those that got us here...

Acknowledgements

Every day in Moe's dojo (the Mojo!), for me, was/is another attempt to emulate Moe Stevens and Chris Parkerson. They have always set the example, and offered as transparent an explanation of their craft as they could muster.

Thank you, Moe and Chris!

By proxy, I have been indirectly influenced by the instructors that played a major role in developing Moe and Chris.

Moe's father Merritt Stevens was the senior instructor for all the Ohio based teachers that made up the core of an organization that eventually became known as Tomiki Aikido of the Americas (TAA). He inspired a style of martial arts called Kachido Aiki Jujutsu and his presence in Tomiki style Aikido is still strong even now, twenty plus years after his death. I never met Merritt in person, but I know him through his many students that credit him with shaping their art and through the many legendary stories of what it was like to train with him. Thank you, Merritt!

Chris is an extremely eclectic and skilled martial artist, to the degree of being a master of many arts. I benefitted greatly from his ability to compare and contrast various styles. It was impossible to stay in my comfort zone. With such a high degree of skill and a background as complex as Chris', it should bear significant prestige that; above all, Chris credits much of his ability to the instruction he received from John Clodig.

John is the source of Yanagi Hara Ryu Aiki Jujutsu. There is a good bit of history in John's story, and as with all good history lessons, I understand it involves some politics. So, in effort to commit no offense, I urge you to seek John's lineage for yourself. Suffice to say that John's background can be traced back to the influences of Yoshida Kotaro's Aiki Jujutsu.

It was only after presentation of a signed copy of 'Six Precepts' to John by Chris that I learned of the great influence he played in our studies.

Through Chris I discovered that John gave praise to the 'Six Precepts' in that John deemed some of the content to be interpretations of concepts received from Chris Parkerson as taught by John Clodig (in the study of Yanagi Hara Ryu Aiki Jujutsu) and that John requested that I cite influence.

It was almost four years later that I finally met John for nine hours of training and was able to experience first-hand just how much influence John had in Chris' art. What I was exposed to was way above the level we were working at. There was much to decode when we went back home. I am still trying to wrap my head around that event.

At the time of this writing, John resides in California and is actively documenting the details of Yanagi Hara Ryu Aiki Jujutsu. If John does share with the world directly, I strongly suggest that you visit with any materials offered. If you find any overlap between his content and mine, I assert that John has a significantly higher awareness and I am a hack with nothing more than a peripheral, second hand (mis)understanding.

Thank you, John Clodig for developing such a significant talent in Chris and for recognizing your influence.

Now, let us acknowledge others that physically threw me:

For having profound influence on my life and with the highest degree of esteem, Sensei Steve Carlisle, thank you for getting me hooked on Aikido and for shaping my mind by showing me the method to finding 'bunkai'. You instilled in me the foundation for these books, especially the cross-overs to the Chinese philosophy, medicine, and martial practice. The Crucible itself is the embodiment of your instruction that we challenge convention not for the sake of discarding it, but for understanding it, accepting what is wise, appreciating the path it took to becoming convention, and using it in its appropriate contexts to improve ourselves. With heartfelt intensity I say, 'Thank You, Sensei Carlisle!'

Special thank you to Grand Master John R. Malmo and Sifu Tim Wolfe. Grand Master Malmo has instructed me in Kombatan via his GGM Ernesto Presas lineage, as well as in Indonesian, Chinese, and traditional Filipino martial styles. Sifu Tim Wolfe shared with me the inner secrets of his Black Panther Kung Fu. It has been an honor and very literally a privilege to receive this instruction (not many have). Although neither of these instructors were directly teaching me 'Aikido', by tangent, they have opened my eyes to the martial reasons behind our actions and how best to apply Aiki, the control of stability and balance, as a martial concept.

There are many others, although not as skilled as Moe and Chris that have had a direct influence on me and were absolutely critical in developing an understanding of Aiki. Although they do not hold the rank of 'Aikido Sensei', they have taught me much, and I offer them my deepest appreciation...

To name just a few: Shawn, Austin (DJ), Eli, Ed, Jess, Ken, Rob, Gene, Gordon, Will, and Ray are those persons, but none have been as important to my practice of Aikido as Joel Copeland. If Aikido is compared to a chemistry class, Joel has been my lab partner. More than any other, he knows firsthand how many mistakes it took to get even this far toward unraveling the mystery of Moe and Chris. Joel is well past the abilities of most fourth-degree black belts in Tomiki Ryu Aikido, all he needs do is take the test tomorrow to achieve the next (two!) rank(s). The expectations that Joel and I have put on ourselves are much higher than is practical and accounts for why we have progressed so far in skill but not in rank. Joel is very much a Shihan in his own right. I look forward to having him document his view on Aiki. Thank you, Joel!

I hold in my heart great appreciation for all in the Tomiki Aikido of the Americas (TAA) and the many partner international Aikido associations. No matter where my martial arts practice takes me, it always comes back to the lessons I learned in Tomiki Aikido. Thank you!

Special thanks, as always, to those that continue pushing the boundaries of how we approach the Martial Arts, especially in the art of Aikido. Thank you to George Ledyard. I consider you among the greatest of the pioneers of teaching Aiki through action and words. I still (re)view your videos, and every time, I find something new and thought provoking.

Thank you to the many (all across the globe!) that have read 'Aiki Secrets: Six Precepts and the Dynamic COB' and have taken time to contact me. You have helped me more than you can know. Glad to know there are others like me that have to think and talk about their craft in order to gain understanding and develop physical skill. This book might never have happened if not for the encouragement. This book is for you!

Thank you to our military and law enforcement for defending and putting into effect the U.S. Constitution that affords me so many freedoms; especially the first two amendments, which together affirm our right to practice martial arts and to publish books about it! I hope this book finds you and helps keep you safe. I will see many of you at the next seminar!

Thank you for reading this book!

Even bigger 'Thank You' if you bought it!

How to read this book

This book is written in much the same format as 'Six Precepts'.

References to 'techniques' are from the Tomiki Juu Nana Hon Kata.

The first section will cover some 'Basic Terms' before explaining more about 'Where Uke is Weak', 'Where we are strong', and then start putting it all together in 'Advanced Connection Builders'.

The Basic Terms are just that, basic; not quite significant enough to be a foundational concept. They are like short stories that add color to other concepts. Expect these terms to show up again throughout this book.

The rest of this book is a description that, at first, might seem slow, even a bit disjointed, and not at all connected, but the intent is that, when you finish, you come to the realization that each section was very much like describing the different sides of a multi sided object (e.g., a cube). Explaining these concepts was an authentic example of the 'Elephant and the Blind Men' parable. The real intent is to reveal the animal that ties them all together (Aiki), but it is very difficult to understand that holistic concept until you know the parts. Each part is phenomenal and miraculous in themselves, but none as great as the coordination of combining them to a singular purpose. Read with the intent to circle back and find the complementary relationships between each section.

A major tip! This is not a 'story'. Although this book is written in a conversational tone, it is in fact a text book, not a novel. You would not want someone to explain this content to you at the speed of subtext in a typical TV commercial. Savor the words, read a section twice if you lose track of 'my voice', create the mental pictures, imagine yourself in the scenario, relate it to your life experiences, pause your reading to analyze the drill(s), integrate the concepts into your practice, challenge what you read! Then read some more.

Most importantly, enjoy!

With and for the love of Aiki, please let me set the stage...

> Physically (e.g., on the mat, in battle),
> Aiki is the control of stability and balance.

With no further ado, Basic Terms...

Basic Terms

Willowed

If you have practiced Aikido for longer than three months, you must have taken a few good falls and you know what it feels like to be thrown using Aiki; but have you ever had a discussion about how it feels to be the person throwing? What is Tori sensing when Uke is unstable?

At first, as with most things in life, it is difficult to get results. Uke feels like a giant rock or at least like one of those punching bags that will wobble but never really falls down. The reason for this is our lack of control of Uke's stability.

Soon enough, we learn to stop colliding with Uke, and as we gain skill in destabilizing Uke, they become more pliable. Uke's weight seems to fade away; as if they suddenly have less mass. This is a sensational experience for, as Uke becomes 'softer', you can use less and less effort to move them.

Sounds great huh? 'Uke is easy to move' and feels 'weightless', but these words still do not capture what Tori is experiencing.

My favorite way to describe the sensation is that Uke feels 'willowed'.

By that, I mean that Uke feels very similar to how a weeping willow tree branch feels if you pull downward just a bit. The branch wants to go upward, but will just as easily be pulled further downward if you apply your weight to it. If you allow it to, a willow branch can help raise you up, or support you a good bit as you continue to stretch it closer to the ground, at least until it reaches the point that it detaches from the tree and falls to the ground. The tensioned willow branch behaves a lot like an unstable Uke, and when the branch detaches, it feels exactly like Uke does when their balance is truly broken and they fall.

Uke feels 'willowed' when they are unstable. See if you agree.

By the way, in a most eerie coincidence, the day I explained my view of 'willowed' to Chris, his face changed, as if shocked. He then proceeded to tell me how to say 'willow' in Japanese. It was then my turn to be shocked.

Basic Terms

Bounce it off the Rock

We often refer to the Center of Balance (COB) as 'The Rock'. Why? Because the COB often feels like an immovable object and because the nickname has one fewer syllable.

'Bounce it off the Rock', refers to 'bouncing' Uke off of your COB, and bouncing Uke off of your COB has many uses.

We are going to investigate two 'bounces'. The first bounce drives Uke directly backward and it relies completely on something Moe coined as the 'hip-pop'. The second bounce ends with the same hip-pop, but the hip-pop is delayed and comes at the end of a hip wheel turn.

Both bounces are best described within the context of a wrist grab.

Let us first start with the bounce directly back toward Uke, with an additional purpose of beginning our discovery on how we generate 'power' versus leveraging 'strength'.

As the previous sentence implies, we are going to bring Uke directly to our COB and then use our hips to drive Uke backward.

That sounds like a pretty straightforward task, and it can be, but in order to gain a sense of the 'power' within our hips, we are going to isolate the power source that 'bounces Uke' off our COB.

We create this isolation by maintaining a weighted forward leg stance and not ever shifting our weight. Eliminating a weight shift here leaves our hip muscles as our only engine to create force. (You can get back to adding the additional power of the weight shift later, after we experiment.)

Let us begin.

Stand with a weighted forward right leg (left leg back and at most, mildly weighted, maybe 90/10 with 90 percent of your weight on the front foot)

Extend your right arm for Uke to grab.

Use a 'reverse S' and a centripetal arm to lead Uke toward your COB.

Bounce it off the Rock

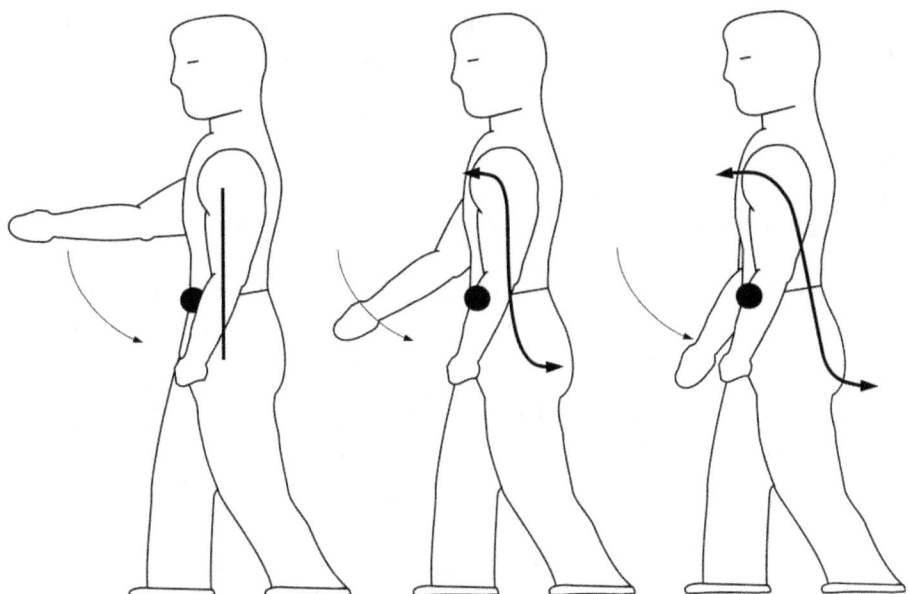

Figure 1: Right-side forward weighted, arm dropping to the COB; three moments in time

It is subtle, so look at the above pictures closely. The primary difference between left and center positions is a slight spread of the shoulder blades [shoulders move forward] and the reverse CAM in the hips (butt starts to stick out as if trying to sit). I kept it subtle within the pics, because 'real world', it can be even more subtle. Unless you are aware of what you are looking for, you will likely not notice this in Sensei's actions. Please also note that the right most picture completes the arm drop and settles into the reverse-S, weight still positioned on the forward right leg.

We should feel chambered and ready to bounce Uke backward. (Checklist of our condition: we are still fully weighted on our right leg; shoulder blades spread; our posterior back a bit, much like we were getting ready to sit; our right arm against our body, right hand in front of the right-side inguinal crease. You might think of your arm in the position it would be in immediately after having dropped open the zipper of your pants.)

We bounce Uke backward by thrusting our hip forward, forming a 'forward-C'; i.e., with your hand wresting on your inguinal crease, you very literally use your hip thrust (transition from reverse to forward CAM) to push the Conjunction Point (CP) back toward Uke. (quick reminder: CP is where you and Uke are in physical contact with each other, in this case: Uke's grasp of your wrist)

Basic Terms

Remember to keep your shoulder blades spread, your collar bones parallel to the floor and against your rib cage, or you risk getting pushed backward instead of Uke.

The hip-pop can easily unbury your hand even against a much larger opponent. It takes very little effort (overall) and is the start of the discovery of how to generate power versus relying on strength.

In review, this 'hip-pop' bounce moves the conjunction point (i.e., the place where Uke has grabbed) directly to our COB, and then, using a 'direct touch' from our 'Rock', it bounces the Conjunction Point (CP) directly away (and back into Uke). Take note of the direct touch of our COB (see Figure 2 below). It is the direct touch to our COB that distinguishes this version of 'bounce it off the rock' drill from the next example.

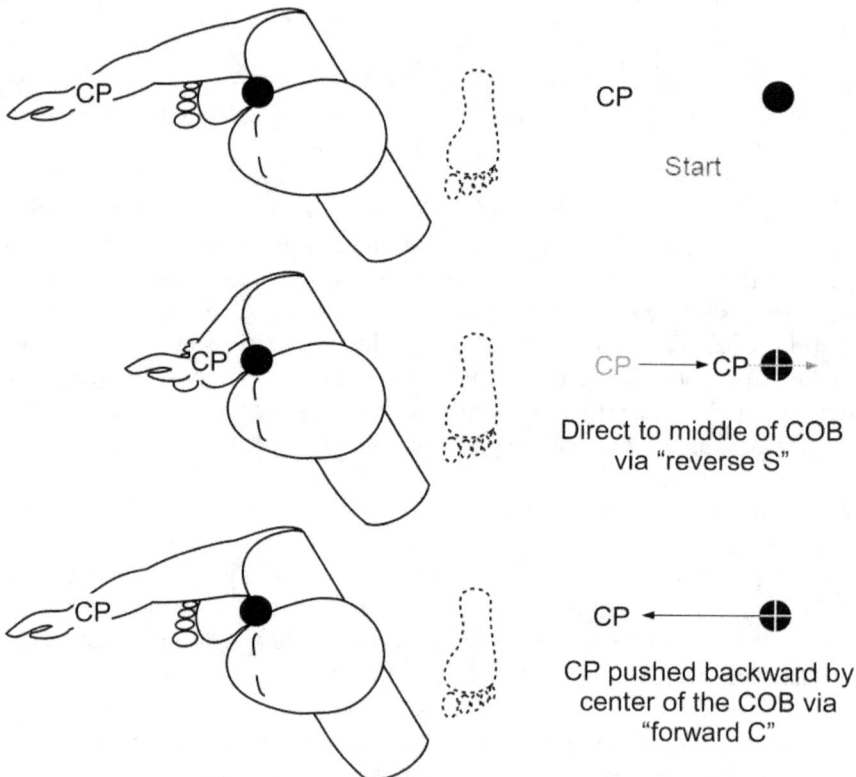

Figure 2: The 'hip pop' directs the Conjunction Point directly to the center of the COB (via reverse S), and then away via forward C

Bounce it off the Rock

Got it? Great! That first drill is a game changer that keeps you from trying to bicep curl Uke. Just keep in mind that the direction you pop back toward Uke (in relation to Uke's COB) is just as important as the hip-pop; i.e., bouncing directly at Uke's COB will move Uke differently than pushing back at the sides of Uke's COB (or even above/below Uke's COB!).

With 'leading Uke into a certain direction' in mind, let us explore the second drill.

In this second example of 'Bounce it off the Rock', we allow Uke's energy to assist us in moving through a hip wheel turn. We add the 'bounce' (the hip-pop) at the very end, when we are facing the same direction that Uke first came toward us; i.e., we do a 180 degree turn and hip-pop.

How do we allow Uke to assist our hip wheel turn?

We leverage Uke's pressure as a 'Mostly Direct Touch' to our COB.

In other words, we allow Uke's pressure to lay 'almost directly' upon our COB. (from a horizontal perspective.) In the case of the drill we just completed, we would be letting Uke touch our COB just a little left of center (our left). Let us try it.

Start exactly as we did in the first drill, but this time, as we allow our arm to drop centripetally, we bring the pressure slightly to the left of our COB. The 'mostly direct' pressure should have two effects: it begins to rotate our body leftward (counter clock-wise if viewed from above) as it drives us backward, shifting our weight backward toward our left leg. (See Figure 3)

You might describe the motion as Uke helping us to 'roll' our COB horizontally from 'post right to post left'; (i.e., as we shift weight from right foot to left foot)

If we maintain a slight reverse CAM throughout the hip-wheel turn, we can finish with a hip-pop (transition from reverse to forward CAM); which will 'bounce' Uke's grip upward and motion Uke past us. As your hand reaches full outward extension, you should consider quickly shifting into a reverse-C or you may experience Uke pulling you out of your stance.

Note the synchronicity of the COB and the hand. Moving them together is the relationship of the 'elephant arms' drill and provides us stability and power throughout the motion.

Basic Terms

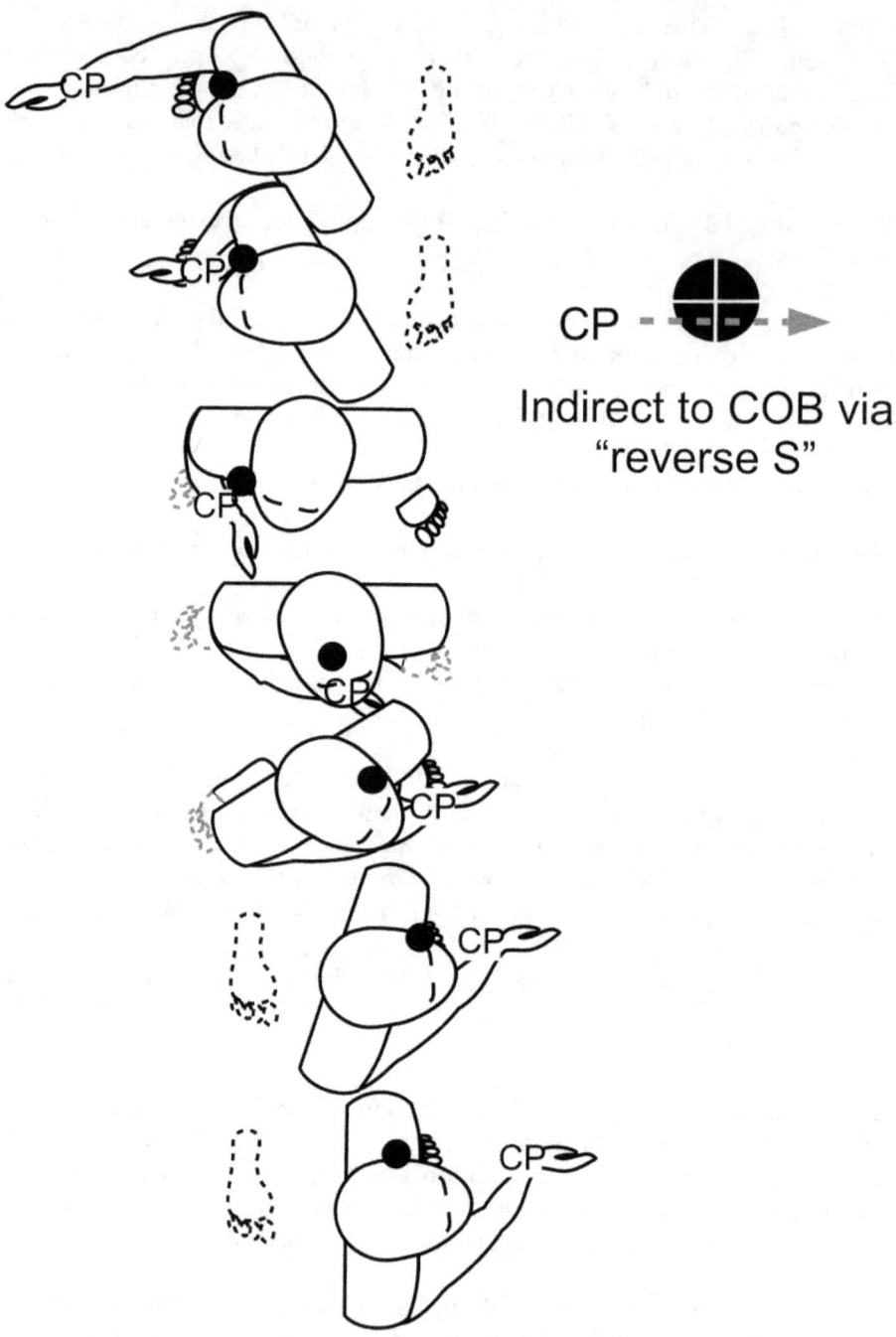

Figure 3: The Conjunction Point pushes 'Mostly Directly' on our COB and pushes us through the hip wheel turn; hip-pop Uke at the end of the motion

Three key lessons exist in these two bounces. First is the obvious emphasis on the hip pop and how we generate power from our hips as we attract and repel Uke from our COB (aka 'The Rock'). Second, we learn what to do if we find Uke is directly upon our COB (hip-pop!). Lastly, we begin to understand how to use Uke's momentum to assist us in turning/stepping should we find Uke's pressure hitting us indirectly. Eventually, we can begin to guide and control how Uke's pressure is received by our center, and thus better control the outcome of the situation.

Hmm...

Hands and center in synch, we generated power from our hips and transmitted that power through our hands. We also touched upon how to use Uke's momentum to move us in an advantageous way, but in the second drill, as we performed the hip wheel turn, did we remember to use 'Front Wheel Drive' and move 'down, over, and up'?

Great segue to....

Drop step, walking in a pool

Repeat with me, 'down, over, and up', 'down, over, and up', ...

Got that? It is important! It is the foundation of generating power when stepping and it keeps us 'grounded' instead of uprooted.

Those of you with Tai Chi experience are already vividly aware of this method of moving. Those that read 'Six Precepts' should recognize it as 'front wheel drive'. Why give a new perspective on this concept? It is just that important! (and I have found that some people feel this explanation of front-wheel drive makes it easier to understand how to apply the principle.)

Here is a great experiment to cement some physical learning. All you need is a pool of waist deep water. (Fear not if you have no access to a pool, at the end of this section we will perform a version that uses a wall, no water. For now, go get wet or please use your imagination.

Let us start by walking inefficiently, so we know what works less optimally.

Wade into the pool waist deep, start with your feet together; then try moving across the pool by moving your torso in any direction other than down; be sure to try and kick a leg outward if you think it will help your momentum. (Spoiler alert! It will not help.)

Basic Terms

You should find that you have very little ability to use the bottom of the pool to move you. Your kicking leg meets a good bit of resistance, you move slowly, and you should be sensing that you are almost floating.

Your actions do not work well because you are not 'grounded'. At best, you are waddling (if you simply shifted your weight laterally) or what I consider even worse (martially worse), you are jumping as you shift weight. (Much like astronauts might do as they walk on the moon. It is fun! It is just not martially sensible under most conditions.)

Let us stop exerting so much energy and try something different.

What is the first thing you should do when you get grabbed? That is correct! Bend your knees and lower your center of balance (COB).

What is the first thing you should do when you want to walk across a pool in waist deep water? That is correct! Same thing! Bend your knees and lower your center. (After all, the water has grabbed you.)

This time, with Figure 5 in mind (skip a head and take a quick look), start with your feet together, bend your right knee, and 'sink' weight into your right leg. This is the 'down' in 'down, over, and up'. As you sink that weight, create a forward CAM in your hips. The CAM of your hips should begin moving the un-weighted left leg forward almost on its own. The CAM will help add force as you push the un-weighted left leg outward and down through the water to catch your step. You might benefit from sliding the ball of the left foot along the floor of the pool as you extend the left leg.

Now plant that forward left foot at the normal stepping distance and use that forward left leg momentum to pull you forward and down even further. It should feel subtly like a 'lunge' step. You might even use a 'scratch at the floor' action with your planted left foot to help pull your body forward and down to get your hips swinging (left foot stays in place, it only feels like you are scratching with the left foot).

You should soon find yourself at the lowest point of this 'downward lunge' as you reach a near even-weighted-ness between your feet.

Super important: do not settle into the bottom of the step! Wide spread leg stances (e.g., horse stances, lunge stance) are only transitional poses for Aikido practitioners. We only stay there for an instant. Instead, from this lowest point and onward, we continue to 'swing' ourselves upward and rise onto the forward left foot in a very particular way…

Drop step, walking in a pool

Keeping the forward CAM in your hips, that forward left foot becomes your new post as the weight shifts. Push down on the floor with your forward left leg, and eventually transition all of your weight to the forward left leg. As the weight shifts to the forward left leg, relax and drag your back-right leg in a manner that closes your stance again. (Drag the ball of the right foot along the bottom of the pool.) Do not push off with the back-right leg!

I practice this in the pool all the time. You will definitely begin to see a difference in the power of your walk in no time. Practice this kind of stepping long enough to remove any desire to hop from place to place.

This small nuance in walking also applies to running and is one explanation why some football running-backs run over linebackers, while others simply get uprooted and bounced backward. (Learning to drop as you start to push instead of jumping as you push also explains a difference between an average and a hall of fame caliber offensive lineman, how a baseball pitcher generates force 'from the stretch', there are countless examples. Let us move on.)

Walking in (under?) water sounds so 'Tai Chi'. Now you know how.

For those of you with no access to a pool, try this the way it was originally taught to me by Sensei Carlisle. (He taught this on my first day of class! As far as I know, he teaches it to everyone on their first day of class. It is just that important! Bonus: this method adds another awareness! Side-sway)

Those of us without gills use contact with a wall to help us judge the 'down, over, and up' of our step.

Stand perpendicular to a wall with your right shoulder against that wall. You can substitute a hip or elbow if that is easier for you, but it is best to bring both feet as close to the wall as you can without creating a serious lean. This means keeping near even weight on both feet.

From this starting position, we take a step forward using our left foot ('away from the wall' foot) while maintaining light contact between our right shoulder and the wall throughout the step.

Simple, but there are two measures of success that we should be maintaining awareness of: 'not bouncing against the wall', and 'smile, not frown (or flat)'; (i.e., swing over, do not leap over)

We should not bounce against the wall when we step (Side-sway). We do not need to lean our torso rightward to step forward with our left foot.

Basic Terms

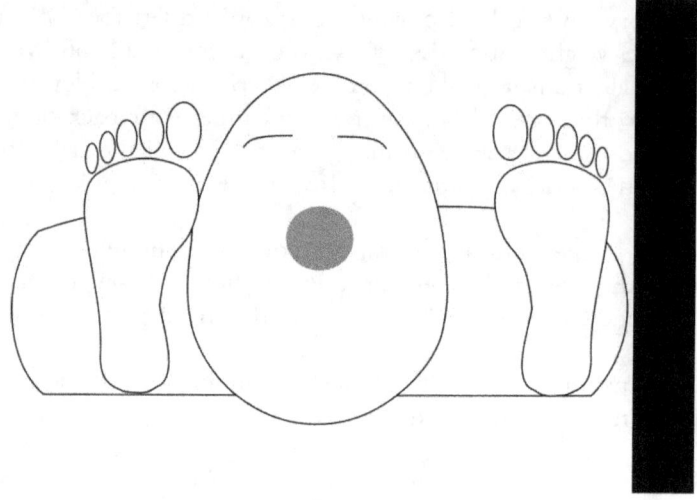

Figure 4: Setting up against the wall to analyze our stepping

The most efficient thing we can do at the start of the step is to lower ourselves directly downward onto our right leg by bending our right knee as we CAM our hips forward (the 'down' part of 'down, over, and up').

The CAM will thrust our left leg outward to catch us just prior to ending the downward movement. Left foot in place, we quickly finish the downward swing of our torso and then start rising upward onto the left leg.

Measure your success by sensing your shoulder tracing that 'smile' (upward cup or a bowl) arc along the wall. As above, so below. Shoulders and hips will be in synch.

There are three common areas of issue to be wary of.

You have already read the first. It is super important to not 'bounce off the wall'. That bounce means your first inclination is to move to the right and off the line of the attack, but then back onto the line of attack as you move forward or left. Many pugilists are trained to take advantage of the vulnerability in this reflex.

Trust me, your Sensei is not rocking back and forth when stepping.

Once you have eliminated any desire to tip rightward before moving your left leg outward; try again with your left side to the wall, and be sure you do not have an issue in either side.

Drop step, walking in a pool

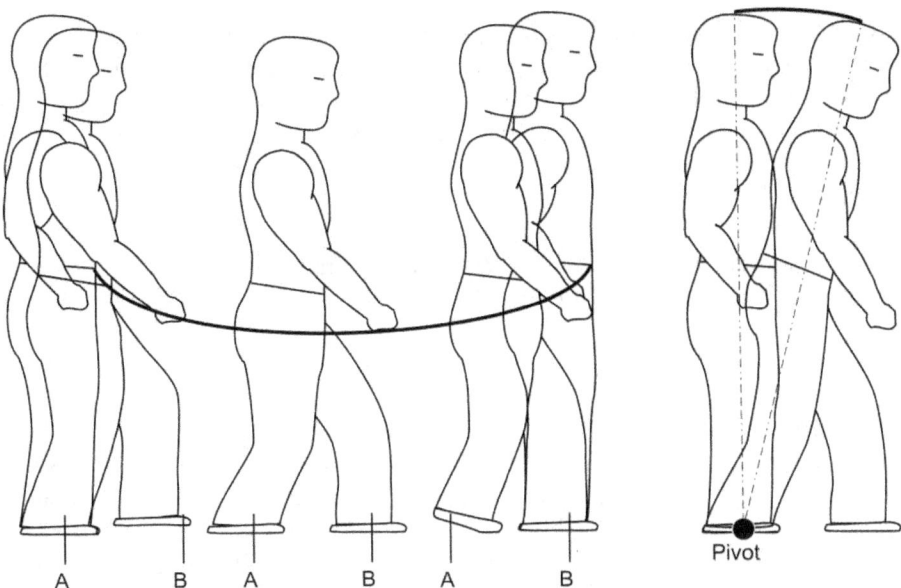

Figure 5: Left: Down, over, and Up (shown in five points in time). Good! versus Right: toppling forward. Bad!

The second potential issue to be aware of: toppling forward is just as bad as bouncing against the wall. (Toppling = right side of Figure 5, super bad!)

We want to keep our torso integrated as we move forward, not topple as a tree would. The forward CAM of the hips helps a lot with this goal.

Boxers know the 'toppling' as leading with your head/face. It is not desirable. Your eyes tell you that you have moved, but in effect all you have really done is destroy a lot of the structure that would help you generate power. Do not confuse toppling with a forward lean. There are arts that favor a forward leaning stance as a natural stance (e.g., Pentjak Silat Bukti Negara). There are martial reasons for why they prefer the forward body positioning, but their stepping is an integrated movement of their entire person (torso), not a toppling effect. (Their shoulders and hips, although not directly above/below each other, are still in synch.)

The third issue is less an issue and more of a guideline. Be sure that the 'up' at the end of the swing does not result in locking out your knees unless you are ready to stop moving.

You can glide directly into the next step if you maintain a slight bend in both of your knees.

Other fun: side note: every so often, someone in the 'against the wall' starting position feels as though they cannot step. They feel they cannot move because their 'away leg' is too heavily weighted and therefore unable leave the floor to allow the step. It can be a bit amusing when this happens. The person will bounce their shoulder against the wall, all the while claiming they cannot step.

Do not despair if this happens. You are not trapped. You will find your escape by simply bending the knee on that weighted outside leg and falling away from the wall. You should be able to catch yourself or simply use some Ukemi.

Lastly, for extra practice and a chance to take 'drop stepping' to the extreme, take this experience to a wave pool. Walk into the waves! Want to be the first one to walk into the deep end? Try synchronizing your 'drop steps' with the wave troughs. You should feel the water making you 'light' and carrying you along. Reverse it, walk backward out of the deep end. (There is a pool in Kissimmee Florida near what used to be 'downtown' that is perfect for this drill.)

Silk-Reeling and the Drop Step

Most of you are familiar with a drill commonly known as 'Silk Reeling'.

I have been shown two versions of the drill, one where the circle is drawn parallel to the line of your shoulders; i.e., across the front of the body (depicted on the left side of Figure 6), and another where it is drawn outward, perpendicular to the body. (See right side of Figure 6)

On the surface, these are simple exercises, but there is a great deal of nuance for both the meditative and martial aspects of this move. Thus, there are more than a few ways to mess things up.

I reserve any comment about the meditative practice, but as for the martial aspects, as I learned to incorporate drop stepping and lever arms into silk reeling, I was surprised to find a very common flaw in my coordination that significantly reduced my ability to generate power.

I am not alone in this issue, so let us explore.

The moral of this story is: do not allow your arms to guide your hips; i.e., always use the drop-step (i.e., move your torso 'down, over, and up'), even when lifting your arms.

Silk-Reeling and the Drop Step

Sounds easy does it not? Well, for some of us, not at first. Please let me explain.

Let us start with, "Why use weight shift in silk reeling?"

Standing static, equally weighted on both feet as you trace a wide circle with your hand might be useful for meditation, but it also greatly enhances the likelihood to rip out your shoulder or lower back in an active martial situation.

Thus, it is pragmatic to practice silk-reeling drills for martial benefit using weight shift or hip turns.

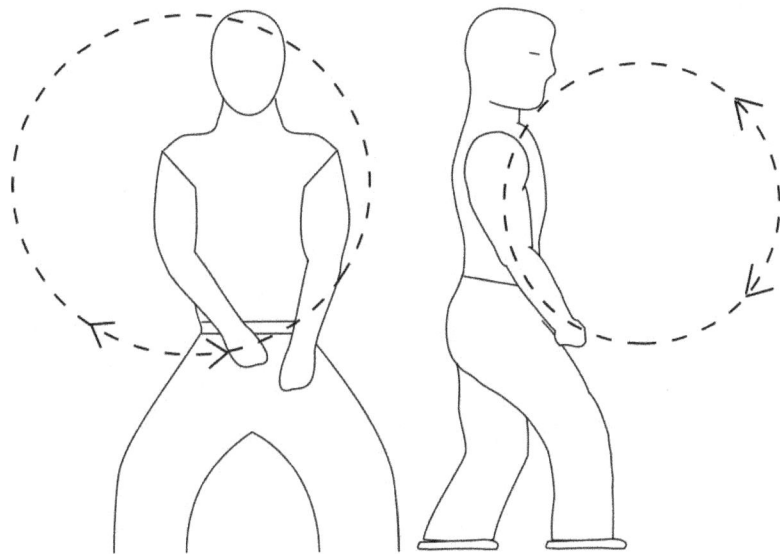

Figure 6: Two versions of silk reeling; parallel and across the body (left) and outward, perpendicular to the body (right)

Let us stick with solely weight shift (no hip turns), ala the 'drop stepping' we just detailed, and let us start with the context of the parallel version on the left side of Figure 6 (before we go tackle the perpendicular on the right)

The good news for those that have read 'Six Precepts' is that incorporating drop stepping with the version of silk reeling depicted on the left of Figure 6 (sideways/parallel to the shoulders) turns this exercise into not much more than a modified 'elephant arms' drill, but the nuance it creates, especially as we trace the top of the circle, is greater than you might expect.

Please pay attention to the detail as we progress.

Basic Terms

The left side of Figure 6 is the point in time where we are at the middle of the sideways weight shift; i.e., the lowest point of the swing of our torso from one side to the next. AKA the bottom of the 'smile'. We do not start here.

Instead, we will start the 'parallel to our shoulders' silk-reeling drill with our weight on our right leg; left leg to the side slightly wider than shoulder distance. (See the left side of Figure 7)

Our first action is to begin a weight shift from right to left foot as we trace the bottom quarter of the wheel with our hands. ('Smile' in hips and hands)

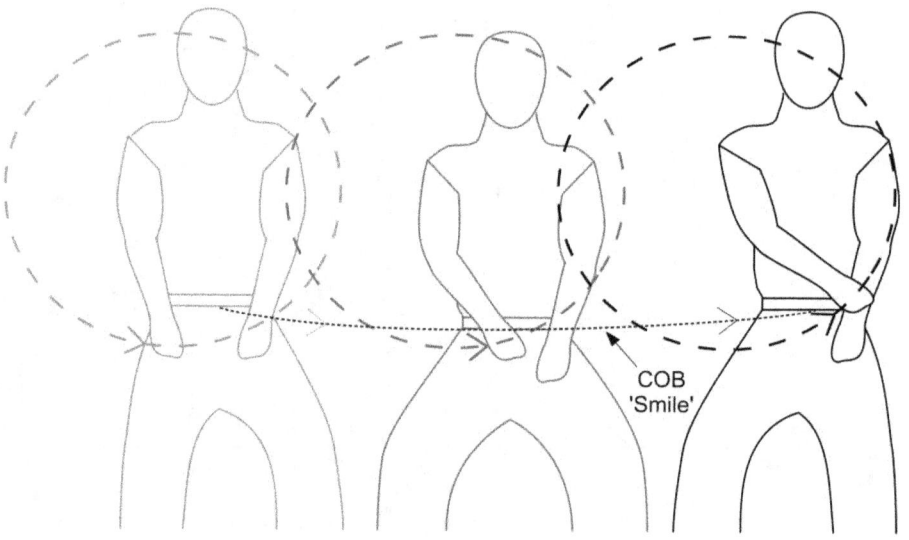

Figure 7: starting in 'Posted right', shifting weight to our left leg 'down, over and up' (COB 'Smile') as we rotate our arm along the lower portion of the wheel

Easily accomplished, for as we drop step with a leftward, sideways weight shift and allow our hand to follow our COB, our hand naturally traces the lower portion of the 'wheel' as the entire body (hand and all) dips along the smile arc.

Are you experiencing flashbacks of the 'elephant arms' drill?

You should be.

One handed this time, but the theory is the exactly the same.

Silk-Reeling and the Drop Step

As we bring most of our weight to our left leg, we raise our hand upward through our newly positioned Sagittal Plane (posted left) and, without pausing, we reach the point at which we have lifted our hand through the horizontal half-way point (right-side Figure 8), and now turn our attention to drifting back to the right leg with the intent to draw the upper portion of the silk reeling wheel; (i.e., Figure 8: paint a frown with our hands as our hips re-paint a 'smile'.)

This is exactly where things get tricky for some of us: our hands go 'up, over, and down' as we move our hips 'down, over, and up'. Performing both at the same time befuddles a lot of people (me especially!)

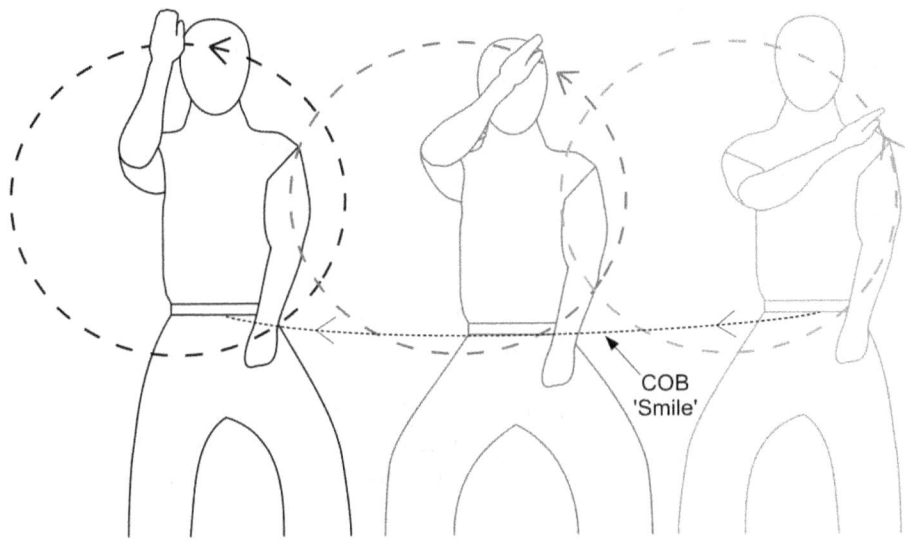

Figure 8: Right to Left! Shifting weight to our right leg 'down, over and up' (COB 'Smile') as we rotate our arm to trace the upper quarter of the wheel (depicted above as desired; without toppling/jumping)

This 'hand moving one way, hips moving the other', can create a common awkwardness that often results in giving up on the swing of the hips and produces a sideways 'toppling'. Worse yet, it can result in an almost hopping motion where the hips follow the arc traced by the hand (sideways 'moon walking', like an astronaut hopping from side to side on the moon.)

It is crucial that your drop-stepping become independent of arm motion. Not 'leaving the floor' and 'staying rooted as you step' is foundational to practically every technique, so if you are finding difficulty, take the time to correct the situation.

Basic Terms

With any luck, you do not have the tendency and are at ease with dropping your torso as you raise your arm (and thus, ready to learn how your arms take advantage of the rise and fall of your torso, as generated by your legs.)

Let us move on and look at the perpendicular to the body silk-reeling exercise. (Right side of Figure 6)

This 'perpendicular to the body' wheel lends itself to bringing together hip-pop and drop-stepping practice.

We will work with just one arm/hand, but this version of silk-reeling also allows us to engage both hands/arms simultaneously. Once you find proficiency at this drill, I suggest switching hands and then try with both; then switch which leg is forward for even more variation. (I can hear you thinking, "Thank you, Captain Obvious!" Ha! I deserve that!)

The motion of our torso throughout this drill is fairly simple, we are not going to do much more than swing back and forth in a 'down, over, and up' fashion while maintaining a 'forward leg' type stance.

In other words, instead of taking that full step, where we come up completely on either leg and bring our feet together, we will keep our legs apart, one in front of the other at the distance of a normal step.

It is not always easy to keep your feet in place as you continually swing your hips forward and back from one leg to the other. It may require your back-foot heel to come off the floor as you shift forward, just be sure to keep the ball of your back foot in place.

Once we get our hips swinging, let us trace some circles with our hands.

Just as in the parallel version, when we allow our arms to stay dropped, the swing of our hips in the 'down, over, and up' fashion will literally create the bottom of the circle for both our hips and our hands. Our hands do little more than stay in place. It does not take a lot of concentration to do almost nothing with our hands. Easy and boring!

Let us add some spice to the (easy) bottom half of the circle. Let us shake it up with hip-pops, some Aiki-Age/Sage, and a nod to Shihonage (#14).

You could incorporate a hip-pop at almost any point when moving forward in the 'lower half of the circle' portion of this drill, but it is quite effective at the very beginning of that weight shift.

Silk-Reeling and the Drop Step

For example, starting with a fully weighted rear leg, and with a reverse CAM position in your hips; you can hip-pop right away to drive yourself downward into the swing toward the forward foot (while tracing the bottom of the circle with your hand(s).

Practicing the hip-pop as an opening to a forward weight shift should feel very familiar, as it is present in a great deal of Aikido techniques.

We can also get fancy by allowing Uke to help us through the move.

It requires positioning our arm in relation to Uke's grip in such a way that Uke leans on us as we redirect that pressure in a manner that drives us forward within the 'down, over, and up' motion. How? Aiki-Age/Sage!

Get into a cat stance with your left leg as the weighted back leg; right foot forward and lightly, if at all, weighted. (see left side of Figure 9)

Reach outward a bit and allow Uke to grab your right wrist. Touch Uke's center (compression connection), and then shift the fulcrum of the lever up your arm such that Uke is pushing the teeter-totter of your arm downward.

If you are doing it correctly, Uke should feel as though they are rolling down your arm; toward your wrist. (More about creating the sensation of rolling for Uke is detailed later in this book as we explore 'Rolling'. Do not get too frustrated until you have explored 'Rolling', your 'Transmission', and 'Rear Cuts' (together they can create an Aiki-Sage; i.e., review this section later with new perspective.)

Using Aiki-Age or Aiki-Sage outside of the traditional Age/Sage drills is not native to all, so let us be aware of that use, but keep the focus on the hip-pop.

As you shift weight, your hip will come forward to meet your hand at, or a bit before, the midpoint of your step; i.e., hip meets the hand at the middle of the 'down, over, and up' swing to the forward right leg.

Once your hand has met your hip (as in the right portion of Figure 9, ala just as though you dropped your pants zipper), use the momentum to affect a hip-pop that drives Uke backward as you complete the forward-most portion of the bottom half of the wheel.

This is very similar to the earlier 'Bounce It Off the Rock' experience, but this time we hip-pop in the middle of a forward step. Adding the weight shift generates a lot more force for the same amount of physical exertion.

Basic Terms

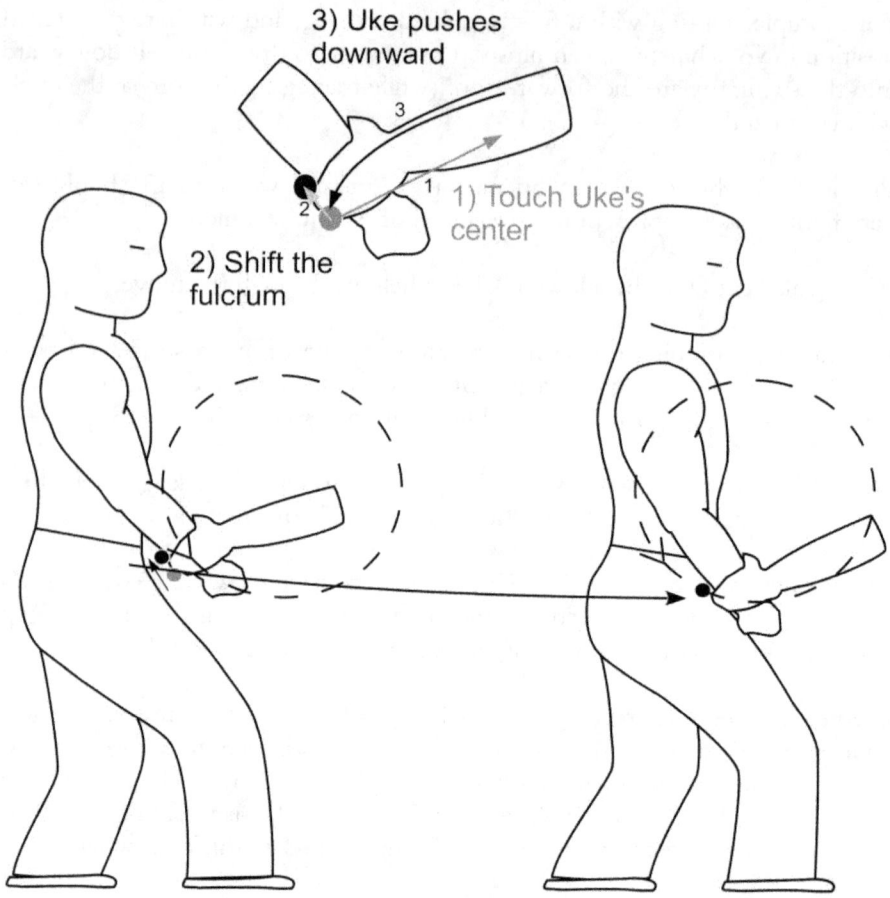

Figure 9: (left) Starting to trace the lower half of the circle, touching Uke's center, shifting our fulcrum (transmission!) to use Uke to motivate our motion forward; (right) the mid-step point at which we hip-pop

[Consider using that 'middle of your step' hip-pop as a way to unbury yourself to complete a hip-wheel turn, ala Shihonage (#14)!]

I realize that I just threw a lot at you there.

Hopefully you got fancy with the hip-pops and letting Uke move you through the motion, but regardless, after having completed the 'down, over, and up; smile with hips and hands', we should have our weight on our forward foot, hands chest high in front of us... (Right side of Figure 10)

Ok, back to the basic drill.

Silk-Reeling and the Drop Step

We complete the drill by swinging our torso to the back leg while tracing the top half of the circle with our hands, but again, for some of us, easier said than done. (see Figure 10. Caution! View it right-to-left.)

Top of the wheel can be tricky. Same issue as before, raising our arms as we drop our hips is commonly an awkward motion. Many allow their hips to follow the action of their hands and will try to rise up instead of dropping and swinging their weight to their back leg.

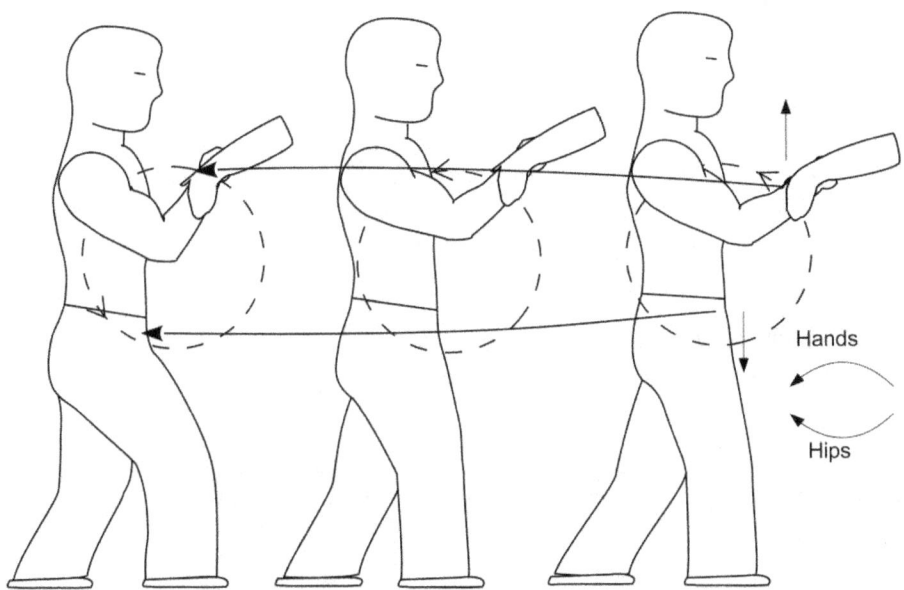

Figure 10: <u>Working</u> <u>right</u>-<u>to</u>-<u>left</u>! Drawing the upper half of the 'wheel'; our hips move 'down, over and up', our elbows drop as we raise our hands. The left most pose is near the halfway point of the swing (no 'up' yet); weight nearly even on both feet, ready to swing up onto the back foot (left foot). It would have taken two more images to show the rise onto Uke's back foot

Doing so results in 'toppling' or using 'rear wheel drive'; (i.e., 'walking on the moon', hopping from leg to leg.) and we lose a lot of power.

If you find you have awkwardness, focus solely on transitioning seamlessly from one foot to the other, and once you have the 'smile' in the hips autonomic, you can then begin to integrate tracing the upper portion of the wheel (tracing the 'frown' with your hands). You can do it!

Once you do remove any difficulty, be sure to try applying Aiki-Age as a way to let Uke help you move down-over-and-up on return to a weighted back leg.

(Again, come back and read this section again after you have come to understand 'Rolling', 'Transmission', and 'Cutting')

The important lesson in these 'Silk-Reeling' exercises is clearing up any difficulty in raising your arms as you drop step. On the surface, it does not seem like much a hindrance, but this skill is integral to cutting with a sword.

Just about all Aikido techniques are somehow related to cutting, thus it is super important to fix this issue if you have it.

Flying the Plane

Let us go a bit easier on this explanation. Less detail, more ambiance.

The 'Flying the Plane' visualization assists us in reducing the over-use of our biceps, and brings us closer to 'pushing' through our techniques.

In this trick, we apply to our forearm/hand a very important concept related to airplane motion: 'a plane only propels itself forward'.

Sure, a plane can rise or fall relative to the ground, but for the Bernoulli principle to function, the plane must move forward to keep air moving from front to back across the wings. (No wise crack comments about Harrier jets please!)

In Aikido practice, the rising and falling of our hands is very similar, in that, if we are efficiently using our hands to transmit the force generated by our legs and center (i.e., 'moving from our hips'), our hands are often moving forward through the finger tips (or you might think 'through the wrists') as they rise, or backward through the tip of our elbow as the hands drop.

This concept was especially important for me. I, like many others, reflexively associate power with the sensation of performing a 'bicep curl', but power comes from the COB, not from the bicep.

To counter my unconscious desire to feel the bicep at work, I had to come up with a visualization to trick my bicep into a more passive state.

When I think of how my hand should move through a technique, I often picture my entire forearm and hand as though it were the fuselage of a small airplane. The airplane is everything from the end of the elbow, all the way out to the fingertips.

Flying the Plane

Where a helicopter might simply rise directly upward, (liken that to using your bicep to raise your hand directly upward) a plane must instead move forward in order to gain altitude. So instead of flexing our bicep to make our hand rise, let us allow the elbow be the rear jet engine that pushes our hand forward, and much like a plane, our hand 'takes off' into a rise.

Note that this 'take off' is how we lift a sword. The sword blade must run itself along an object if it hopes to cut it. Every motion should enable cutting. Lifting a sword as if setting the hook when fishing does not cut.

To further support the cutting motion, we will find that the forward thrust of our hand/elbow combination is made most comfortable when we either move our entire body forward (step) or by twisting our hips.

When stepping or twisting, be sure to take notice of how the forward CAM affects your elbow. With your hands down, sword in a ready position (pointing at the space between Uke's eyes), one leg forward ('Cat stance'), try moving from neutral spine to a forward 'C' and sense how your elbow is thrust forward by the spread of your shoulder blades and reinforced by the forward CAM in your hips. In essence, it feels a bit like a hip-pop without having to literally have the arm in contact with your hips. This practice begins to build our awareness of how we tie our 'hand rise' to our hips.

Dropping our hands/sword (e.g., cutting downward) is akin to flipping the plane around and now the nose of the plane is our elbow. I often find myself relating it to painting with an extremely big brush. A brush the size of a broom! (that broom analogy will make even more sense later!)

If you are dropping a sword using this plane analogy, your elbow is going to flex a little bit, but use 'unbendable arm', 'at rest tension', to maintain nearly the same amount of elbow bend throughout the cut. Imagine the jet engines in your hands pushing toward your elbows, sense the elbows leading as you engage your rear shoulder muscles (rear deltoid) and your 'lats' (latissimus dorsi muscles). (This applies primarily in the beginning of the cut.)

You may notice that, when using the 'Flying the Plane' visualization, it becomes a bit easier to keep the elbows dropped and to your sides (instead of out and turned like you are flapping your wings; i.e., dancing the 'chicken dance'.)

Elbows dropped, even when extending your arm, is typically the martially correct thing to do and it will best allow the hips to drive the elbows.

This analogy works wonders with empty hand techniques as well. Leverage the visualization as you practice your hand blade motions (especially when practicing the Tomiki Aikido drill called 'The Walks' [Onsoku/Tandoku])

In summary, we become efficient at channeling the energy in our COB when our elbows move with the expression of our hips and chest (the movements of hips and chest are also known as the reverse and forward 'S' and 'C') The 'flying the plane' imagery reduces the desire to use the bicep, which in turn allows the energy of our hips to be channeled into our hands.

In any case, I hope this short anecdote helps those that suffer, as I did, from OABD (Over Active Bicep Disorder).

The Timing of Tick-Tick, Whoosh!

Have you ever been thrown by an ocean wave? I have. Here is something I noticed while playing Uke for nature.

I call it 'Tick-Tick, Whoosh' and it represents one of the timings Moe and Chris use to enter into an attack. Apparently, Mother Nature uses it too!

I noticed the timing as I stood in the water, facing out toward the ocean in a rear-weighted cat stance.

The waves were not tall; the water, at its lowest, was about knee high; at peak, the tallest waves were hitting me just below my solar plexus.

I was in a cat stance, weighted on the back leg, letting my lead foot lie lightly on the sand as the water receded to form a wave. My intent was to let the receding water pull my lead foot forward and cause me to step into a wave (and then I would jump a bit to rise above the wave. Best laid plans…)

The receding water did not force my step right away. It took to the point that the wave was starting to swell for it to generate enough pressure to move my lead foot outward; and then it happened, a peculiar, yet strangely familiar event timing occurred.

The small undertow pushed the un-weighted lead foot into the wave base; I followed its lead and allowed that push to guide me into a drop-step forward into the forming wave.

The Timing of Tick-Tick, Whoosh!

That quick, complete drop step reminded me of 'tick-tick' ("one, two"); i.e., the one second (actually, two-successive half seconds) that it takes to throw an attack at Tori (when playing Uke of course).

There I was, safe and sound, having completed my forward motion, I had met success; but only for about an additional half a second! (the 'three!')

Unfortunately, during that third half second, I was ill-aware that the wave was pulling me in and fully forming to meet me. (This wave, for some reason, seemed larger than the rest. [excuses!]) As the wave formed, it drew my calves forward a bit; thus, my body (and COB!) moved forward to the frontward edge of my stance. I was willowed and could not jump!

With the wave fully formed and ready to pounce, it came crashing in on me in that final (fourth) half second with a great whoosh! ('four!')

I got pummeled and as I shot backward, salt water shot up my nose!

I went back to class to feel the Aikido version and soon came to understand the parallel. Moe and Chris were using the first two half seconds ('tick-tick') to allow me to step and to position themselves as they stepped around me and redirected my attack. The third half second felt much like a pause to me, but Sensei was completing the shift of my COB to the edge of my stance, willowing me even further as he prepared the throw. The final half-second, the 'four', manifested my fall as Sensei knocked me down as if he was a wave crashing; i.e., whoosh!

The tick-tick ('one, two') opened up an additional awareness. I realized that Sensei was actively influencing me with both halves of that first second. I mean immediately, as in, from the very moment Sensei started to re-position; (i.e., redirecting Uke begins at the 'down' in 'down, over, and up')

My kata practice was changed forever. I had not been applying directed intent to influence Uke from the initial 'down' in the step used to reposition myself off the line of attack and into a superior martial position. I had been ignoring the fact that the relationship with Uke starts the second we step onto the mat and does not end till we step off the mat. Sure, I was aware of attacks and I was avoiding strikes, but the problem was that I was waiting to complete the first step before influencing Uke (wasting the 'one, two'). As such, I was not succeeding at throwing Uke much either.

All that from a nose enema at the beach!

Basic Terms

Link your shoulders

This is going to be old news if you have read the Tai Chi classics, but just in case you have not, I wanted to be certain I did not lead you astray with the descriptions of the 'S' and 'C' in 'Six Precepts'.

The lesson is: you have to keep the shoulders dropped (i.e., collar bone parallel to the floor) if you hope to transfer the energy of your COB into your arms. (When standing of course.)

The reason I am going out of my way to mention this, is that, all too often, when I describe the action of spreading apart or closing the gap between our shoulder blades (when describing the S and C positions), the person I am trying to help raises their shoulders. (Their collar bones [clavicles] go slanted in relation to the floor)

Raising the shoulders is most often the martially inefficient thing to do, so instead, keep your shoulders dropped and 'linked' to your torso; i.e., keep your collar bone parallel to the floor.

To help relate this something you might do in other areas of your life, let us look at lifting weights; specifically, the dumbbell fly exercise.

Bringing the two weights from the sides of your body to touch in the middle is an act of spreading your shoulder blades apart (just as you would in the forward C and backward S) It might help to imagine touching your elbows together in front of you as if seated on a cable-type weight lifting machine versus lying on a bench with free weights.

Conversely, consider the action of a seated Lat Pull (Rowing action) and how pulling toward you closes the distance between your shoulder blades. These exercises are very difficult to perform with weights if you do not keep your collar bones parallel to the floor.

These are not perfect descriptions, as they still leave some room for other error, but the form required for weight lifting is the same for Aikido.

Do not expect to generate a lot of power through your arms when you tilt your collarbones; i.e., raise the shoulders.

In almost all situations, keep the shoulders dropped and 'linked' to your torso as you separate or pinch your shoulder blades together; i.e., while performing the S's and C's.

Cheating the Arc (Cheating the Circle)

'Cheating the Arc' or 'Cheating the Circle' are terms we use to describe a common error made when attempting to throw Uke while relying on Pin and Spin.

In short, it means we are no longer moving the Conjunction Point (CP) orbitally in relation to Uke's COB. That can be bad if our intention is to solely 'spin' Uke's COB. What happens instead of spinning Uke's COB is that we end up ruining the Pin by moving Uke's center away from or through the edge of their stance.

Let us add some detail so we are both on the same page with this. It is important to learn how to identify when we are cheating the circle, for this is also the introduction to moving circularly, and the topic of circularity is going to come up again and again.

Three questions will reveal what it means to 'Cheat the Arc': 1) 'How do you Pin Uke?', 2) 'What is so important about the Pin?', and 3) 'What is this common mistake that un-Pin's Uke?'

First, Question 1: 'How do you Pin Uke?'
The Pin in 'Pin and Spin' is when we move (emphasis on 'move') Uke's Center of Balance to the edge of Uke's stance.

Next Question 2: 'What is so important about the Pin?'
The Pin de-stabilizes Uke.

De-stabilizing Uke is the highest goal if you wish to disrupt Uke's ability to deliver force. The Pin will 'Willow' Uke; i.e., it greatly reduces Uke's stability. Our ability to 'temporarily physically dominate' Uke depends primarily upon being significantly more stable than Uke. Being stable is more important than being stronger, bigger, quicker, etc. This is why we avoid the Pinning condition in ourselves, while at the same time; we create the Pin in Uke.

Once we are there, we cannot afford to let Uke out of this predicament.

Lastly Question 3: 'What is the common mistake that un-Pin's Uke?'
In 'Pin and Spin', once you have pinned Uke's COB against their stance, your goal is to leave Uke's COB pinned as you spin it. Please note that the word 'spin' is motion, but in this case, does not have the same meaning as 'move' (in the sense of horizontal/vertical; X, Y, and Z dimensions)

Basic Terms

The two movements we want to avoid in Uke's center are moving away from or through the edge of Uke's stance (a.k.a. away from or through the 'glass wall' of Uke's stance). Both actions ruin the 'Pin'.

That is easier said than done right? Agreed!

Here are the indicators that tell us which of those two mistakes we have made.

The first movement is called a 'breach' and as the name implies, it means Uke's COB travels through the edge of and outside of Uke's stance. It forces Uke to step, which means we have to reset and move Uke toward a new 'pin'.

The second type of loss of pin is attributed to allowing Uke's COB to lose contact with the perimeter and shift back into 'the field' of Uke's stance. We call that 'relaxing the pin' and it results in Uke gaining stability right where they are; i.e., Uke does not step, the willowing fades and Uke feels solid again. (Cough, (re)read 'Six Precepts' if none of this section just made sense to you or it did not feel like retrospection)

Good overview of 'The secret to Aiki', right? So where are the arcs/circles?

Once we have successfully Pinned Uke's COB, the spin is created by manipulating the part of Uke's body we are in contact with. (The literal point at which we are in contact with Uke; i.e., the CP, 'Conjunction Point')

Under typical conditions, in order to spin, but not move Uke's COB, we need to move the conjunction point along a path where all points on that path are equally distant from the COB.

Take a moment to trace all the paths around and equally distant from Uke's COB and you will ultimately trace an orb (or sphere). You might liken it to the valence shell (outermost shell) an electron travels in around the nucleus of an atom. The nucleus being the COB and the CP is the electron.

A circle is just a subset of that orb. An arc is a section of that circle.

To perceive the circles, imagine that orb as a globe and consider the equator or any two longitudes directly opposite each other. The circle that the equator and longitudes create are a lot like a hula hoop, or even more relatable, the outer edges of a clock with the 'Earth's core' (or in our case, the COB) at the center. (You might think of the lines on a common beach ball as

Cheating the Arc (Cheating the Circle)

the edges of a clock face which is sitting inside the ball. The clock divides the inside of the beach ball into halves.)

Since an arc is just a subset of any of the 'clocks' we just imagined, we define an arc by moving along the edge of the clock from one 'hour' of the clock to any other.

Figure 11 should help, but if we are talking clock faces; i.e., complete circles, why do we bother to notice the arc's in lieu of complete circles?

An interesting aspect of 'Pin and Spin' is that, in the practice of knocking Uke down, we never really get to use a full circle.

Uke will fall long before you trace a full circle with the conjunction point, so the word 'arc' is typically more appropriate when describing a technique, but since arcs are a portion of a circle, we often interchange the words 'arc' and 'circle'. (This is a very circular discussion! and puns are never funny.)

So, what then is 'Cheating the Arc/Circle'?

'Cheating the arc' is when we move the conjunction point off of the arc. Said in a more visual way, it means that we have allowed the CP to lose contact with the surface of the orb; by either allowing the CP to 'take off' and away from the COB, or diving in toward the middle (toward the COB.)

You could think of 'driving inward' as intending to move from 12 o'clock to 5 o'clock, but instead of moving through all of the numbers in between, simply 'cheating', by going directly from 12 to 5. Result: Uke gains stability.

The 'take off' (going outside the surface of the clock) is a circular 'breach' because it takes Uke's COB with it (akin to the linear 'breach' described above and in 'Six Precepts'). This typically forces Uke to step.

The diving toward the middle is the circular counterpart of allowing Uke's COB move into the field of the stance (akin to the linear 'relaxing the pin').

Both of these actions result in a force that pulls or pushes on Uke's COB; i.e., once Uke is Pinned, pushing or pulling on Uke's COB will most often move Uke's COB in an X or Y position and cause the breach or relax the pin. (or even far worse, we lose 'connection'. No connection, no Aiki!)

It sounds easy to simply move the CP along an arc, but the truth is, it is rather difficult. Here is why.

Basic Terms

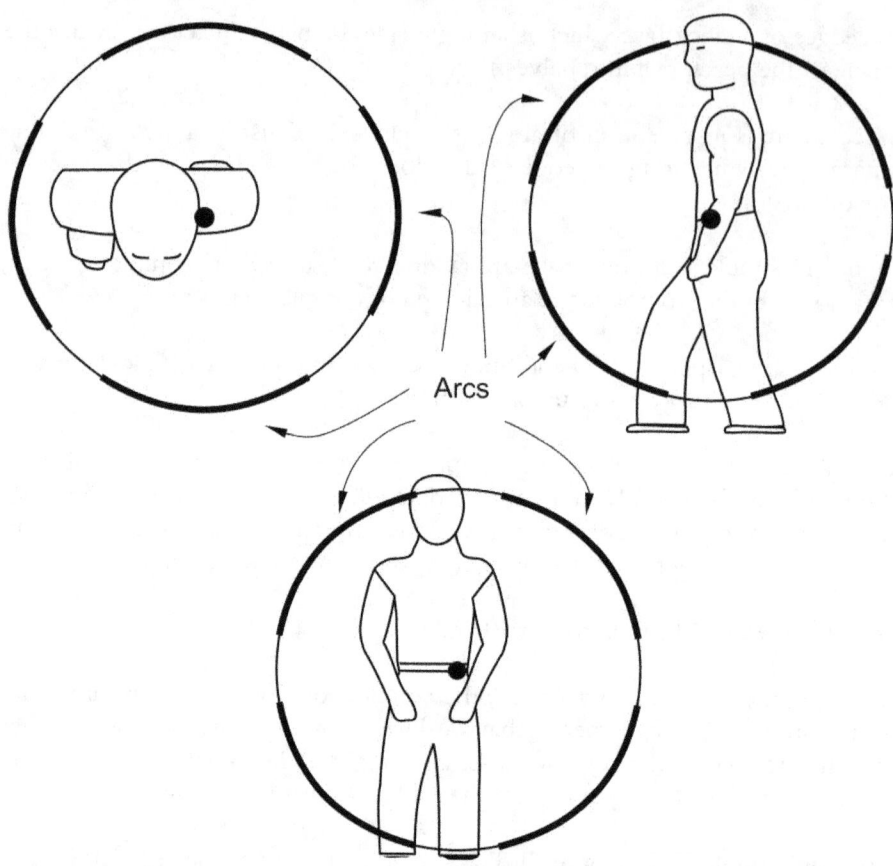

Figure 11: Representation of a few 'Circles' and a few of their inherent 'Arcs' that surround Uke's COB. The picture above shows three views of arcs/circles created when Uke 'posts left'. All three circles above are parts of the same 'orb'. (four arcs are highlighted within each circle)

When applying a technique, the radius of the orb is most often the distance between the center of Uke's COB and your point of contact (the CP).

That radius is established by the rigidity in Uke's body; either by Uke electing to (or not to) tense up, by the application of joint locks, or simply by the structure of Uke's body (e.g., tendons, ligaments, bones, joints, etc.; i.e., Uke's normal, natural range of motion, which is different for everyone).

This is why I did not draw the 'hula hoops' in Figure 11 as though Uke was neatly inside of a bubble. The orb used to throw Uke is rarely large enough to

Cheating the Arc (Cheating the Circle)

encompass Uke and almost never resides in Uke's physical center (just below belly button). The orb is dynamic! It moves, shrinks, and/or grows!

It also should be obvious that, since rigidity in Uke's body is often variable within the progression of a technique, so too are the arcs we use to influence the COB. The overall size of the orb and related arc/circles changes as the rigidity in Uke's body changes. Thus, as a technique is performed, the circles/arcs begin to shrink or grow, and instead of being spherical, the arcs tend to resemble the edge of a French curve.

It sounds complex because it is. It truly is amazing that we can ever throw Uke, but fear not! You have a friend in all of this complexity. Your best friend is 'the Pin' of Uke's COB against the edge of Uke's stance.

The Pin makes our goal of spinning Uke's COB much more manageable.

Besides significantly reducing Uke's ability to generate force in resistance to our intention, the Pin keeps Uke's COB somewhat stationary, because it limits Uke's ability to move. Therefore, the arcs tend to stay in place as well. Besides, without the Pin, the Spin has almost no Aiki effect. This is why one of the most important secrets in creating Aiki is to keep Uke's COB pinned; (i.e., keep Uke willowed; unstable).

It is a good practice to work out in teams of three persons, and together, identify where the optimal arc or arcs exist. (Always have at least one black belt in the group!)

The plan is to have all persons take turns assessing where the 'seam' of the arc exists by (re)viewing from three key perspectives: as Tori, as Uke, and as an outside spectator.

Many, many times, even with an in-depth third person perspective and an understanding of what the correct motions feel like as Uke, I play Tori and the motions I think will take me along the arc are grossly out of proportion, position, and/or direction.

The best of Uke's will help lead you along the appropriate arc; i.e., help you direct the CP along the path where Uke is most vulnerable and has the least amount of ability to resist the throw.

I do not mean throw themselves, but rather, help you assess the arcs and help you follow them, especially through the changes in rigidity (and resist respectfully when you do not).

Basic Terms

That Uke will be asking for and sharing insights using words as much as actions. Work with and be the Uke that can consistently lead others to and through the arcs (Thank you, Joel!).

If you cannot find the Uke that is able and/or willing to help, know that any time you feel Uke becoming less willowed; i.e., they step, or Uke feels like an immovable object, you are almost certainly 'Cheating the Arc'. Change your direction and feel for Uke's vulnerability. It takes a while to find the arcs using this trial and error method, but it can be done.

We are finished discussing the arcs, so hold onto that concept and work with it for a bit, but for those that can dig a little deeper yet; I want to highlight a subtle, but important nuance: the difference between 'spinning' and 'scratching' Uke's COB against the glass. Both will knock Uke down, but they are graphically different. Spinning Uke's COB involves arcs, scratching does not. You should recognize 'scratching' when you see it.

The 'secret to Aiki' is abbreviated as 'Pin and Spin', but the expanded version can be stated as "Pin and 'Spin or Scratch' the COB against the glass". (The secret to Aiki was originally conceived/perceived as 'grinding the pearl [COB] against the glass', scratch or spin, it was all a 'grind' to me.)

'Scratching' Uke's COB against the glass is a 'mobile' pin. We move Uke's COB, but it never leaves the edge of Uke's stance; (i.e., Uke's COB stays 'pinned' throughout the X/Y horizontal and vertical Z up/down motions)

With perimeter of Uke's stance representing 'the glass', 'scratching' is like moving the COB as though it were a bar of soap being used to write upon that glass. We do not spin the soap as we write; we smear the soap across (or upward, downward, or along) the glass, leaving traces of soap as we go.

The best example of this is in the 'secret of the 20' (described in next section of this book; adding clarity to the '20' description in 'Six Precepts') where our focus is shifting Uke's COB horizontally against the wall of Uke's stance.

There is no spin, so arcs do not apply here. Hence, 'Cheating the Arc' has no relevance.

Scratching offers you the freedom of 'moving' Uke and not losing that willowed sensation. Almost all techniques can use either a spin or a scratch. For example, my favorite way to throw Gedanate (#4) is to use a scratch along either the front or back of Uke's stance (secret of the '20'). The scratch can be just as helpful in Gyakugamaeate (#3) if you just want to 'drop Uke to the floor' in front of you (by foregoing the spin of the CP and just scratching Uke's COB downward on the glass)

Become aware of the important distinction between spinning and scratching Uke's COB against the wall of their stance, so you can look for it and not be confused when the technique calls for 'Pin and Scratch' and 'Cheating the Arc' does not apply.

Peeling Tape Off the Wall

'Peeling Tape Off the Wall' is a way to describe how we manipulate Uke's spine in a manner that creates vertical pressure on Uke's COB.

This methodology is also helpful when trying to compensate for how the spine modifies the 'arcs'. (You may have already discovered that the arcs become less like circles and more like French curves as you remove the slack from Uke's spine. 'Peeling Tape Off the Wall' is a great way to remove the slack from Uke's spine.)

In 'Six Precepts' we introduced the spine as a wobbly stack of plates.

Well, what happens when the top plate begins to protrude further than the rest of the stack? It falls of course. In the case of the spine, since all of the vertebrae 'plates' are attached to each other, like dominoes, or maybe better described as the cars of a roller coaster, each subsequent vertebrae follows the next, shifting each other out of alignment, putting tension on the connective tissues, and making it easier to influence Uke's COB.

One way to quickly describe this pattern is 'Peeling Tape Off the Wall'.

Picture a strip of painter's tape stuck on the wall vertically, ceiling to floor, and the method with which a painter might remove that tape.

From the top, if the tape is pulled directly outward at a ninety-degree angle from the wall, it has a tendency to lift paint and cause damage. Additionally, the painter would have to move a rather long distance away from the wall to remove all of the tape. Pulling the tape directly outward and maintaining a parallel relationship between the tape and the floor will eventually remove the tape all the way to the floor, but the tape will not touch the floor until the very last bit of tape is removed. Many a Tori has tried to move Uke's spine similarly, moving Uke outward, but never down, and as such, Uke steps and is rarely, if ever, thrown in this manner.

Basic Terms

The more appropriate way to remove tape from a wall, is to peel downward, at around a thirty-degree angle, against the length of the tape. This reduces the likelihood that paint underneath will lift off the wall. Of course, the tape in your fingers will move away from the wall, but certainly not as quickly as it would if you were trying to move perpendicularly from the wall. The added and major benefit is that the tape begins to be drawn toward the floor.

The parallel here is that we should treat Uke's top vertebrae in much the same way as the top of the tape. By treating Uke's spine as though it were the tape we are peeling off of the wall, we create an indirect downward vertical pressure on each subsequent lower vertebra and influence Uke's spine to bend much as if Uke was doing yoga; folding one vertebra at time to touch the floor. The key result is the compound and constant indirect vertical pressure on Uke's spine/COB. Uke is bent toward the floor and not led across the room.

This analogy is not quite as scientific as most of the analogies in our previous sections, but can prove useful when we find ourselves moving Uke 'outward' instead of 'downward' (and you might have learned a little about home improvement! Bonus! No extra charge.)

Rolling the Sleeves

We have to remember that, except at the 'north and south poles' of Uke's 'orb' (sphere), in order to maintain connection, Uke and Tori COB's need to be (or at least have the intent to be) moving away or toward each other. If not, there is no connection. No connection, no Aiki.

The easy mistake to make when performing a technique using Uke's arm (e.g., Oshitaoshi (#6) aka Ikkyo) is to abandon connection by ceasing to maintain a pressure that moves Uke's upper arm (Humerus bone) into or out of the shoulder socket. (Side note: into the shoulder socket is an expression of compression connection, outward is extension connection.)

Typically, this loss of connection happens because, instead of focusing on moving the arm in or out of the socket, we begin putting pressure on bones that will not flex (e.g., the forearm bones), or start applying pressure on Uke's elbow joint. You could break the bone(s) you are pressing upon, or if your intent was on Uke's joint, you might even destroy Uke's elbow, but you still will not have Aiki. (Your Uke is likely less than happy too!)

In 'Rolling the Sleeves' we find a way to improve our focus on keeping the extension or compression connection flowing.

The description is the key, 'rolling the sleeves'. Instead of trying to manipulate Uke's arms, try instead to slide/push/roll Uke's sleeves down or up Uke's arms as part of the technique you are performing. It does not matter if Uke has sleeves to actually move. The intention is the secret. It is the friction on Uke's arm that allows us to keep moving Uke's arm into or away from their shoulder socket.

You should quickly find that it is not always enough to push the sleeves upward, it helps even more to put a bit of twisting into the action. This is exactly how we follow 'the snakes' (described in 'Six Precepts').

The Plank

'The Plank' is an attack to Uke's knees, but not for the purposes of breaking their knee; i.e., The 'Plank' is nothing like kicking Uke's knee.

The Plank is used to make Uke's knee most uncomfortable. This helps us move Uke in such a way as to make Uke want to step; which means we can more easily draw Uke out of a rooted stance.

The Plank is best described as a wooden plank under Uke's foot, as if you glued a piece of rectangular hardwood flooring to the bottom of Uke's foot.

The corners of this plank are as far in front of the foot as where Uke's knee would land if Uke left the ball of their foot in place and went to a knee. In the reverse direction, it is practically the same distance behind as it was in the forward direction. You might think of it as having the distance of where the back of Uke's knee would be if Uke kept the heel on the same spot and were sitting on the floor (as if seated, legs straight, stretching to touch the toes).

Like I said, practically the same distance forward and backward.

The width of the plank is about two to three inches on either side of Uke's foot. The wider the foot, the closer to three inches on either side, the narrower the foot, closer to two. (See Figure 12)

Why are the points of these rectangles so important? To be blunt, if the ball of your foot was stuck to the ground (much as if you were wearing a cleated shoe on grass) and your knee was to actually go near the forward corners of the plank, your knee would be in extremely high risk of coming apart. Do not even try it. The same would be said for your foot being rooted (flat on the floor) and having the body weight moving toward the back corners.

Basic Terms

Your body will sense this without any special training and it will be very enticing to shift your weight off of the knee in danger; most likely in the form of a step, at minimum, as a repositioning of the foot.

Knowing this, we can leverage this reflex to coerce Uke into uprooting and moving out of a rooted stance. All we need do is guide Uke to the corners of the Plank.

The most common use of the Plank is what we call 'Up on 2'.

What is the '2'?

We numbered the four corners of the plank. For no special reason, we started numbering on the front corner, nearest the no-line and then worked our way around and to the back. (See Figure 12 again)

The number '2' corner is particularly useful when attempting an Oshitaoshi (#6) (Ikkyo) on Uke.

With Uke in the right foot forward stance (ala Figure 12), lead Uke's right elbow to rise over 'the R2' via a 'Football' (i.e., Lotus) curve.

This action will encourage Uke to shift weight in such a way that Uke's left leg/foot will kick the right foot out of place; and the left foot will replace the right as the support. (See the 'teeter' position/pose described in the end of this book's 'Basic Terms' section)

Uke is left teetering on one foot (their left foot), with their right elbow above their head, and the near side (right side) of their body stretched.

Be sure to raise the elbow as it approaches R2, so as to avoid getting hit and to assure that Uke ends up with right elbow directly above their weighted left foot.

You will find that it helps immensely to 'Roll Uke's Sleeves downward' when taking Uke 'up on 2'. It is especially helpful to mix in that bit of twist along 'the snakes' as you apply extension connection.

Another practical use of the Plank is when we are literally pushing on Uke's knee toward a corner of the Plank.

The Plank

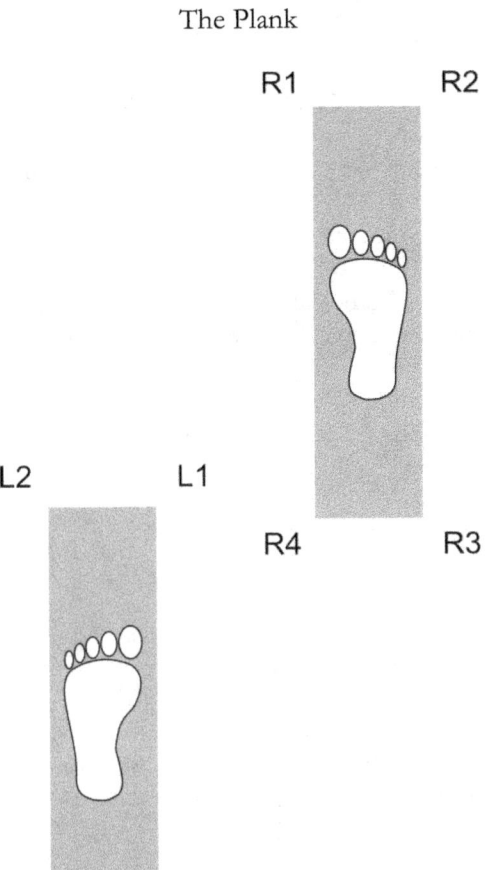

Figure 12: The Plank(s), the numbered corners are the points at which Uke's knee becomes uncomfortable should their weight be directed to it. (foot rooted to the ground; i.e., weighted)

When Uke's leg is straight, pushing on the front of the knee toward either the '3' or the '4' will work fine, it all depends on which direction you want Uke to fall. Typically, this happens when Tori is in kneeling position and Uke is standing. (Hanmi-handachi).

You should recognize using the '3' and '4' as a parallel to how we might hyperextend an elbow on a straight arm.

The point is that you should focus the pressure through Uke's knee and at the corners of the Plank, not parallel to the floor, but directly at the corners of the Plank.

Basic Terms

Rolling the Tack

'Rolling the Tack' teaches us a way to approach spinning Uke's COB such that we can keep Uke's COB pinned while we move them from leaning forward to leaning backward (or vice versa).

Ever try to move Uke from leaning forward to leaning backward? (You might have if you have ever tried to perform Irimi Nage.) It does not work well if we try to simply pick Uke's head straight up and then move Uke into a back bend. It might work if you manage to keep Uke's COB pinned, but most often, simply picking Uke's head straight up will put Uke back into the middle of their stance and back to being stable.

Instead, let us roll Uke like a tack!

Ever see a tack roll around?

With the point of the tack touching the floor, the top of the tack actually spins like a wheel. Eventually all points along the edge of the tack's crown will touch the surface of the floor. A key aspect to notice is that the point of the tack does not move, rather it stays in the same spot on the floor and spins. (Granted this only happens when the length of the stem and size of the top of the tack are a certain ratio. Work with me.)

The tack mimics how we want to move Uke, in that the tack is completely committed to leaning on the floor at all points along the way as it moves from 'leaning forward' to 'leaning backward'.

Uke can be made to do this too! We can swing Uke's torso in such a way that we leave Uke's COB immobile in the X and Y dimensions as we 'roll' Uke's head from forward to rear.

Uke's COB is going to be represented by the point of the tack. You can view Uke's shoulders as the top of the tack (the crown). By pinning Uke's center to the side of their stance, we can imagine how we 'Roll Uke like a tack'; i.e., how we make the 'crown' of Uke's shoulders rotate as we maintain vertical pressure on Uke's COB ('Peeling the Tape') and shift Uke from a forward to a backward lean.

The maintenance of the vertical pressure and keeping Uke's COB in place, (locked against the side of their stance) is the key here. Pin and Spin! Uke remains burdened, never regains stability, and we reposition Uke for a technique. Aikido and office supplies!

Bonus thought: you could learn a good deal about this analogy by using an umbrella. Use an open umbrella, set on the ground, with the handle resting on the ground. Get into a kneeling position, and 'roll the umbrella'.

Hmm, having trouble spinning that umbrella because it would have to roll over you? You might notice the same issue when rolling Uke. Do not get in Uke's way. If that still does not help, you can think of 'Rolling the Tack' as a rotational 'Peeling Tape Off the Wall'; i.e., Peel the tape from the Maypole'

The Snow Shovel

This is easily one of my most favorite drills to share, as it is easy to do and very undeniable. You are going to get some wide eyes in your dojo with this one.

The 'snow shovel' teaches us a method to 'uproot' Uke from a strong stance when we cannot get to their no-line or 20; (i.e., how to move Uke through the strongest part of their stance.) This drill is taught using a jo.

Have Uke get into a cat stance, as if Uke had just thrust a jo staff at your belly; test Uke a bit by pushing on the jo staff in a few different angles to be sure Uke is trying to maintain this stance.

Now, grab the end of the jo and try to move Uke out of the stance, specifically, try to push the tip downward. Uke should resist the action.

I could not do it (at first), and I will lay you a bet you did the same as me. My first experience: Uke stood there in a deeply rooted stance, the jo staff bent a bit, but Uke's hands stayed pretty much where they were at the start.

When this happens, it is primarily because we are allowing the tip of the jo to move in an arc with Uke as the center/pivot. (See Figure 13)

Now, have Uke assume that same stance, but this time have the tip of the jo touching a wall; preferably a wall you can make marks on, because chances are you will make a mark, for I am going to ask you to grab the end of the jo closest to the wall and, while keeping the tip against the wall as you progress downward, push the tip all the way to the floor; (i.e., scrape the wall with the jo staff tip, all the way to the floor.)

The only trick here is to keep the tip of the jo against the wall, nothing more.

Well, that and learn to drop your weight.

Basic Terms

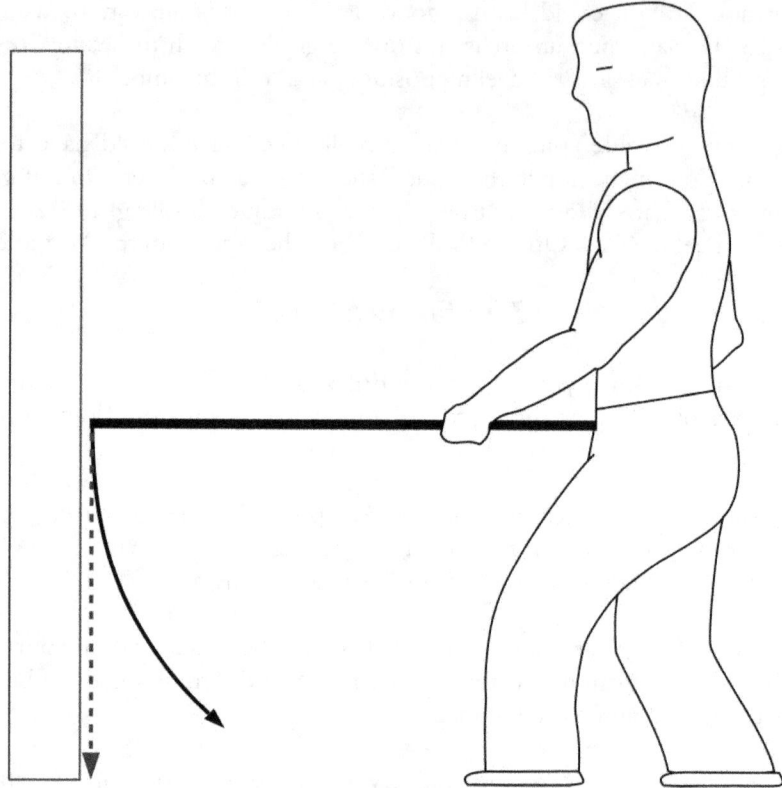

Figure 13: Uke's holding a Jo staff; Pushing down in such a way that drifts/arcs back toward Uke does not work, but moving down along the wall gives the desired effect.

Uke is going to have to let the hands move away from their body or Uke is going to be pulled out of that stance in a hot second. Ask that Uke does their best to keep their hands in place so you can fully experience the feeling of uprooting Uke in a unique way: along the strongest line that Uke has. (No 'no-line', no '20', just lifted out of their stance. Using the no-line and 20 with this technique is just candy!)

Why do we call this the 'snow shovel'? It snows where I live, and I imagined this scenario as a case of someone holding a snow shovel similarly to how Uke holds the Jo in Figure 13 and what would happen if a large amount of snow would drop down from a roof and onto the snow shovel.

That sudden weight applied to the shovel would drive the shovel directly downward and launch the poor person trying to clear their driveway. The snow would follow gravity and move directly downward; no arcing back into

Uke. Therein lies the secret, that the tip of the shovel (or Jo) has to stay 'resolute' in the sense of X and Y positioning. The weight must influence the tip of the shovel (or Jo) directly downward; (i.e., downward in the Z dimension; the tip of the jo scratches along an imaginary wall)

The lesson here is an option available to us when we have extended Uke as far as Uke can go. For example, when we stretch Uke's arm outward as far as it will let itself reach.

With Uke fully extended, we can use the path that takes our conjunction point directly to the floor, and consequently, Uke gets uprooted.

Crush Box, Stretch Box, the Teeter, the Back Teeter, Twisted, and Checked

This is the last of our 'warm-up' toward the bigger learning and this is arguably the best of the warm up so…

Quick! Get a blank piece of paper and a pen…

Are you back?

Now, write down all the words you use to describe the poses Uke will be made to assume just prior to a throw. (Go ahead, take a minute, I will wait.)

Back so soon? Let us look at that list.

Are you staring at a blank piece of paper? My expectation: most will. We typically spend all of our time watching what Sensei is doing and it can leave us completely blind to the nuance in the condition of Uke's overall posture. Exceptional benefits arise when we take the time to focus on Uke's posture.

Welcome to 'the other side' of the throw, where we analyze, in detail, Uke's posture at the point of and potentially even leading up to the throw.

This section describes five poses that are effectively reactions (postures) we all naturally try to avoid as we become burdened; (i.e., as we bear significant weight; e.g., as Uke when coping with Tori leaning on us).

When combined with 'Pinning', these poses leave Uke extremely unstable and susceptible to collapse. That is why Uke does not want to be there and exactly why we are going to put them there.

Basic Terms

Now, those paying close attention counted six items in the title above and noticed that I said 'five poses'. Good eye!

The last item is more of a 'condition' than a pose. You will see what that means in just a bit, but first, please give me a minute to sell you on why you should even bother to read this section.

What follows immediately is a description of the common evolution that occurs as we learn to focus on Uke's condition. Trust me, we will talk about the poses, but it is important to know why these poses are important and what using them looks like…

It begins by improving our personal (individual) awareness.

We learn to identify the poses and start paying attention to which pose Uke is in immediately prior to, and potentially even leading up to the throw.

The goal is to use this awareness (awareness developed as Tori, Uke, and observer! Teams of three!) as a benefit toward mimicking Sensei.

That sounds rather obvious, as if it were common sense, but if we are not trained to put detailed analysis on Uke, a lot of important feedback goes unnoticed. For example, we could easily mistake a crush box for a stretch box (or worse yet, not even be aware of either!) and wonder why 'the throw worked yesterday' or 'last time' but nothing is working today or when playing with a different Uke. Instead of chalking it up to 'bad days' or 'good days', we can put our head in the game, apply a bit of knowledge, and, through an understanding of Uke's pose, reveal to ourselves where we may be cheating the circle, relaxing the pin, etc.

We are mastering this first phase when we can identify each pose and can sense 'the Pin' that creates the willowing in all of them.

This beginning phase has a group effect as well. Once these poses are shared with others and are generally understood, the common reaction when playing with others is, "Hey! Sensei seems to always use Crush Box to perform that particular technique, but you just did it with a stretch box!"

"Which is correct?"

The answer, in truth, is that both are correct.

Here is one way to explain it:

Crush Box, Stretch Box, the Teeter, the Back Teeter, Twisted, and Checked

"Over time, Sensei has demonstrated mastery of that technique in more than just these two situations. Sensei likely showed you one before the others because Sensei knows what would work best for you."

This too is the truth, and is some good verbal Aiki side stepping! The good news is that your training partner has demonstrated awareness; and you might just have become aware of something you did not know you were doing differently than Sensei (but hey! You were doing it successfully!). Awareness is the path to appropriateness. (Your training partner also indirectly complemented your abilities in more than one way!)

We evolve to the second phase when we become aware of the nuance within and between the poses. More than one benefit will arise.

We begin to understand that each pose has its own special characteristics; i.e., the poses do not all work the same way. This gives rise to knowing why and how we keep Uke in a particular pose as we move to deliver the throw/technique.

Eventually, it evolves into awareness of the path between poses and all the nuance necessary to successfully make that trip. At that point, it becomes glaringly obvious that some techniques require Uke to be locked into one pose and only that pose; yet still other techniques can leverage more than one. (Same technique, but different pose(s))

As we break from static kata and delve into free form sparring, we find that the poses are also a method to keep Uke in place as we flow from one technique/throw to another; i.e., the poses are 'linkages' between techniques. Same pose, different techniques (e.g., Crush box allows us to perform both Aigamaeate (#2) and Ushiroate (#5)).

On the flip side, we begin to see ways to flow Uke into other poses and techniques as the situation decrees; (e.g., multi-person randori where we realize we need to shift Uke and change the intended technique in order to perform a particular throw and simultaneously keep others at bay. True improvisation.)

It is also in this second phase that we find solace from frustration and break from boredom as we switch from focusing solely on what our body is doing to focusing on how Uke is or is not reacting.

The change in focus often hastens us to break through plateaus in our training. The upside is that the 'high level' goal of 'shaping Uke' takes the attention off of our body and often allows us to 'get out of our own way'; i.e., focusing on Uke can help us 'put it all together' within ourselves; for instead of nit-picking precepts, posture, judging ourselves, etc., we get right to the task at hand: controlling Uke's stability.

On the downside, shifting focus from the traditional 'what am I doing?' to 'just get Uke to pose!' can help, but of course, it can also create whole new kind of frustration on its own!

Neither approach is the real point of Aiki.

The end goal is not to focus on your own actions, nor is it to focus upon Uke's reactions, but to master the relationship between the two. It is extremely difficult to manage a relationship without a deep understanding and awareness of both sides, hence this focus on Uke's pose.

At the far end of the second phase, (and I do not know that we ever actually leave this final part of the second phase), you develop increasingly finer grain control in the amount of willowing that you create in Uke (e.g., sensing just how 'crushed' the Crush Box was and how to add or subtract from it.) The seemingly endless goal at this point two-fold: to reduce the amount of effort it takes to move Uke into a pose and how much willowing to create in Uke (that we do not telegraph our intent before it is necessary.)

Finesse and graceful action will appear more often and expect three things to happen: first, making more throws than ever before, second, hearing "Wow, that was a good throw!" from others, and lastly finding yourself saying "That felt effortless". May all three be your everyday experience!

One final note of 'why learn the poses?'. At a master level progression, (the third phase, and I speak of this only because I have seen it, not consistently nor intentionally experienced it) we learn to stop trying to 'put Uke into a pose' and eventually start creating the conditions that 'influence Uke to assume the pose themselves'. I do not mean Uke-do: (UKEDO - oo-kay-doh! - the art of Uke doing the throw for you), but true Aikido, where we align with 'The Tao' and Uke breaks their structure against it/us.

Dream big! Ride the spiral to the end…

Apologies for the long digression about the path, but what good is a tool if you do not know why it was forged?

So, let us get into it. Five positions and a common condition.

The Boxes: Crush Box and Stretch Box

These five poses are not the only forms Uke wants to avoid when unstable, but these are foundational, or at least the ones that I felt I should detail in this book. There is room for you to find others and share. For instance,

I leave it up you to name or research the moniker we use to describe the two positions Uke assumes when willowed to the front of their stance, ready for a face first, forward throw; i.e., when Uke is on the balls of their feet, trying to stop short before falling forward (ala stopping short of running off of a cliff.)

The primary difference between these two poses is position/condition of Uke's hips (one is 'butt out', the other 'belly out'). In the 'Mojo' we actually use the same multi-word phrase to identify both, but the difference, as Chris tells me, is that one is traditionally very indicative of Kito Ryu influence. I did not coin the phrase we use for the forward-directed willowing, and I know extremely little about Kito Ryu, so I mention these two poses in passing, as I did not feel I could detail these poses sufficiently. I urge you to ask Chris! Tell him, "Bill asked me to ask you about a hub."

Remember to use the following five poses (and the 'checked' condition) as a key indicator of your and other's success at achieving the same effect as Sensei. Expect to use these names often as they are an efficient way to communicate corrections and/or confirm success.

The Boxes: Crush Box and Stretch Box
The first two terms, 'Crush Box' and 'Stretch Box', describe a 'box'.

The box is Uke's torso; the hips and shoulders forming the four corners.

As you might guess, the 'crush' or 'stretch' describe a distortion in the natural shape of the torso.

These box distortions are a sideways flexing of Uke's spine. Basically, a side stretch; where one side's shoulder is brought closer to or 'crushed toward' that same side's hip. Conversely, as one side of our torso box gets 'crushed', the other side of the box is 'stretched' as the shoulder moves further away from the hip.

You might be asking "Can Uke be stretched and crushed at the same time!?!", and the answer is, "No, absolutely not", and the distinction is extremely important.

The key to identifying which type of box distortion we have applied to Uke is in locating Uke's COB. If Uke is posted (i.e., weighted) on the 'crushed' side, it is called 'Crush Box', and if Uke is posted on the 'stretched' side, it is called 'Stretch Box'.

Let us take a closer look...

Crush box

To show 'Crush Box', I borrowed a picture that was originally drawn to be included in 'Six Precepts' but did not quite make it (see Figure 14). I used this picture for a couple of reasons.

First, because it was already drawn and I was lazy! Just kidding, actually, although it is true that this picture was drawn years ago, I shared this image because this picture caught my notice as a way to expand your awareness of 'the Pin' as we learn these poses.

Some will look at the coming picture and notice that Uke's center is not quite at the side-edge of the stance (i.e., not at the blade of the foot). That might lead us to believe that Uke is not willowed, but in this case, Uke is actually pinned to the front of the stance; i.e., Uke's COB is shifted to the ball of Uke's right foot. (I guess you could also imagine Uke also being pinned to the back of the stance; i.e., pinned on the right heel).

The point here is that the 'Box' positions are not themselves the pin.

The Boxes are the equivalent of a spinal 'joint lock'.

You can lock Uke's wrist and not have Aiki, the same goes for their spine. The flexion of the spine weakens Uke's ability to deliver force, but if not done correctly, it will not create willowing; thus, we can use a 'Pin' instead.

The point is that your perspective can make it all too easy to miss where Sensei has Uke pinned (e.g., frontward instead of sideward, see figure 14).

Notice that in crush box the Sagittal Plane crosses the same side shoulder and hip; the spine is bent, aka 'Crushed', toward the COB. (Go and review the 'salt shaker' in 'Six Precepts'.

Were you aware of and creating the 'pin' when you practiced the 'Salt Shaker' drill?

You were! Awesome, as I did not highlight that in the text, but the diagrams show the Crush Box moving to the pin. The 'Salt Shaker' drill works much better with a pin.)

The Boxes: Crush Box and Stretch Box

'Crush box' is a real crowd pleaser for those that prefer to keep Uke's COB close to them. You can call it the true sense of 'in-fighting'.

Most of the times that I have stopped to take notice, Tori makes Uke assume a Crush Box during the application of an Irimi Nage, and thusly, in many of the techniques that have Uke running around Tori.

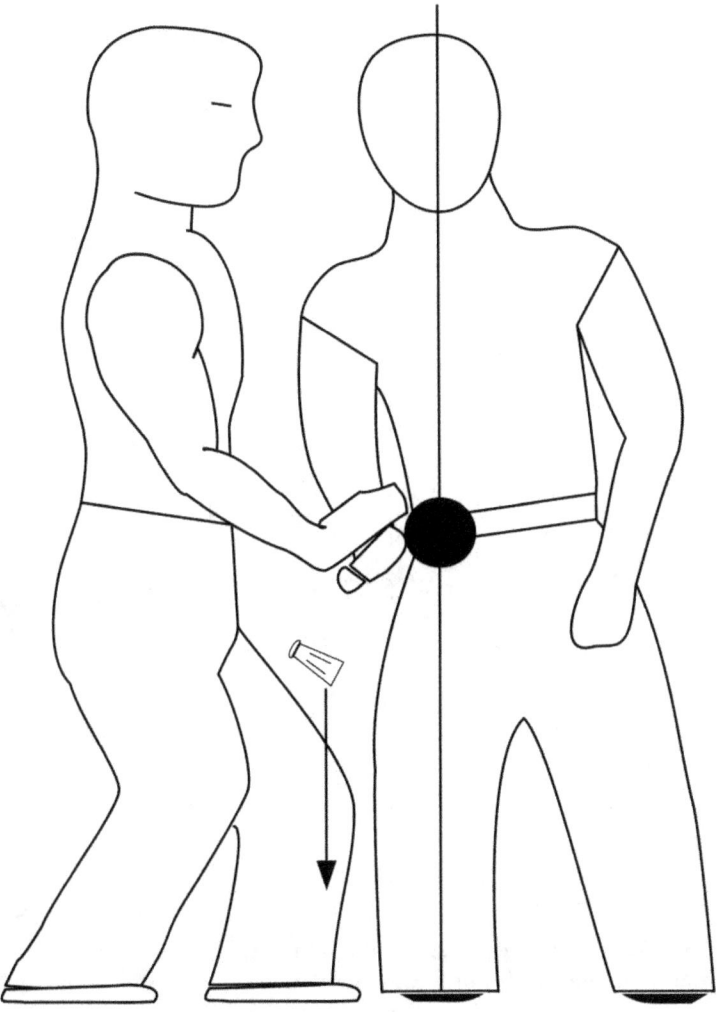

Figure 14: 'Crush Box' in Uke; Tori has Uke weighted on the right side and has Uke's spine flexed such that Uke's right shoulder is closer to Uke's right hip. You might notice this picture is similar to the 'salt shaker' drills. (Here Uke is pinned forward and onto the ball of the right foot; not the side-edge of the foot/stance as in the diagrams in 'Six Precepts'.)

Stretch Box

Stretch box is likely my favorite, as it feels a bit 'roomier' than Crush Box (Personal space!). Below are two examples of a left-side Stretch Box in Uke. (Right side of Figure 15 lacks a Tori applying Oshitaoshi (#6) [Ikkyo]).

Even though Tori is controlling different arms and at different heights, it is still the same Stretch Box (from Uke's perspective).

We often find ourselves on the left side of Figure15 as a follow up to the right-side Crush Box in Figure 14 (See Figure 16); i.e., we maintained the crush-box bend in Uke's spine, and shifted Uke's weight from right foot to their left.

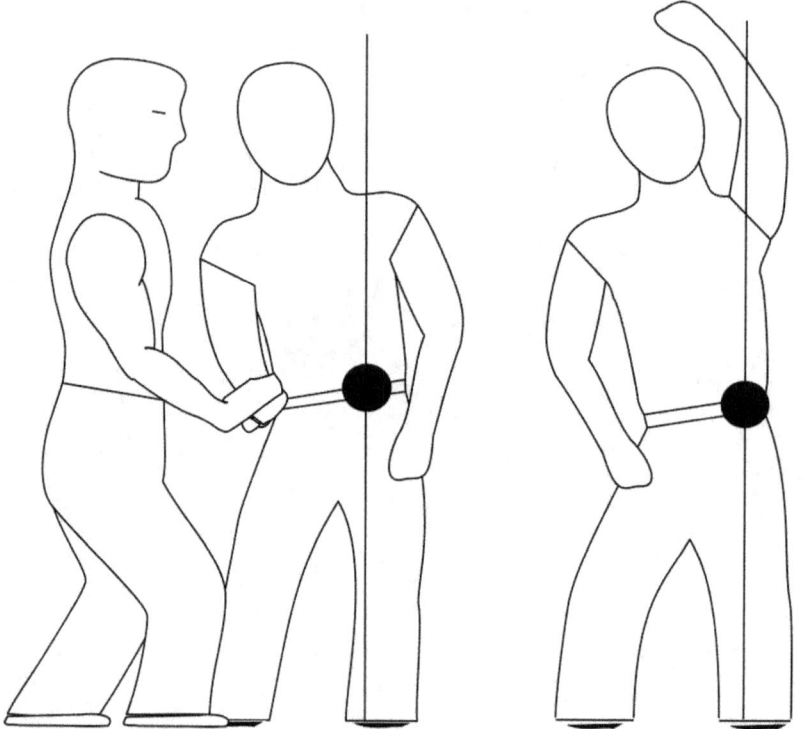

Figure 15: Uke in left-side 'Stretch Box'; On left, a follow up to earlier 'crush box' explanation. On right, only Uke shown, in typical pose delivered as part of Oshitaoshi (#6) (Ikkyo)

Remember 'Cheating the circle/arc'? If you are not aware of the shift in Uke's COB and the differences between the 'Stretch' and 'Crush' box poses, that little shift can create some big frustration for Tori when trying to follow an arc that spins, but does not move Uke's COB. Subtle huh?

The Boxes: Crush Box and Stretch Box

Some key 'notables' about the right side of Figure 16 (same image as the left side of Figure 15).

In this particular stretch, Uke is posted left, and Uke's left hip and shoulder moved apart (subtly); i.e., Uke's left side is stretched; exposing and spreading the ribs. Uke's sagittal plane still does not cross the body, instead, it passes through the same side hip and shoulder.

The same lesson about 'the pin' applies: we do not necessarily 'pin' Uke's center to the outside blade of the left foot. We can but, in both sides of Figure 15 and 16, Uke is again pinned to the front (at the ball of area of the left foot).

How about the right side of Figure 15? Seen it before, huh? Looks like part of Oshitaoshi (#6) (Ikkyo) right? (Imagine Tori attached to Uke's left arm)

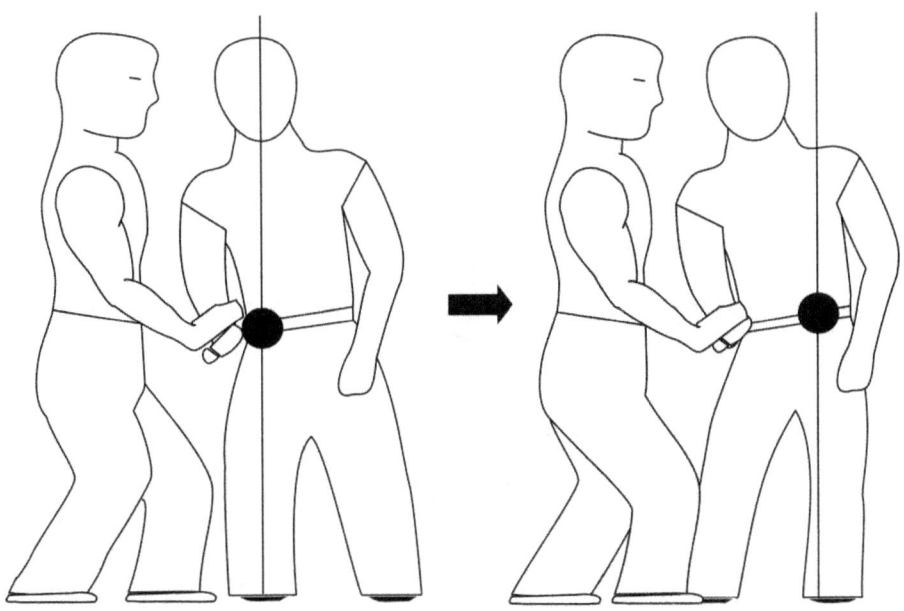

Figure 16: starting on the left as in Figure 14 'Crush Box', shifting Uke to their left side as in Figure 15 'Stretch Box'; we could just as easily shift back

In the right side of Figure 15, we took control of Uke by raising their left arm and drawing Uke's COB toward the edge of Uke's stance.

Oshitaoshi (#6) (Ikkyo) will often start 'in the stretch', but unless we are doing a 'Tenkan' (turning) version, we will likely move Uke's COB to the

Basic Terms

other hip (by 'scratching' Uke's COB along the '20') and create what we call the 'Teeter'...

The Teeters: Teeter and Back Teeter
We crushed and stretched Uke's box, so now we tip the box so that it 'teeters' on a side of its base. You will catch on quickly if you have ever poured a box of cereal.

The (side) Teeter
The (side) teeter is Aikido's version of pouring cereal out of a box with the intent of getting all of the cereal into a bowl. (Tipped on the narrow side of the box) We sometimes called this condition 'Tipped Box', or even 'The Sprinkler'.

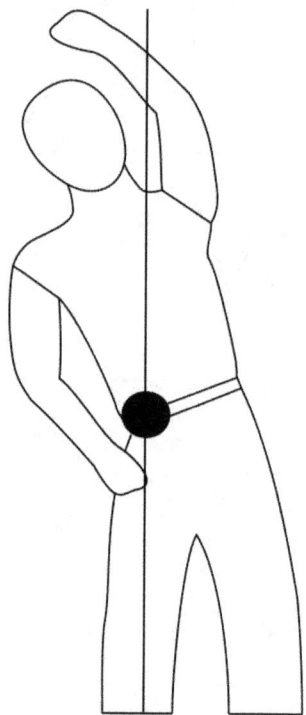

Figure 17: Uke in a 'Teeter'; opposite shoulder and hip are nearly aligned on the posted side (vertical line between them cuts through the COB)

How do we Uke get here? Remember the stretch we achieved in Uke in that right side of picture in Figure 15: 'stretch box'?

We can maintain that same stretch in Uke as we shift Uke's COB to the opposite hip, but the stretch in the ribs is not the key ingredient.

The Teeters: Teeter and Back Teeter

What makes the Teeter drastically different than 'the boxes' is the relationship between opposite corners of Uke's box and their COB. (See Figure 18).

An important point to make on this position is that Uke does not necessarily have to be completely on one foot. Some throws rely on keeping Uke's hips facing a certain direction. In these cases, you will be seeking to maintain friction between the floor and Uke's less weighted foot.

"Where does this happen" you ask? Mentally walk through 'rolling Uke like a tack', and you can find the Teeter with drag on Uke's unweighted foot. If that visualization is too new for you, you could analyze what you feel is a demonstration of a perfect Irimi Nage.

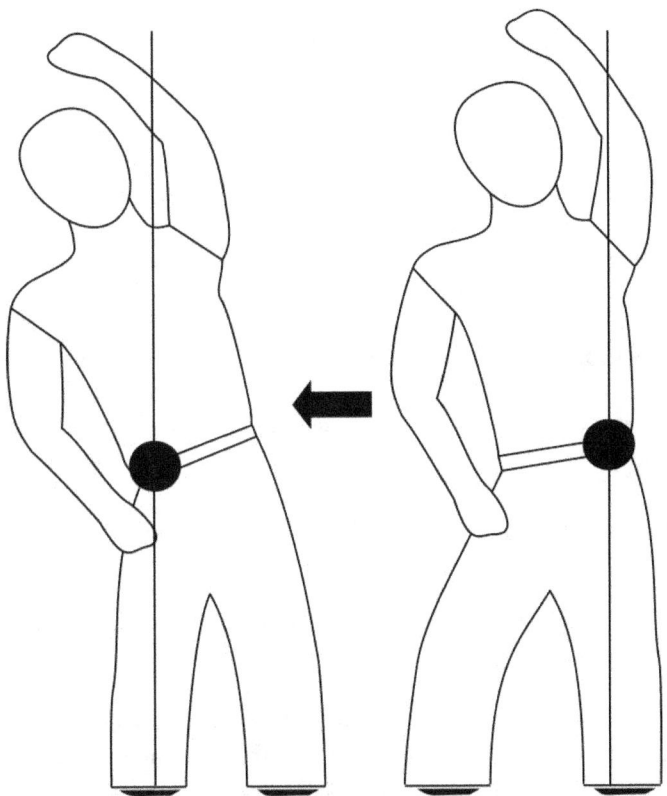

Figure 18: Shifting Uke from 'Stretch Box' to a 'Side Teeter'

In the demonstrations of Irimi Nage that I most revere, it is clear that Uke is placed into a 'single-foot' Teeter only at the very end, immediately after a good 'roll the tack' session!

Until that final throw is delivered, Uke's feet are typically both on the ground, with Uke's weighted-leg's knee bent just enough that their 'unweighted' foot has some drag on the floor.

Uke's hips stay square and Uke cannot slip out. (You might notice Irimi Nage often performed as a Crush Box until the very last second where it transitions to a Teeter. Go ahead, check it out on the web. I will wait.)

Be aware that the key element in the (side) Teeter is that Uke has opposite shoulder above the weighted hip.

If you find Uke's shoulder is above Uke's belly button, it typically means you have not moved Uke's COB over far enough, or you have moved Uke's COB away from the edge of their stance (you have 'relaxed the pin' and Uke has gained some sense of mobility and stability). Fix it by getting that shoulder over the opposite hip.

Back Teeter
The next to last of the poses we will describe is the 'back teeter'.

The back teeter is the position we wish to see Uke display in Shomenate (#1) or in Gyakugamaeate (#3) (and others, e.g., Ushiroate (#5)).

It will look as though Uke is leaning back as if trying to do the limbo.

This pose seems very basic in nature and that it would be easy to knock down an Uke in this position, but it is much more complicated than it looks.

In fact, it is arguably the most complicated of all the poses we have discussed thus far.

Three key understandings will help simplify your attempts to throw Uke when Uke is in this pose.

First, that in this pose, Uke is always double weighted; second, that there is a big difference between being pinned on the 'front of the stance' (rectangle) or pinned on the 'rear of the stance'; and lastly, that being pinned on the front of the stance can be further complicated by two conditions we call 'belly weighted' and 'tail weighted'.

Let us dig in!

The first awareness, that Uke is double weighted, is very important.

The Teeters: Teeter and Back Teeter

It is not a back teeter if Uke is on one foot.

Being 'on one foot', bending backward, is instead a rather awkward Teeter. You would expect that Uke would be destined to fall, but Uke is in a good position to wriggle out.

Single foot stances are prone to make Uke's COB 'breach' the wall. The base of the stance is only the size of the triangle of the foot. You do not have to push Uke's COB far to make Uke breach and step.

So, if you are going to try knock Uke down in Shomenate (#1), it helps to keep Uke double weighted.

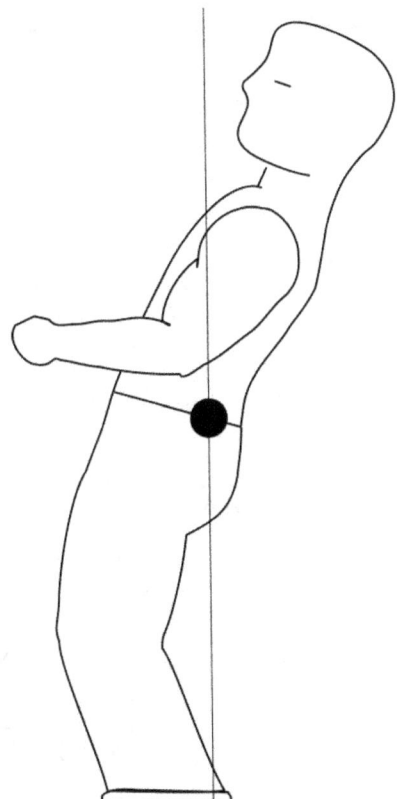

Figure 19: Uke in 'Back Teeter'; Uke is double-weighted

The second concept, 'which side of the rectangle Uke is pinned to' (Front or Back side?), is of great importance when trying to spin Uke's center.

Basic Terms

The key reason most attempts to throw Uke in a back-teeter fail, is that most Tori (especially me!) will cause Uke's COB to breach the stance.

The cause is that, instead of keeping Uke's COB 'pinned and spinning' by moving the CP along an 'arc', Tori instead 'cheats the arc' and ends up moving Uke's COB horizontally; causing a breach and Uke steps out.

[In the case of Shomenate (#1), this means that Uke gets pushed backward and gains stability; i.e., Uke starts to walk backward.]

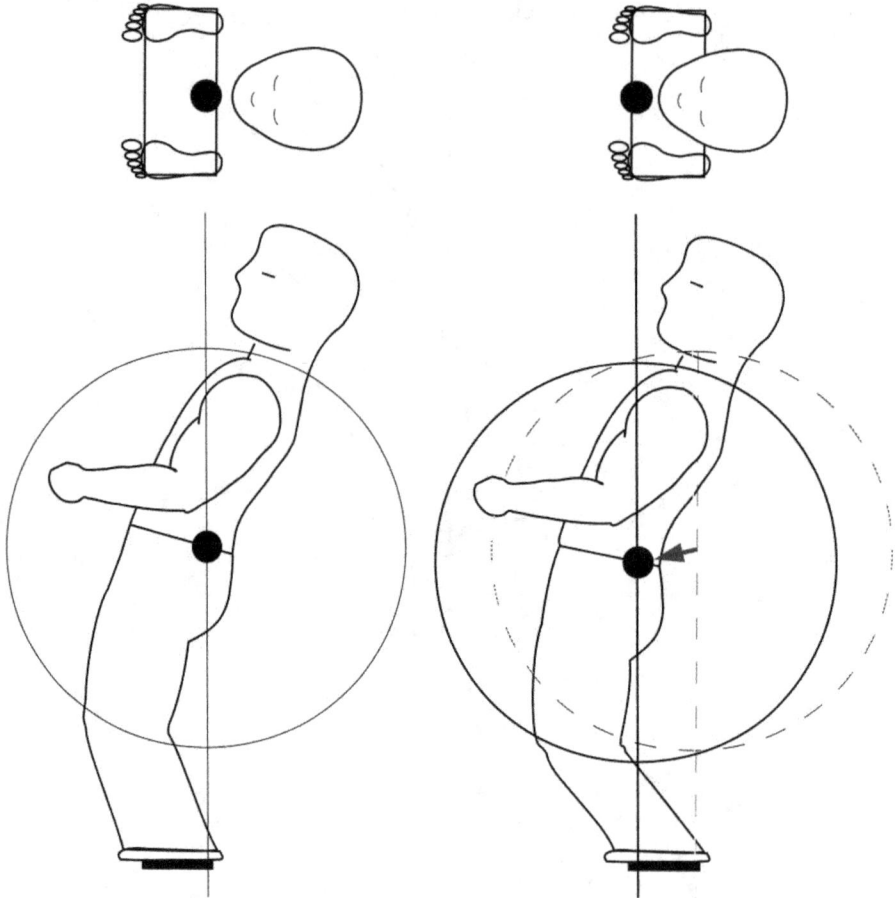

Figure 20: Top-down and side view: Uke in a 'back teeter', (left) pinned to back of the rectangle; and (right) pinned to front of the rectangle. Arcs shifted

Let us focus on the arc in a back-teeter. Look closely at Figure 20.

The Teeters: Teeter and Back Teeter

The differences here are subtle and that is why you likely did not notice this small detail in real life practice.

Take note of the difference in the position of the COB in relation to the 'rectangle' of Uke's stance (front or back side), and just as importantly, where the arcs are.

The arcs formed when Uke is pinned on the back of the rectangle (double weighted) stance (left side Figure 20) do not reside in the same space as the arcs that show up when Uke's COB is pinned on the front of the rectangle. The center will have shifted as much as eight inches or more depending upon the size of Uke's feet, the arcs will follow along as well.

This truly subtle shift radically changes the dynamic of the throw.

Ever have Uke melt like butter one time and then offer a great deal of resistance the next while in this back-teeter pose?

If so, it is highly likely that you were 'on the arc' when it worked, and when it did not work, at minimum, you were cheating the arc, if not completely on some other path.

Now, it is true, pinning and spinning is not easy in any situation, but it can be especially tricky for a back teeter (even if you see the differences we just spoke of.)

Here is why...

Back Teeter requires a spinal lock. No two Uke's are exactly the same. The variances between when different Uke's spines are 'locked' in a backbend can be great, and that is a major complication. It is almost impossible to spin Uke's COB in this 'back teeter' position if we have not taken the slack out of Uke's back. So, what do we do?

Let us not despair. Let us leverage our new awareness and turn it into a drill that helps us gain proficiency at finding the arcs within a back teeter.

(Before we start: Warning: Do not test the upper limits of resistance Uke can generate in this pose. Avoid back strain at all costs.

Staying bent backward with Tori pushing down on you will hurt your back and you only ever get one spine. Do not chance it! Protect yourself, perform Ukemi.)

Find an Uke you are compatible with and have them alternate their pose between the back teeter pinned to the front and to the back of their (rectangle) stance. (Just like Figure 20)

While knowing exactly which of the back-teeter pins Uke is in (Front or Back side), keep Uke pinned there and experiment by initiating a technique along the arc; e.g., Aigamaeate (#2) or Gyakugamaeate (#3).

You may find that, at first, you need Uke to lead you to the place where they feel the fall developing. That is the path of the arc. Learn to feel for it yourself, step back, and take a look at where it is, watch others perform the drill, jump in and play Uke/lead Tori to your arc.

I suggest it is easiest to start practicing with an arc that is in the same line/direction as Uke's 'no-line'. Try other arcs later. Eventually try other techniques, but for now…

This practice is tricky at first, but you are on the right path as you discover it is increasingly effortless to spin and topple Uke. (Getting easy? Quick! Mix it up! Get a different Uke!)

Even with all this excellent insight and assistance, there still exists a place to get stuck here and it involves that last nuance I spoke of: belly versus tail weighted-ness in Uke. (Third key understanding)

Let me guess…

I will bet you are having more issues finding an arc when you are pinning Uke on the front of the rectangle than when at the back.

For many, the first issue is that we are trying to spin Uke in a 'backward' direction, and that can easily lead to relaxing the pin on the front of Uke's stance; i.e., we move Uke's COB from the front of their rectangle into the field of stance. That truly is an issue, but one I think you can easily identify and deal with. Let us look at the less obvious culprit that complicates this back-teeter position: 'belly' and 'tail' weighted.

The focus here is in the position of Uke's COB in reference to the coronal plane (Front/back) of Uke's body. You see, although we have pinned Uke's center to the front of Uke's stance, if we want to consistently make Uke fall, we have to assure that we have Uke's COB stuck in the tailbone.

The Teeters: Teeter and Back Teeter

If we leave Uke's COB in the front the body (in the belly; aka 'belly weighted'), there is going to be a bit of slack in the lower spine and Uke can leverage a shift in the hips to gain the wiggle room necessary to step out.

Study Figure 21 until you are vividly aware of the difference(s).

Experience this for yourself. Try a bit of a 'limbo'. Bend backward a bit, put your weight into your belly, and position that weight at the line of the balls of your feet. (Left side Figure 21)

This is where the good limbo dancers put their attention. It is a lot more forgiving on the lower back and that is why most Uke will naturally assume this pose. Our bodies do not want to be tail weighted in a back teeter and our brain/reflex knows it (even if we do not realize it consciously)

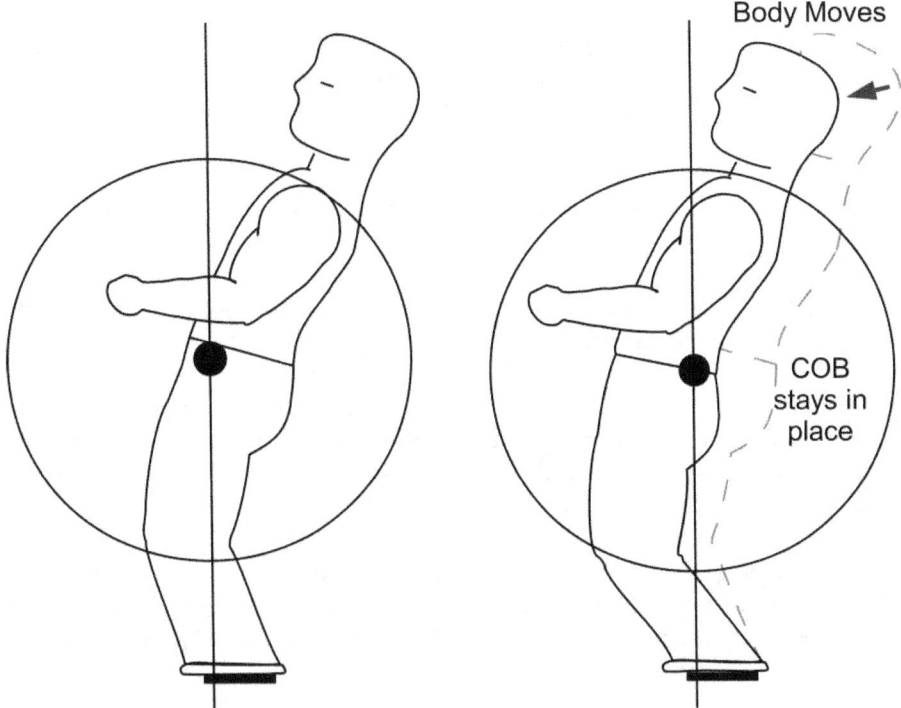

Figure 21: Uke in a back teeter, pinned to the front; 'belly weighted' on the left, and 'tail weighted' on the right. Note the arcs are shifted in relation to Uke's body.

Basic Terms

Uke changing the tilt of their hips when shifting from belly to tail weightedness does not translate much in the way of a vertical shift and it does not do much in the way of bending their knees. As Tori, you might not be able to sense that anything has happened, but that little shift moves the 'arcs' in a drastic way. Study Figure 21 again and come to the realization that the orb dynamically shifted to a completely different reference within the body.

Tori can be tricked into a false sense of success by sensing the willowed Uke in a belly weighted condition, not realizing that Uke still has the tail weighted option to escape to. Non-compliant Uke's will use this shift from 'belly' to 'tail' to avoid the throw. Although this action is a bit of a sacrifice, in that it moves Uke into a more vulnerable position, it shifts the arcs enough to make the actions of a non-aware Tori incongruent with the arcs. The common result is that Tori ends up pushing Uke out of the stance and likely into a back step.

Take your time and play till you understand the subtle nuances in Uke's back-teeter position as they are subtle and will significantly determine which arc you must follow for a pure spin (and achieve a true Aiki throw!)

Twisted

Another shape that needs no introduction and thusly almost no explanation, 'Twisted'.

Put simply, it is a twist in Uke's spine, and it can be used solely on its own or to spice up a 'Box' or 'Teeter' distortion.

Your torso is a lot stronger when your shoulders are above your hips. The reason is that your buttocks and abdominal wall can best work together to support your frame if your body is aligned. The last thing you want to do is to bear a great deal of external weight while your spine is twisted.

With that thought in mind, I will ask that you remember to keep your frame aligned, but feel very free to misalign Uke. Placing even a modest amount of weight on Uke's frame, while twisted, will cause Uke to fall. Apply liberally.

Checked

Here is the short version so you can get out quickly before the explanation gets a little esoteric and I go on a little rant.

Put succinctly: 'Checked' means you have removed Uke's desire to harm you; i.e., Uke is 'Willowed'

Checked

If you are good with just that explanation, and figure you already know where this explanation is heading, skip the rest of this and proceed to the next section of the book. This is the last of the 'basic terms' and technically, it really is all you need to know to before we get into the real meat of this book,

but…

if you want to explore the reason we use 'desire to harm you', instead of 'ability to harm you', keep reading, but I warn that you might find it reads a lot like an Aiki blog you would find on the web (flame on! I am going to have a bit of my own fun as I write this. I have a warped sense of fun when it comes to writing.)

Last chance to skip this…

OK, if you are still reading, it is your fault and you lose the right to comment on the part of the book that got a little didactic: 'intended to teach, particularly in having moral instruction as an ulterior motive'. (Yeah, I found that word in the thesaurus and had to consult the dictionary too!)

As we learned in the introduction of these poses, 'Checked' is more of a 'condition' than a pose.

Each of the poses we just described offers a good bit of temporary physical dominance over Uke, but if we are not careful enough to create 'willowing', Uke might still be able to strike you. That typically means that Uke can hit you with some body part other than the one you are directly manipulating/impeding. That is not safe.

So, let us explore the 'Checked' concept by asking, 'Where is the safest place to be when engaged with Uke in a confrontation?'

Most persons I have asked have told me something along the lines of, "Get to the side and slightly behind Uke, like you were in their back pocket; this way, Uke has only the near side of their body that they can attack with, and that near side is limited in its usefulness"

That answer gets about half credit.

It is a well thought out answer, but the question "where is the safest place to be…" is really a trick question that helps reveal where you are in your understanding of Aiki(do) theory. (and might reveal a lack of awareness of the usefulness, to Uke in this case, of Karate's Naihanchi katas!)

Basic Terms

The answer we were seeking to "Where is the safest place to be when engaged with Uke in a confrontation?" is: 'In control', even though 'Control' in this sense, is itself a bit of a misnomer. (After all, control is just an illusion. Right? I mean, it must be true that there is no such thing as control. I overheard a tortoise tell that to a bear-cat under a tree. [I will watch any movie about martial arts. What can I say? I love Kung Fu!])

The more precise words for 'Control' are 'temporary physical dominance' and that certainly exists, especially if we make the rules.

The two rules are: opportunity and motivation.

The ultimate 'control' (temporary physical dominance) of a situation would be to remove opportunity; i.e., 'be elsewhere' and never end up in the confrontation in the first place.

This first, basic tenet of behavior modification: 'completely remove the opportunity to perform the action stops the action' is absolute and indisputable in its effectiveness.

The down side is that, with zero opportunity for an altercation, it also means there is zero opportunity for any other magical things in life to happen. (Boring! Is not this why we were created in the first place? 'Someone' resolved that 'us' hanging around with 'other than Us' is infinitely better than being alone. The miraculous part is that when Wu-chi moved and became Tai-Chi, unity remained unblemished. Nothing was separated, instead all things were made a part, not apart; given identity and an opportunity in the grand design... I digress. Didactic right?)

So... since we intend to live amongst other humans, and every other human can be an Uke, Uke will have some amount of opportunity to hurt us. Life just got risky. How do we mitigate this risk? Motivation!

Some will argue that if we can 'slip a punch', then we have effectively created the same condition as 'not being there', and are 'in control'. We are on the right track, but not getting hit is still not a pure Aiki condition.

All martial arts block and/or reposition to avoid an attack, and, if this is true, why then, would we not add 'Aiki' to name of every art?

The reason is because Aiki does something rather extraordinary, and not every art takes advantage of this focus: Aiki does not rely solely on physical position or restriction; Aiki attacks the root of the problem: Uke's motivation.

Checked

To prove this point, let us raise the stakes and set a goal that, with Aiki, we can be standing anywhere, even directly in front of Uke and be completely safe and 'In Control'.

Before you call me insane, know that this goal is achievable, and in the 'Mojo', we call this condition 'Checked'. So, what exactly is checked?

Remember that quick explanation at the start of this chapter? ('Checked' means you have removed Uke's desire to harm you.)

Well...

Where there is a will, there is a way. That means Uke's weapons are not the real danger, it is Uke's intent that makes them dangerous. (Weapons do not actually exist. Only tools do, Uke is intentionally using a tool as a weapon.)

Barring accidents, (if you believe in accidents; i.e., there are no accidents), Uke is not going to harm you until they make up their mind to do so.

Your mission is to steal Uke's mind and that is nothing short of incredible.

"Why?"

Altering someone's motivation; i.e., 'changing their mind', is a very daunting task. No one can 'force' someone else to do anything.

Literally (and in this case I do literally mean 'literally'), you could threaten someone, their family, their finances, etc. but that is all influence, not control.

Try to 'force' someone to walk across a room and it will not happen unless they decide to walk across the room. They only ever move when they have decided that the influence is great enough to motivate them, but that is their decision, not yours. Control of others = illusion; (and peach pits do not produce apple or orange trees; that turtle was right!)

What can we possibly offer Uke that could influence them from mindful focus upon attacking us to a condition of neutrality?

"Take away Uke's stability!"

That is correct!

Nothing focuses the mind more than the 'near death' experience of falling.

Basic Terms

Quoting Sensei Merritt Stevens, "The brain will sacrifice the body to protect itself".

I suggest that, not only will the brain sacrifice the body, your brain will also sacrifice intent and focus solely on the base chakra: self-preservation.

In this sense, Aiki is 'mind control', an instant hypnosis.

The lack of stability consumes Uke's mind and, even if they have been training how to strike when in the midst of Ukemi (and who really does that kind of training?), you have nothing to fear once you have truly 'willowed' Uke, no matter where you may stand in relation to Uke's position.

Voila!

'Temporary Physical Dominance', opportunity is there, but multiplied by 'zero intent to hurt you' because Uke is busy saving themselves = safe, you are 'In control'.

It is through physical Aiki (the control of stability and balance) that we achieve the highest order of ethical resolution to physical violence brought about by Uke's 'intentional aggression'. (the focus is on 'intentional')

Both parties leaving the engagement, not only unharmed, but no longer focused on the offensive. (Didactic! [Come on. At least smile for me, that was funny!])

'Entering into the attack' requires that we establish this kind of 'control'.

Now, even though we know that 'willowing' Uke has Uke 'checked', it is still in our best interest to use a 'belt and suspenders' approach to staying 'In control'.

Please do not rely solely upon removing Uke's intent.

Be certain to limit Uke's physical options at the same time.

'Double-check all of Uke's weapons' by physically trapping or locking Uke's extremities and...

 you might even position yourself in Uke's 'back pocket', because that is actually a very martially advantageous position to be in. (even if it is only a half credit answer!)

Advanced Concepts

Let us get started with the advanced concepts. This section will be more rewarding if you have read the basic terms, as the basic terms are really not all that basic. We touched upon many of the advanced topics, just left out some of the details. Well no more! Time to dive in!

Where 'Six Precepts' focused on the roots of the physical theory behind why Aiki works, in this book ('Codex'), we get past the 'why it works' and begin to shift focus toward practical theory of how to use some Aiki 'tools'.

There is a lot to convey in these next chapters, and the hardest part was not the content, it was how to present it. There is no one 'thing' that defines Aiki. Aiki is the sum of many moving parts. Not all of them are physical.

So, to do our best, we will investigate some of most important parts of Aiki by loosely repeating the pattern used in 'Six Precepts': 'Where Uke is weak', 'Where we are strong', and 'Advanced connection builders'. You will find a lot of food for thought in each section, but please remember that Aiki is where these parts come together.

In 'Where Uke is weak', we focus on Uke in two key areas: 'Weakness in Uke's stance' and 'Aiki Joint Locks'. The weaknesses in Uke's stance become explicit as we advance the 'No-Line and 20' concepts by perfecting the '20' before we proceed into completely new ground; revealing where Uke's stance is weak when Uke is standing on one foot, and when Uke is in a three-point stance. One foot or three, the weaknesses are related and we call them the 'Three-legged Dog Spots'. From there we (re?)define and clarify a very often misunderstood part of Aikido: the Aiki joint lock. Are you not sure what an 'Aiki Joint Lock' is? Expect to never have trouble with joint locks again and the ability to use them without hurting Uke (unless you want to).

We then turn the focus on 'Where we are strong' and look at methods to improve our overall power and effectiveness.

We start with the atmosphere of 'Presence' (not everything in Aiki is physical!) and continue with the 'intangible' by learning how to use 'Tori's Hypotenuse' as a way to multiply our overall power. For all of our food lovers, we will cover the duality within our techniques called the 'Fork and Knife', introduce ourselves to 'internal power' through 'Fascial Grasping', further the 'internal power' study as we solidify ourselves restfully in Zhan Zhuang (we practice standing to better learn how to move!), and finish maximizing our ability to be powerful by revealing the key coordination/skill that starts us along the path to putting the power of our hips into our hands; it is called 'the transmission'.

In the final section, we explore 'Advanced Connection Builders'. Our goal is to reveal the rest of the path to putting the power of the COB into our hands. We start with 'advanced elephant arms' to find a new friend: 'The Albatross'. We then progress to expose the link between levers, calligraphy, and swordplay as we explore the 'Broom and Shovel' as a method to temper our 'sword hand'. Our punches, sword cuts, and jo strikes are also blocks, so we take what we learned of the brooms and shovels and apply them as our shield. We reach our final destination in the path to putting power into our hands (and learning how to cut!) when we explore 'Rolling'.

If having 'Aiki hands' was not enough, we will end our discussion with uncovering the true foundation of 'Circular Aiki' as we bring to light our 'Ba Gua Zhang' roots in something we call 'Orbits'. May you never step straight again after you learn how to become a catapult! (Trebuchet)

Along the entire way we will be pointing out some of the roots of 'internal power'. Where I feel confident in my opinion, I will point out cross-overs to the Chinese arts of Hsing-I, Ba Gua, and Tai Chi Chuan. Much of my understanding of Aiki comes from Sensei Carlisle putting me on a path to thread together and understand the progression between those Chinese arts (arts that I admittedly, only tangentially, even cursorily, study; for I apply their insights to Aikido practice, not the other way around). I (much) later learned that the progression stems from evolving through 'Zhan Zhuang', to Hsing-I, then Ba Gua, and finally reaching Tai Chi, but I am getting ahead of myself. More detail revealed as we progress.

Keep your eyes wide, mind open, try each concept as you understand it. Most of all, take it slow, none of these concepts are trivial by any measure.

Here we go...

Where Uke is Weak

The Secret of the 20

The power of constructive criticism! Since the publishing of 'Six Precepts' it has come to my attention that I should have included a bit more detail on the mechanics and usage of 'The 20'.

The critique came, more than once, when discussing how to use the '20' when performing both Gyakugamaeate (#3) and Gedanate (#4). The question was, "Why did you not explain the '20' like that in 'Six Precepts?'"

That is a fair point. Please allow me to make up for that miss…

The '20' is invaluable to us when we need to move Uke's COB from side to side. The ability to move Uke laterally is critical to many Aikido techniques as it is the source of how we control Uke's weighted-ness (Shifting/controlling Uke from singular to double-weighted, and vice versa)

For example, the '20' simplifies the actions of throwing Uke by 'leading them forward and out over Uke's forward-foot's little toe' or 'driving Uke backward on-angle over their rear-foot heel' (in both cases, when Uke is in a forward foot stance, of course)

Issues often arise when we are challenged to move Uke to and through these 'toe' and 'heel' points. Many beginners are stumped and often cannot get Uke to shift weight at all, even we 'more experienced' Tori will find times where Uke puts up a great deal of resistance to our intent (and then the next time none). Let us get consistent.

All we need know is that the easiest way to move Uke's COB from side to side is to exploit Uke's 20! "How do we exploit Uke's 20", you ask?

In building awareness of the '20', we must first call to mind what we know about Uke's No-line: that the No-line is a singular line that runs between Uke's legs and bisects the field of Uke's stance (splits the field into left and right sides). (See Figure 22)

Uke cannot resist as we move their COB 'forward' or 'backward' along the No-Line path because the No-Line is an inherent flaw in being a biped. (No third leg to keep us stable)

Where Uke is Weak

The net of the situation is that there is only one 'No-Line'; it is easy to find and requires only a little exertion on our part to use.

To get the same ease of use out of the '20', we must realize that the '20' is actually two lines and using them requires 'willowing' and 'scratching' (as described in the end of 'Cheating the Arc' in this book's 'Basic Terms').

The two '20' lines are the 'front' and 'back' of Uke's stance. (See Figure 22)

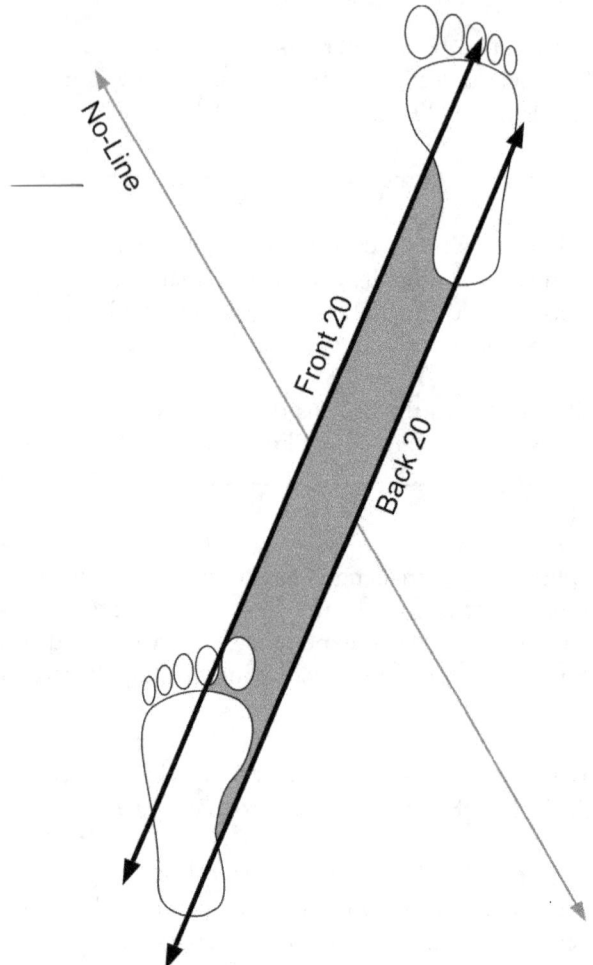

Figure 22: Secret of the 20: Uke's COB has to 'Scratch' along the path from Uke's 'heel to heel' or 'ball of one foot to the other'. Use the No-line to get Uke's COB to a '20' line and then keep it out of the gray area ('the field') as you move Uke's COB to one side or the other

The Secret of the 20

The 'front 20' is the line that runs between the balls of Uke's feet, the 'back 20' is the line between the heels of Uke's feet; (i.e., chop off the back of 'the diamond' shape from 'Six Precepts' and re-imagine Uke's stance as a parallelogram to find the 'back 20' running between Uke's heels.)

If we want to throw Uke in a typical 'forward and over the little toe' fashion, we have to first pin Uke's COB to the 'front 20' and keep them there as we 'Scratch' their COB along the 'front 20' path.

How do we first get Uke pinned against their Front 20? By moving Uke's COB along the No-Line! Simply shift Uke's COB forward along the No-line until it reaches their 20. Easy to do no matter how big Uke is.

Once Uke's COB reaches the 'front 20', they are willowed, and we discover that Uke is easy to guide sideways provided we keep Uke's COB 'scratching along the 20'; i.e., we do not allow Uke's COB to drift into the middle of Uke's stance (a.k.a. 'the field' of Uke's stance - Gray area of Figure 22)

Here is a quick drill to help illuminate the point.

Let us start to build contrast by first focusing on what does not work well.

Have Uke assume a right foot forward stance with their COB in the middle of their stance; (i.e., 'rooted in' a bit) Now, have Uke's give you their right arm and lightly pull away from Uke along the line formed between the arches (middle) of Uke's feet. Uke should still be very stable and difficult to move. (Figure 23A)

When we try to move Uke in this way, we are trying to drag their COB through the middle of their stance (trying to move Uke through the gray area; i.e., 'the field' of Uke's stance).

Except along the No-Line, moving Uke's COB through the 'field' of their stance is not easy. (Read that last sentence again, it is very important to fully 'own' that understanding)

Let us try again, this time, ask Uke to shift their weight to the front of their stance; (i.e., ask Uke to pin themselves forward; get on the balls of their feet). Follow along with Uke and shift your pull forward to align/overlay with the line formed between the balls of Uke's feet. (Figure 23B)

You will find Uke moving sideways along the front 20 with the same kind of ease you have when moving Uke forward or backward along Uke's No-Line.

Where Uke is Weak

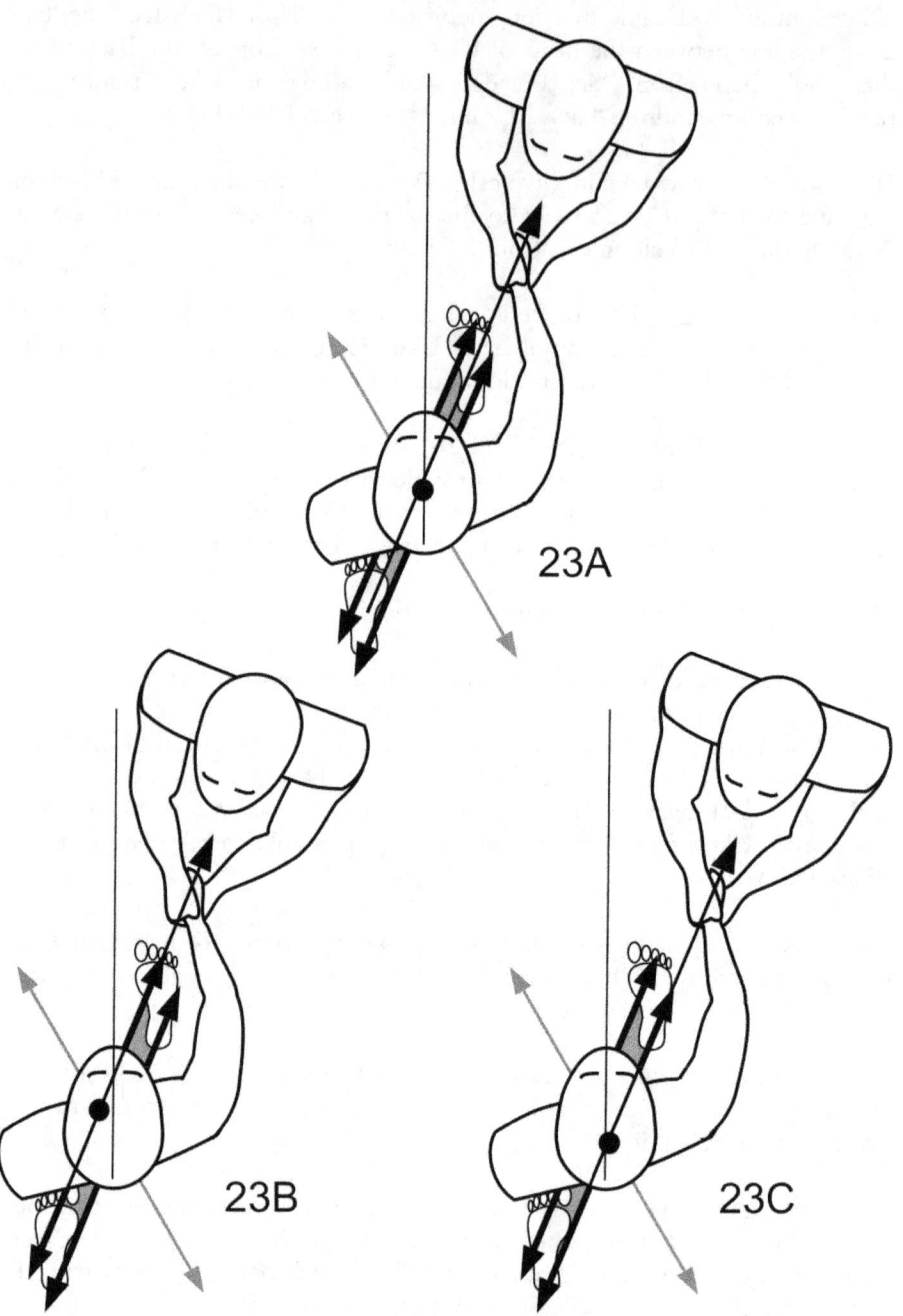

Figure 23: Finding the two '20s' using extension connection

The Secret of the 20

Expect Uke will not like it and most Uke will immediately un-Pin themselves (i.e., only allow a short scratch), but you now have graphic and direct experience of the 'Front' 20. (Hopefully your Uke will comply and let you feel scratching until the point that they are on one foot; i.e., a 'full length scratch'. Even if they do not, you will get to sense longer 'scratches' when you are leading; i.e., as you practice your techniques)

To find the 'Back 20', reset Uke's stance and set up with a pull; this time have Uke shift their weight (COB) to the line of the heels. Again, shift your pull slightly to coincide with the line that cuts across Uke's heels and voila! You have found the 'Back' 20. (Figure 23C)

Subtle stuff huh? A game of inches really.

Aiki (Aikido) is very much a study in placing weight and/or the correct pivot in just the right spot to gain mechanical and structural advantage!

One quick note before we move on to compression connection: did you notice the path Uke used to Pin themselves? (Hint: It is a super safe bet that Uke moved their COB along/through their No-Line to reach their '20')

Try pinning yourself to your '20' lines (i.e., go to the balls or heels of your feet) and experience your own natural tendency to use the No-Line line to get there. Improve awareness of your own No-Line by moving your COB toward and away from either 20 in various stances. Practice in a mirror to literally 'see' it in yourself, so you can begin to 'see' it in others.

Let us get back to it.

Now that we have tried the 20's using extension connection, let us try it with compression; (i.e., we move toward Uke rather than away).

Have Uke assume that right foot forward stance again, but this time, with hands at their sides. Put your hands on Uke's shoulder and take a moment to try pushing Uke away from you, toward their left foot instep.

This push will be a bit trickier if Uke's torso is not aligned with their '20'. It hardly ever is, so you will likely have to rely on parallelism with Uke's '20' and a bit of finesse to make the push (and not spin Uke).

Pushing Uke 'toward the instep' (i.e., pushing Uke's COB along the line between the arches of Uke's feet) equates to us trying to push Uke's COB

Where Uke is Weak

through the field of their stance (Figure 24 or ala Figure 23A, but this time moving in the opposite direction; i.e., via a push)

You will soon discover how rooted Uke can be. Even a small Uke can offer massive resistance when their COB is in the field of their stance.

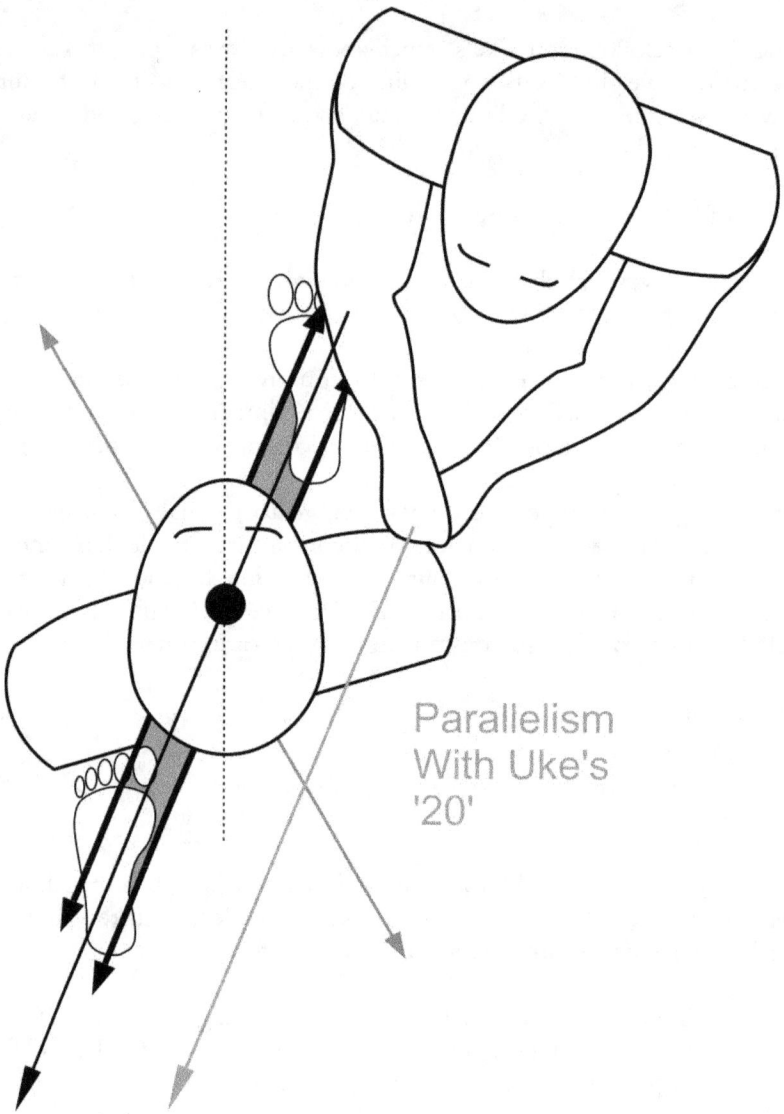

Figure 24: Pushing Uke in the direction of the '20' but through 'the field' (gray area) of Uke's stance (Uke can offer plenty of resistance)

The Secret of the 20

Let us find greater success by using Uke's front 20 this time.

Put your hands on Uke's shoulder again. Get ready to push into Uke along the Front 20 (line across the balls of Uke's feet), but instead of asking Uke to shift their weight forward to the front 20 and into the Pin, enhance your practice by leveraging 'Parallelism' along Uke's No-line to move Uke to the pin yourself; (i.e., move Uke along their No-Line, then along their '20')

Start slow and feel for Uke's COB following the path through the No-Line until it comes to rest against their front 20 line, and then 'into the 20' path.

You should find that the path is never two straight lines with a sharp turn.

You will start sensing the beginning part of the 'footballs' as you round off that corner by simultaneously shifting Uke's COB through their No-Line and '20'; eventually aligning their COB with their 20, where, from that point onward, you will then 'scratch' Uke's COB along the '20'. (See figure 25)

Reset for the drill and try compression on the Back 20 (again, you, not Uke, should be the one that shifts Uke's COB to the back 20 via the No-line).

It is a short trip to get Uke's COB to the 20. Be sure to sense the 'rounding' as you moved from No-line on into the Back 20.

Easy to do even with big people is it not?!? (No one can resist the No-line!)

This 'back 20' compression version of the drill is the root of what I consider the 'basic' Gyakugamaeate (#3). You will know exactly what I mean if you re-try this 'back 20' compression drill, but this time with your arm across Uke's chest (instead of on Uke's shoulder).

There is a good bit more to know about Gyakugamaeate (#3) than just exploiting Uke's 20, but you can get a lot of mileage out of this quick drill.

Push against Uke's hip and you have the root of the basic Gedanate (#4).

Food for thought:
- Which techniques use the extension versions of this drill? (Do not rule out Gyakugamaeate (#3) and Gedanate (#4)!)
- Can those techniques be done on both the back and front 20's. (answer: Yes! Of course!)
- Have you been using the 20's already, but without cognitive awareness? (likely!)

Where Uke is Weak

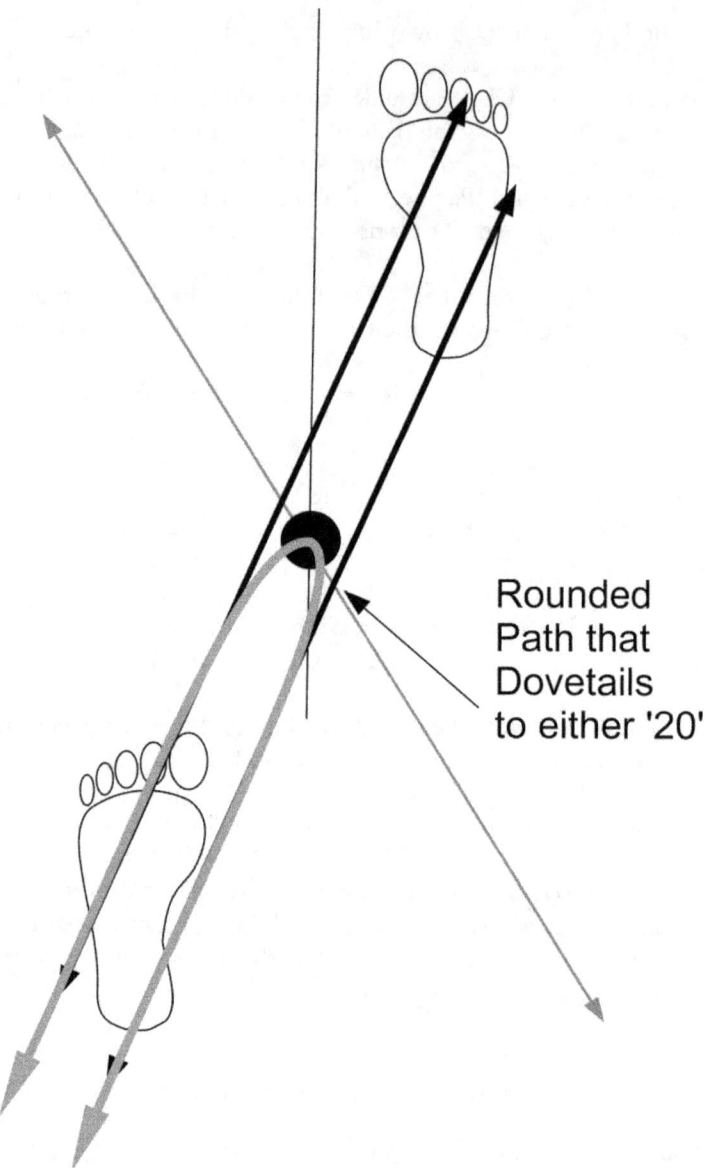

Figure 25: The rounded path that forms as we move Uke simultaneously along the No-line and '20' paths (until we pin Uke's COB and continue to move solely along Uke's '20')

- Have your students/partners? (varies)
- Could you coach them now? (of course; it is likely that you could have even before you read this explanation too! Pfft!)

With a small bit of practice, you will soon be a master of Uke's 20, same as you likely already were with their No-Line. Remember to feel for the rounded path used to bring Uke to their '20's and you may even begin to find you can move Uke from one '20' to the other at will. (That is the goal!)

Now, you may have noticed that moving Uke into the corner of their stance has high likelihood of putting Uke on one foot.

That changes the game quite a bit.

So, let us talk about what happens when Uke is not on two feet...

(Three-legged) Dog Spots

The good news is you are now an expert on the nuance of application of the No-line and 20. The 'footballs' should make excellent sense now too.

The bad news is that the No-line and 20's disappear when Uke is predominantly weighted on one foot (let us define 'weighted on one foot' as Uke supporting three quarters or more of their weight on one foot).

Where do the No/20 lines go? They vanish! How do we draw a 20 line between Uke's feet when one of those feet is not really committed to being on the floor?!?! The same issue applies to the No-line.

That fact is kind of scary to the person that solely has expertise in leveraging the No-line and 20's. Uke is on one foot a lot. I mean a lot!

Think about it. Most of us never really stand double-weighted.

Here are just a handful of ways in which we find ourselves on one foot.

The next time you are in line waiting for anything, stop and take notice of how people stand. Take notice of yourself! How long do you really spend in that line with equal weight on both feet?

When sparring, ever get into a cat stance? (Smart move!)

How about when stepping? When stepping, you are supported by only one foot (75% of your weight or better) for most of the action.

What does Uke often do to avoid being thrown? Step, of course!

Where Uke is Weak

Putting Uke on one foot is a martial goal! A single foot stance can be most unstable; i.e., We can improve the willowing if we reduce Uke's base to just one foot. (The trick is learning how to keep Uke from stepping.)

Let us additionally factor in that, as Aikido practitioners, we sadly spend too little time training to use our legs/feet to perform trapping, blocking, striking (kicking), and sweeping.

Practically all of that action requires a predominantly single weighted stance.

Let us always assume that our opponent is practiced in those actions and ready to attack us with their legs, thus they are putting themselves on one foot intentionally!

When I take the time to take notice, the only time I typically see persons constantly in a double weighted stance is when engaged in drills that do not account for active sparring situations; e.g., during Hubad-Lubad or push-hands drills.

In active combat, we are constantly shifting our weight. (and then there are those that are constantly hopping while sparring. Disconnecting from the floor completely is strategy for some.)

So, let us trust that Uke is going to be primarily weighted on one foot at some point in our experience, so we had best become experts on where Uke is weak when on one foot.

So enter the 'Three-legged Dog spots'!

There are four 'Three-legged Dog Spots' that represent the directions in which Uke is most vulnerable when on one foot.

The name implies that these 'spots' are derived from the behavior of a quadruped (e.g., dog) that is missing a front leg.

Ever see a dog that is missing a front leg try to stop quickly from a speedy run? They do not stop; they fall forward into a roll.

Why is that germane to Aiki? Our brain will commit us to reverting to a quadruped before it gives up the fight for self-preservation; (i.e., unless you practice Ukemi, your hands will reflex-ably go outward to break your fall, your brain would rather you crawl than fall; break your arm before breaking your skull).

(Three-legged) Dog Spots

Since we fully intend to have control over at least one of the Uke's appendages, the analogy of a 'Three-legged Dog' fits, but do not get too smug that we yet understand this flaw in the human anatomy.

The reason these 'Dog Spots' work in a human has nothing to do with why our tri-pedal pooch is challenged by a 'quick-stop'. The dog's issues are momentum based, we instead, will be discovering an issue in the construction of our hips in relation to the overall weight of our bodies...

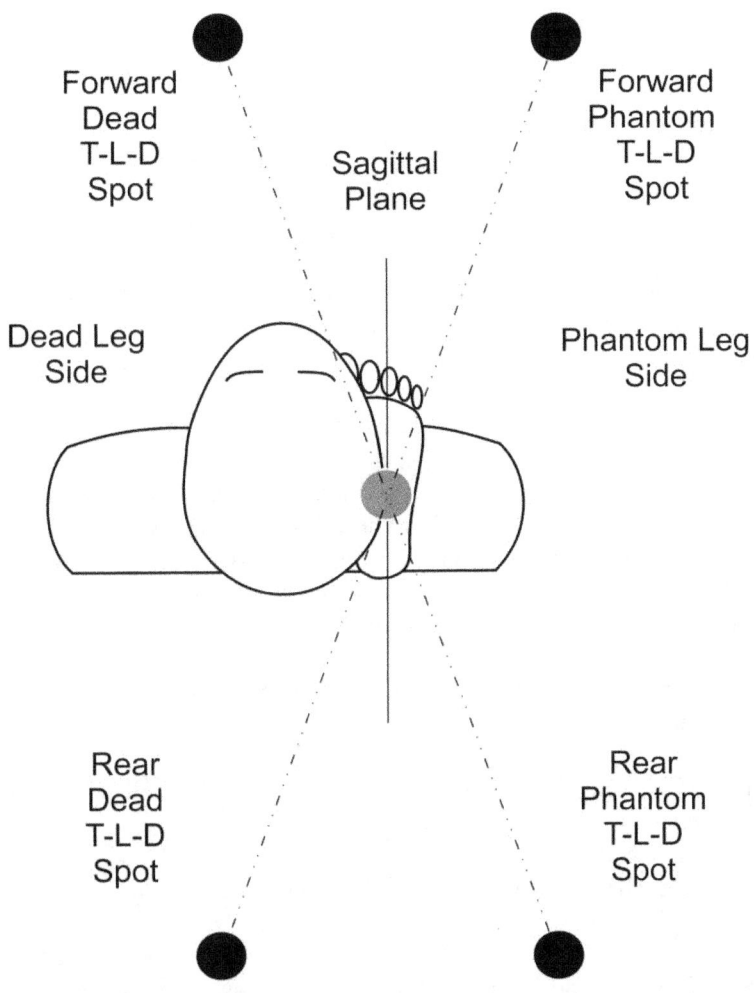

Figure 26: Uke weighted on their right foot and the associated 'Three-legged Dog Spots' (Black spots). There are two on the 'Dead Leg' side (in this stance, left-side) and two more on the 'Phantom Leg' (right) side.

Where Uke is Weak

Take a peek at Figure 26 to find that two of the Dog-Spots appear on each side of Uke's Sagittal Plane (our version of the Sagittal Plane, where the Sagittal Plane follows the COB) and that the COB resides over the weighted foot. (If the COB is not over Uke's foot, Uke falls!)

We call the side with no leg (just Uke's arm; right side of Figure 26) the 'Phantom-leg' side. The other side is called the 'Dead-leg' side (the other leg is there, but you will find it has no committed connection to the floor)

The Phantom Leg side is easier to explain and experience, so let us start there.

The Phantom TLD's

The Dog Spots sit roughly twenty degrees off the centerline of the Sagittal Plane (that is currently bisecting the foot). Each Dog Spot is as just a slight bit further from Uke than the length of Uke's spine (torso).

Hmm... the Forward Phantom Dog Spot is roughly twenty degrees off of the Sagittal plane… That is practically on the same line as the Uke's '20' when in a front foot stance!

Is there a connection? Yes, but there is a catch too!

The very similar paths of the Forward Phantom Dog Spot and the '20' when Uke is in a forward foot stance (ala Figure 25) is advantageous, as it allows us to flow somewhat seamlessly between moving Uke sideways along either of the two-legged '20' lines and on into this new line that forms once Uke is on one foot; but the flow is not key to success…

A little trick must be applied if you want a high rate of success in throwing Uke on either of the Phantom side Dog Spots.

The trick? Uke's hip-kua (inguinal crease) must be locked; (i.e., either fully closed or open)

Let me explain what that means by helping you experience it.

We will start with using the Phantom spots incorrectly (no lock in our hip/kua). There should be little danger in exploring the Phantom spots with an un-locked hip, for we should be able to step with relative ease.

We can then progress to using them in the manner that will offer an irresistible desire to roll (locked hip/kua). Get ready to fall.

(Three-legged) Dog Spots

Without the lock, easy to step and stay off of the floor…

Stand on your right foot with your right hip-kua in neutral. (Neither open, nor closed; your hips and shoulders square with your foot, ala Figure 26; i.e., the diagram we just viewed).

Keep your back straight, hips and shoulders aligned, and bend at the hip to tip your head toward the forward phantom side Dog-Spot (Try touching the Forward Phantom TLD with your right hand).

You should find your body falling toward the Dog-Spot, and that it is rather easy to keep from falling by simply stepping in that direction; (i.e., putting your left leg/foot directly onto the Dog Spot).

Let us try this again, but in this attempt, take a bit of precaution by finding something to lean on with your hand(s) or at least be certain that nothing is in your way if you want to practice Ukemi and fall onto and through that Forward, Phantom side Dog-Spot.

When you are ready to fall or support yourself…

Stand on your right foot again, but this time, close your right hip-kua; (see Figure 27) i.e., keep your right foot in place and rotate your hips clockwise until they can go no further, which will result in your right foot being in a 'pigeon-toed' position.

(You may decide to place your left leg in manner that completes the 'toes touching' pigeon-toed stance, just be sure to maintain the 'single weighted on the right foot' stance. Keep as little weight on the left foot as you can coordinate. Remove any spinal twist – i.e., keep your shoulders and hips aligned.)

Again, with that straight, untwisted spine (shoulders over hips), bend at the hip to tip your head toward the Forward Phantom Dog-Spot.

Yeah, that left leg flies backward pretty quickly, and if you decided to perform ukemi, you pretty much looked like our three-legged best friend trying to make that hasty stop.

The other alternatives were that you used something to hold yourself up, or your body unlocked (i.e., opened) that closed right-side hip-kua to regain the ability to step. The locked hip is the key!

Where Uke is Weak

Go into the pigeon-toed stance again and (safely) try tipping your head toward any direction other than the Phantom side Dog-Spots and you will sense some desire to fall, but nothing near the intensity of the Dog-Spots.

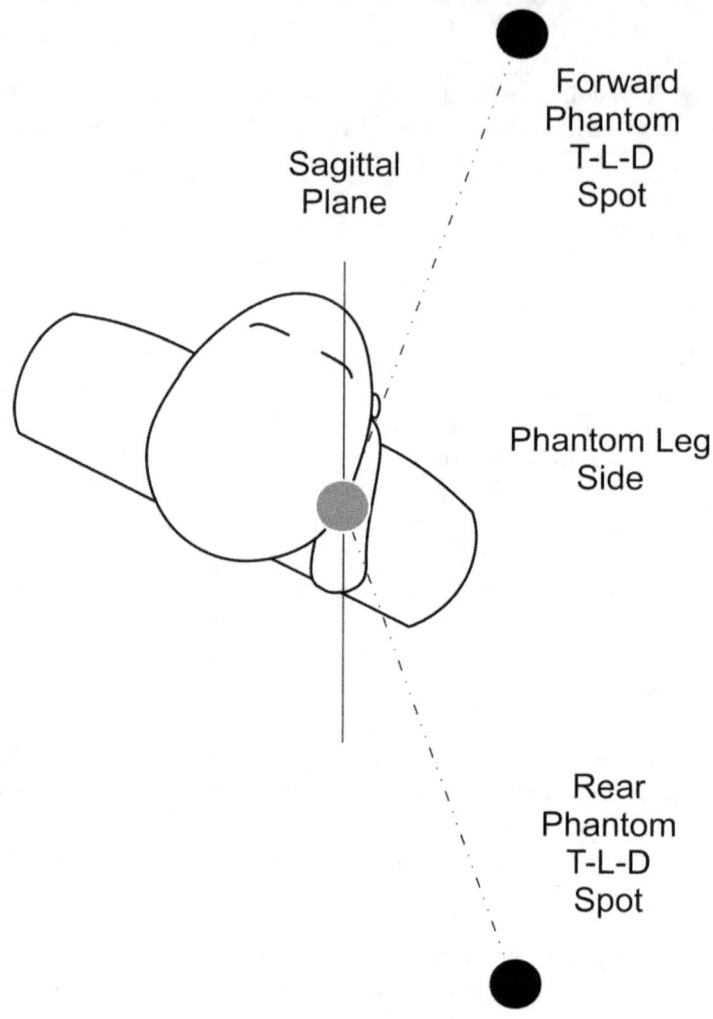

Figure 27: Us weighted on right foot with a 'closed' right hip-kua.

Why does this work this way?

Because a locked hip-kua on the weighted side prevents the floating (non-weighted; in this case 'left') foot from reaching that Dog Spot; (i.e., there is a true physical impediment keeping that left leg from moving forward.)

(Three-legged) Dog Spots

Since that non-weighted leg ('dead leg') cannot go forward, your brain senses your head moving toward the floor and it will trigger your left leg to swing backward to counter balance your torso.

Basically, this drill leverages your reflexes to turn you, from head to left foot, into a propeller that spins at the right-side hip joint.

The counterbalance is not perfect either. Your left and right hip joints are a few inches apart and that keeps your leg from being directly opposite your falling torso. Any lean toward that Dog Spot will force you to tumble.

You just got to love attacking Uke's reflexes! It is irresistible.

We could sense the Rear Phantom Dog Spot by doing this same drill going backward off the heel, but I do not suggest it unless you are truly in an area where you can fall safely.

If you must... The safest bet is to stand with your right side closest to a wall.

Put your right hand on the wall to support yourself. Keep the right foot in place as you turn your torso counter-clockwise, left-ward as far as you can go (spine straight and untwisted, shoulders over hips) and put your heels together in a 'duck feet' stance (toes outward to the sides.) (See Figure 28)

Your right hip-kua should now be wide open. With all of your weight on the right foot, ever so slowly and lightly, allow your head to tip backward toward the rear phantom Dog-Spot.

This method takes very little motion to cause a fall.

If you were wise and successful at being safe with the Rear Phantom Dog Spot test, set yourself up again with the 'duck feet' (toes outward toward the sides), but this time tipping your head toward any direction other than the Rear Phantom TLD and note that it will not have the same effect.

Continue the experiment by performing this drill while alternating between neutral and locked hips, and also leaning toward or away from either of the phantom side Dog Spots. The goal is to have full command of what works and what does not.

Now before you suggest that this is too difficult, or not practical in combat, know that Chris Parkerson absolutely loves to throw persons in a forward dimension and uses the forward-phantom TLD as a mainstay.

Where Uke is Weak

His focus on controlling Uke's hips accounts for much of the aerial acrobatics I have had to endure and he does it at speed!

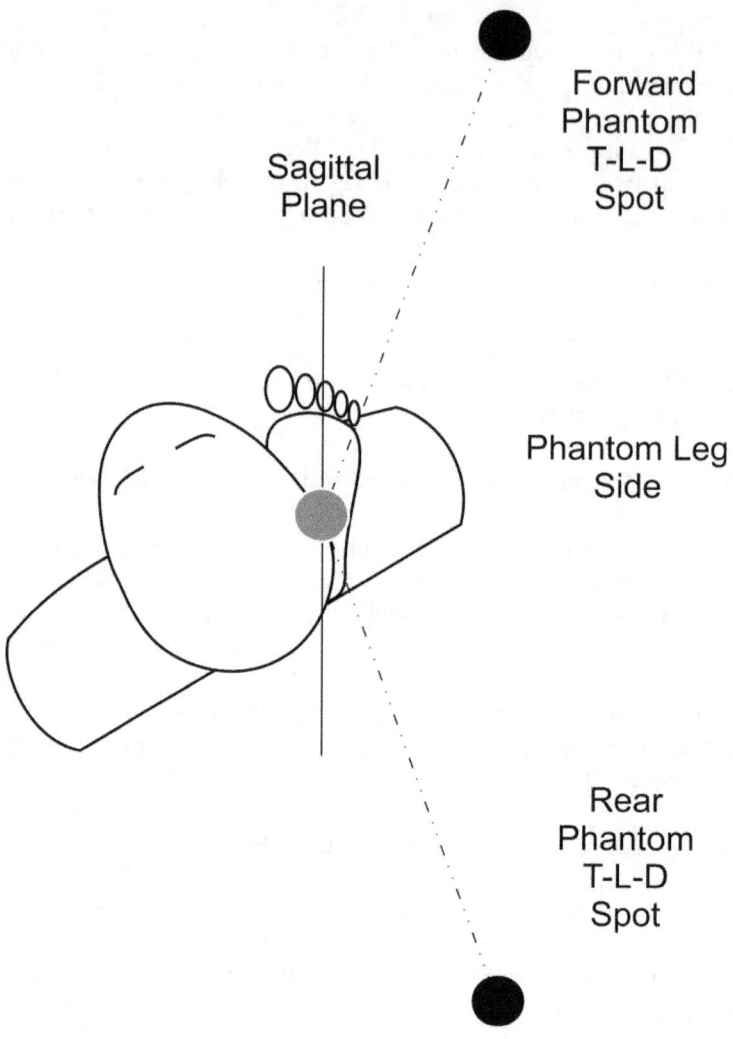

Figure 28: Us weighted on our right foot with an 'open' right hip-kua.

That rear-phantom TLD spot is a source for success in throwing Irimi-Nage.

Not everyone uses the Rear Phantom TLD spot when performing Irimi Nage, but then again, it is not a requirement. It is a cherry on top, one that not everyone knows exists. (Shhh, it is a secret!)

(Three-legged) Dog Spots

The reason most are not aware of these spots is that we are dealing with an extreme subtlety, and if you do not stumble upon it (as Uke!), you are very likely to miss the nuance. Unless your Sensei is actively working with these locks, it is possible you might never have experienced this frailty in our stance at all.

Learn to see past the subtlety of this drill so that you can better apply this skill when throwing Uke. As Uke, it feels like a big action to your internal senses, but looks extremely small on the outside. (Super subtle!) If possible, use a mirror to observe how this solo drill looks to others, then truly observe someone else performing the drill in this alternating way.

The important skill to gain is the awareness of the locked open or locked closed hip joint. Do not let your eyes fool you into thinking you have locked the kua when you have not. Hips and shoulders aligned, no twist in your spine, rotate until you cannot any more. If you do not feel the left leg jetting backward when doing the Front-Phantom TLD spot drill (first one), trust me, you are not locking your kua and/or not keeping it locked.

What you just experienced is the power of locking Uke's hip-kua. It is a very powerful element in throwing Uke and is practically a necessity if you want to throw Uke to their 'Phantom Leg Side' Dog Spots.

On that note, let us discuss an important aspect of Uke's weighted foot...

Grasping Uke's foot

Controlling Uke's hip-kua is a component of something I like to call a 'ground-up' joint collection.

What I mean by that is, similar to the Aiki versions of an inside and outside wrist turns that we are about to learn (Aiki Kotehineri and Aiki Kotegaeshi respectively), we 'grasp' a part of Uke and begin locking all the joints on the way to the COB.

In these TLD drills, we 'grasped' Uke's foot by putting the majority, or all of, Uke's weight onto just one foot; i.e., Weighting Uke's foot 'sticks' it to the ground and thus, for all intents and purposes, we have 'grasped' Uke's foot with the floor.

From there, as we rotate Uke's hips, it will put tension into Uke's ankle, their knee, and, depending upon which direction we are turning Uke, will eventually lock Uke's weighted side hip-kua in either an open or closed position.

Where Uke is Weak

The similarities between 'ground up' and 'Aiki wrist lock' will mean a lot more to you after we explore the next section: 'Secret to Aiki Joint Locking', but I want to point out something that might not be obvious and now is the best time to address it...

if we have done a good 'ground up' joint lock, we have all we need in order to throw Uke; i.e., You could argue that we will not need the coming chapter on upper body Aiki joint collection.

Control of Uke's COB is the sole (physical) goal of Aiki. Once Uke's COB is controlled, let us say, by a 'ground up' collection of joints, the job is complete. Locking any of Uke's upper joints (the arm joints and/or the spine) is just a 'best practice' and merely more icing on the cake.

I point out the lack of necessity to control the upper body as matter of fact, but certainly not as a recommendation. Important benefits exist in adding upper body joint control to a 'ground up' COB control.

Obviously, the safety of having Uke restricted from hitting us is of primary benefit, but the upper body joint locks can also be made to assist us in creating the 'ground up' control. A box, teeter, or twist can be the action that sets up the hip-kua lock and they have the tendency to make Uke's fall look even more dramatic. (Basic terms?!?! Not so basic!)

A danger here is that many practitioner's hybrid the use of upper and lower joint locks and do an incomplete job of either, or worse, both. They make many throws, but it can form gaps in awareness and they miss just as many throws as they succeed at.

Let us save more discussion on the intermingling of 'ground up' and upper body joint locks until after we actually review the upper body joint locks, but before we move on to the Dead Leg side Dog Spots we must address one other aspect that initially causes frustration with 'ground up' locking: keeping Uke's foot completely planted in one spot; (i.e., 'grasping' Uke's foot with the floor.)

When we try to perform 'ground up' joint locking, if we leave Uke's weight in the pinky toe, ball, or heel area, Uke is likely to 'release the floor's grasp' by pivoting like a dancer on that spot.

That pivot removes the tension in Uke's leg. No twist means we lose control of Uke's hip joint and cannot lock it fully open or closed.

(Three-legged) Dog Spots

In order to keep Uke's foot from pivoting, we have to keep Uke's COB in the middle third of the foot. Specifically, in the arch area, just inside, or you might say 'against' the 'blade' or 'side' of the foot.

Just inside the edge of the arch is important (See Figure 29).

If you put Uke's COB over the blade of their foot, they will roll their ankle and/or more likely, simply step out.

Keeping Uke's COB tucked on the blade side of their arch is easier said than done at first. The foot is a rather small, roughly triangular area. There is very little area to work with compared to the double weighted stance's parallelogram; (i.e., compare Figure 29 with Figure 25).

My best advice for keeping Uke's foot pinned to the floor is to focus on compressing Uke's same-side buttock into that middle third of the foot

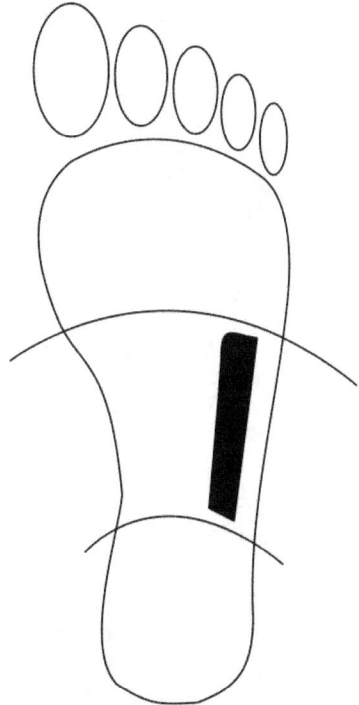

Figure 29: Middle third of the foot, still in the arch area, just inside (against) the blade of the foot. Put Uke's COB 'in the black' to keep it from pivoting

Where Uke is Weak

That line from hip joint to the middle third of the foot is something we call the 'Leg Hypotenuse' (Figure 30), you will see why as you read the later sections of this book; but if this is your first pass through this book, and you have not read the Aiki Joint Lock section yet, know that I mean that you have to have the intent of reducing the distance between Uke's weighted side hip joint and weighted foot arch. Uke may be maintaining a certain distance between the hip joint and arch area, they may even be increasing the distance (i.e., straightening their leg), but Uke must be fighting against your intent to decrease the length of Uke's leg hypotenuse.

Note that I did not say 'make Uke's knee bend'. Making Uke's knee bend is not necessarily going to keep Uke's weight in just the right spot to keep the foot in place.

If you must have the 'knee bend' description, then I would have to interject 'make Uke do a pistol squat', even though I feel even that description has flaws (but hey, that description has helped some people understand.)

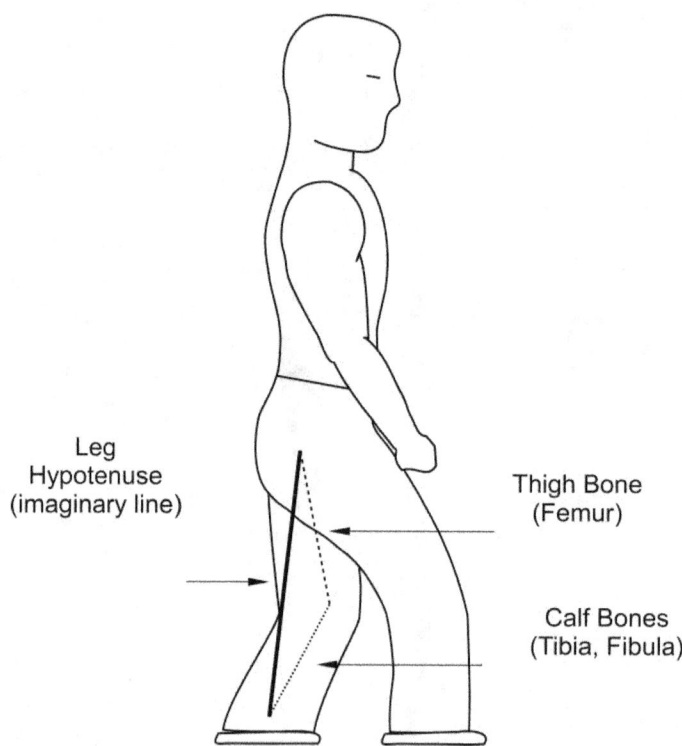

Figure 30: The Leg 'Hypotenuse', imaginary line that completes the triangle with the thigh bone (Femur) and calf bones (Tibia, Fibula)

(Three-legged) Dog Spots

Grasping Uke's foot with the floor and keeping it stationary as you torsion Uke's leg all the way to their hip is somewhat tricky, but you can master the skill. You might already be doing it, just did not put the mental attention to what you were doing.

If you are ready, let us leave the phantoms and 'cross-over' Uke's Sagittal Plane to visit the 'Dead Side' Dog Spots!

The Dead Leg Side Dog Spots are similar to their Phantom Side complements in that they also turn us into a propeller that spins at the hip.

Dead Side TLD's

The Dead Side story is very short. We do not need a hip lock at all, we simply need to keep Uke's COB pinned in the arch area just inside of the blade of the weighted foot.

Convenient, as this is exactly the same method (area of the foot) we just learned about when discussing how to 'grasp Uke's foot with the floor and keep it from pivoting'.

These Dead Side Dog Spots are easy to experience so be ready to support yourself on an object or ready to do some Ukemi.

Clear the area for your fall onto (and through) the Forward Dead Side Dog Spot (or make ready your support object) and put all of your weight onto your right foot as in Figure 31. (It may help to start into a pistol squat.)

The only necessary detail is to be sure to pin your COB in middle third of your weighted foot, just inside of the blade (Figure 29.)

Keep the weight in that special spot as you keep that straight back, shoulders over hips, bend at the hip, and tip your head toward the Forward Dead Side Dog Spot.

If you were diligent in maintaining a pin just inside the blade of your right foot, you should have sensed a compelling desire for your left leg to fly backward and 'start the propeller turning'; thereby putting you into a fall (or giving you the reason to support yourself on something if you chose that option).

Why?

In this drill, your left foot has little to no friction with the floor, so there is little resistance to having that left foot fly backward, but friction is not the only reason for this phenomenon.

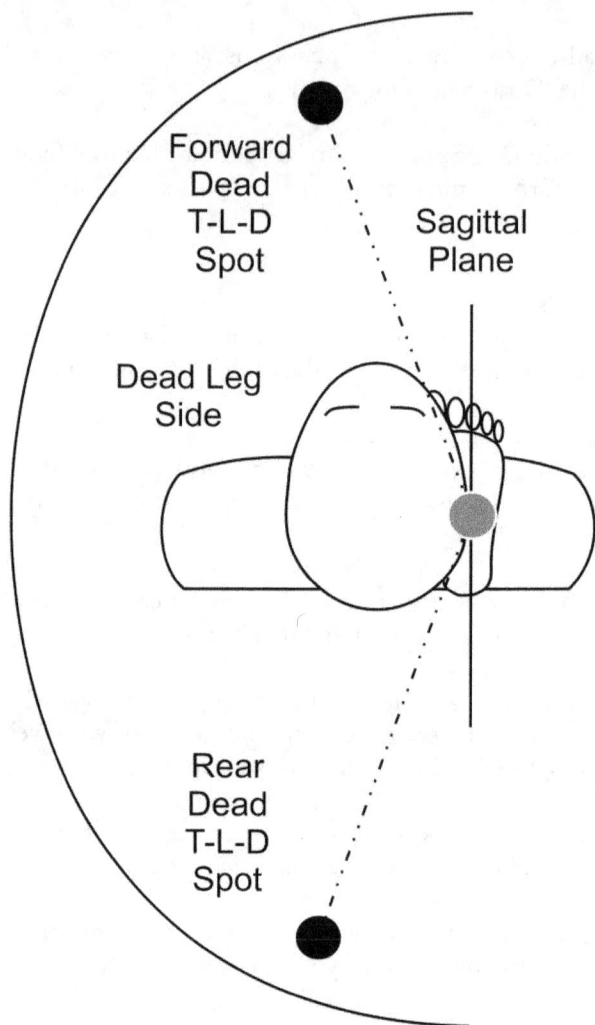

Figure 31: Any direction along the 'Dead Side' has effect, but the effect is greatest at the Dog Spots due to Uke's inability to create an effective counterbalance

We are playing with reflexes again!

Not allowing the weight (COB) to shift away from the Pin is the same as 'not allowing a step' and…

(Three-legged) Dog Spots

Just as before in the Phantom side, the brain senses that it is headed toward the ground and since there is no weight shifting into the left foot (i.e., no stepping happening), your reflexive instinct for self-preservation provides the impulse to put that left leg behind you to counter-balance the rest of the body that is toppling forward.

The problem is, your hips are not shaped to allow the counter-balance.

Interesting huh? Lack of friction, physical restriction, and reflex!

The challenge on the Dead Side is that you are going to tip yourself (and eventually an Uke) in a direction that has a tendency to pull the COB away from that Pin and into the field of the stance. Do not let that pin relax or you will find that you (and Uke) can get some weight into that left leg and basically shift into a double weighted stance; (i.e., you can stand there.)

Clear the area or find your safety support and let us play some more.

Again, place all of your weight onto your right foot arch area, just inside the blade and try tipping your head in any direction other than the forward or rear dead-leg Dog-Spots. You will feel some desire to become a propeller, but the sense of being projected is not there. The angle of the Dog Spots maximizes the Dead Side effect.

For extra fun, place your left foot/leg on the path between your right foot and the Forward Dead Side Dog Spot (no weight on the left foot, simply on the line toward the forward Dead Side Dog Spot). Be certain to keep all of your weight on the right foot and again tip your head toward the Dog-Spot.

If you maintained weight on the right foot (i.e., maintained the pin), you should have felt your left foot uncontrollably moving backward.

Have even more fun with this. Put your hips in any direction you like, left leg in any starting position you like, and repeat the drill. Give that rear Dead Side Dog Spot a try. If you are truly pinning that COB correctly, you will have the same result every time.

This is why we call the Dog-Spots that lay across the centerline of Uke's body the 'Dead Leg Side' spots. Sure, there is a leg, but it may as well not be there. The beauty of the dead-leg side Dog-Spots is that you do not even have to lock a hip-kua. Simply pin Uke's COB to one foot and spin Uke's COB by tipping their head toward the dead-leg side Dog-Spots.

Where Uke is Weak

Before moving on, take a moment and put some thought into how Three-legged Dog Spots can affect sweeps, judo throws, and almost every single Aikido throw you have ever performed. (especially Hikiotoshi (#17))

That wraps up one-legged stances, but what about three-point stances?

Three-legged Stance Considerations

Ever see Uke use three legs? Of course you have! We put them there often. The end of an Oshitaoshi (#6) (Ikkyo) comes to mind.

Have you ever tried to push Uke to the floor at the end of Oshitaoshi (#6) (Ikkyo) and while Uke is in a three-point stance, experience Uke easily shifting/crawling along with you (and even trying to attack your legs?)

Well no more!

Let us start this lesson by paying some attention to how Uke actions that shifting/crawling when we have their arm.

Begin by getting into a crawling position; (i.e., knees and hands on the floor; straight arms and thigh bones all perfectly vertical)

Stay off the floor as you put your right hand over your heart (ala when honoring one's flag or national symbol; e.g., Pledge of Allegiance)

Now take notice of your weight as you shift your left knee toward the position your right hand just vacated; (i.e., move your left knee forward and across your body until under your right shoulder area, or at least as close as you can get.)

If you were forward weighted, your left hand took the lion's share of the weight (and shared some of the weight with your right knee/shin) as you 'stepped' that left knee across and into that forward position.

If you were weighted more heavily in your hips (put your butt over your calves), your right leg did more work to support you, but shared the support with your left hand. The net effect is the same: you 'stepped' your left knee forward and across your body. This is how Uke is chasing you from a three-point stance (after your 'successful' throw.)

Return to the three-point stance and distribute your weight equally amongst your knees and left hand.

(Three-legged) Dog Spots

This time, while keeping your left knee rooted; try to place either your right knee or left hand into the spot vacated by your right hand.

It is almost impossible right?

I am sure some of you did your best to 'hop' your left hand over to that spot, but I hope we can agree that even that action requires a good bit of freedom to make a forceful left-handed push against the floor to jump over there, or you had to stick your butt out over your calves to support yourself primarily with your legs and hips. (In effect, sitting on your knees, ala started into a kneeling/suwari type position)

As Tori, we are not going to let that happen, are we? Of course not! (Well, in my case, I had to stop letting it happen.)

We should understand now that Uke has two options if they hope to chase us when in a three-point stance: 1: to position their weight over their calves (kneeling/suwari position, freeing up their hand) or 2: share the weight between left hand and right knee so they can step that left leg across their body.

So how do we stop them?

If Uke is trying option 1: supporting themselves with hips over calves, simply reapply your arm bar control (Oshitaoshi (#6)/Ikkyo) and return Uke face down to the floor by repeating the technique.

The reason is that, when in the 'suwari' position, Uke is technically on two 'feet' (knees) and has re-introduced a no-line; Uke's left hand will be only mildly committed to the floor.

Bringing Uke's right shoulder to the newly formed No-Line will nudge their left hand out and render the left hand useless as an effective support.

Two options arise. Compress Uke into a fetal position (knees on the floor under their torso) and keep them there; or... 'Strike the match' on the floor in the direction of the No-Line with Uke's right shoulder acting as the match head.

Next, when Uke tries to leverage option 2: stepping the left leg across the body, a very familiar phrase comes to our rescue: two objects cannot occupy the same space at the same time.

Where Uke is Weak

What do I mean?

Well, Uke's knee cannot take the space under the right shoulder if the right shoulder is there first. Written another way, keep Uke's right shoulder lower than Uke's left hip.

Drive Uke's right shoulder directly downward toward the space under their right shoulder; (i.e., push Uke's right shoulder to the floor where their right hand should have been when in a four-point stance.)

Once Uke's right shoulder is lower than the rest of their torso, use Uke's right shoulder to 'strike a match' outward on the floor along the line formed from Uke's left hip toward Uke's 'floating' right shoulder.

Note the line from Uke's left hip to right shoulder is akin to a No-line; i.e., with Uke's left-hand and right-knee substituting for Uke's feet. Uke's left-knee is actually a 'third leg', controlling Uke's right-arm/hand makes Uke a 'three-legged table'.

Allow Uke's body to collapse or risk having Uke tuck their right shoulder and do a forward roll.

Preventing Uke from rolling can be accomplished by locking/twisting Uke's shoulder, but too much shoulder lock and/or clavicle rotation could just as easily work against our goals and add to Uke's ability to roll. (we cover controlling/locking Uke's shoulder in the next section.)

It does take some practice, but the good news is, Uke is not 'three-point crawling/walking' to catch up with us.

Just be sure that you are not supporting Uke by holding their shoulder higher than their opposite-side hip and I expect you are now ready to handle Uke, whether Uke is in a three point stance (pin them in a fetal position, or 'strike a match' to get them on their belly), on two feet (effective use of the No-line and the 20's), or on one foot (Three-legged Dog Spots).

Secret to Aiki Joint Locking

'The secret to Aiki joint locking? Forget the 'secret' to it, what is it?'

Glad you asked!

We will discuss the 'three intentions within joint locking' so you will know the difference between 'what is', and 'what is not' an 'Aiki Joint Lock'.

We will discover why Aiki Joint Locking is best described as 'collecting the joints between the conjunction point and Uke's COB'. In other words, stiffening all Uke's joints from the point of contact all the way to Uke's COB (even through the spine!)

All that joint collecting is for the purposes of manipulating Uke's COB; most times, into a 'Pin and Spin', a 'Box', 'Teeter', or 'Twist', onto one foot for a TLD throw, and other times, to simply dissolve Uke's structure.

Collecting Uke's upper-body joints also allows us to manipulate the rest of Uke's joints between Uke's COB and the floor! ('Ground up locking')

That is a lot to say, so we abbreviate it to: 'collecting the joints'.

In order to make the detail of 'collecting the joints...' simple, we will use examples that center on 'Daito Hands in Uke'. This is complementary to the Daito Hands we use as Tori (as taught in 'Six Precepts'), so it will help if you understand Daito Hands in yourself before learning it in Uke.

You should recognize 'Daito Hands in Uke' as Kotegaeshi and Kotehineri, but how well do you really know them? Expect to find out a good deal more about your good friends than you have ever likely put words to before.

There is a lot to absorb as we learn how to make Kotegaeshi and Kotehineri not just a 'pain in the wrist', but instead an Aiki movement that affects every joint along a path to the COB.

Settle in, go slowly, as this discussion is closely integrated with many of the concepts that follow it. (and those that came before!) Let us get into it!

Three intentions behind locking a joint
There are three intentions behind joint locking: 1) Aiki Joint Locks 2) Pain Compliance and 3) Joint Destruction

Where Uke is Weak

The first joint locking intention, called 'Aiki Joint Locking' is the intention we seek to use in Aikido.

Aiki Joint Locking uses multiple of Uke's joints together to create Aiki (the control of Uke's stability and balance). Sounds simple enough, but the reality is that a good many people have confused Aiki joint locking with our second joint lock intention: Pain Compliance.

In Pain Compliance, Uke jumps out of the situation like a cat on a hot tin roof. The joint under attack is isolated and made to attempt to move in a way that it cannot easily or naturally progress. Moving a joint in ways it was never intended to be used threatens the integrity of ligaments and/or tendons or activates nerves near or inside the joint channel.

In short, Pain Compliance hurts! A lot! Hurting people gives them motivation to move (but it is not Aiki in the physical sense!)

Pain Compliance will offer success in the dojo and often develops a conditioned response in Uke.

The mere risk of joint destruction is enough to encourage our friendly Uke into action.

Unfortunately, these conditioned responses tend to be missing in interactions with persons with adrenaline pumping in their veins (e.g., rage, fear) or other chemical impairments (e.g., intoxication).

The 'it only works if Uke can feel it' aspect is another proof to why Pain Compliance is not an Aiki action. Aiki is the control of Uke's COB in a manner that controls Uke's stability and balance. The state of Uke's faculties is no matter. Although Aiki joint locks create a bit of discomfort, on the large, they do not generate pain (unless you want them to!)

In defense of Pain Compliance, it works more times than not.

The pain in Pain Compliance typically stops shortly after the technique is complete and this is why entire arts are dedicated to creating influential pain; i.e., I like Pain Compliance. I use it from time to time. Learn it. It is extremely effective, but it is not Aiki/Aikido and not the focus of this book.

Our third joint locking intention: Joint Destruction, is the polar opposite to Aiki joint locking.

Secret to Aiki Joint Locking

There are arts that focus on the destruction of a joint; and in damaging ligaments, bone, and/or tendons, creates a difficult to retract experience that will almost certainly require physical therapy and maybe even a bit of traction/surgery to restore normal function. (If it can be restored at all!)

There are only a few situations where we might deem joint destruction admirable, but it is a skill none the less.

Aiki can be used to simplify the destruction of a joint, but Joint Destruction typically does not result in Aiki (control of Uke's stability and balance).

In defense of the efficacy of the joint destruction arts, do you really need to control stability and balance when your opponent's joint is torn apart?

Unless your opponent is in a chemically altered state of mind, they are likely to be writhing on the ground in a state of constant and intense pain or, even if battling through the pain, an entire limb/threat may have been rendered useless for Uke to generate force back at Tori (i.e., you).

Gruesome, effective, but not Aiki/Aikido (not the focus of this book)

Where things get a bit crazy is that these three joint lock intentions: Aiki Locking, Pain Compliance, and Joint Destruction, are not exclusive and that leads to a lot of confusion.

We can use all three simultaneously to create very powerful effects.

For example,

we could use Aiki to lead Uke into a 'willowed' position, then apply some pain compliance to influence Uke to fall, and as Uke falls, brace a joint in such a way that Uke's body is hanging on the other side of that bracing, which would generate a serious amount of force and dismantle the braced joint.

Most times, when these situations happen, they are accidents, but they can be directed intentions.

The point here is not the capabilities, but that you know the three facets of joint locking, and that Aiki joint locking is practically painless

(Although we all know that one person that will wince and make hissing sounds no matter how little pain is involved; e.g., spouses and significant others!)

Collecting the Joints

So how do we create a comparatively painless joint lock that focuses on controlling Uke's stability and balance?

The short explanation is that we 'collect (tension) all the joints along the path from the conjunction point to Uke's center'.

We call this path 'Uke's Chain', as it is a patterned, sequenced, 'chain of events' that offers us a mechanism to directly influence Uke's COB.

When I say 'patterned', it is that, when we start at Uke's wrist, their wrist will always come before their elbow, and the elbow before their shoulder, etc.

In contrast, if we approach Kotegaeshi and Kotehineri as solely 'wrist locks', they only engage one joint and are confined/limited to becoming Pain Compliance or Joint Destruction with no direct control of Uke's COB.

So when we suggest that we are going to turn a 'wrist lock' into an 'Aiki joint lock', we start the lock at the wrist, but we then engage every joint from the wrist, all the way to Uke's COB (and then sometimes through Uke's COB and into every joint along the path from COB to the floor!)

Let us spell it out starting at Uke's wrist, (pay close attention to the words used as this is the outline of what you are about to master.)

Aiki wrist locking misaligns the grasped wrist before tensioning Uke's elbow; it then locks their shoulder in two places, as it twists and/or folds Uke's spine at the joining of the thoracic and lumbar vertebrae, where it places strain on Uke's 'core' muscle structure.

All of that action maximizes our ability to 'move' Uke's COB (through a mix of direct and in-direct touch) into a Pin, which in turn offers us the ability to continue the joint locking through Uke's COB all the way to the floor; (locking the hip (kua), loading the knee(s), and tensioning the ankle.) It can also be the avenue to simply over-burdening Uke's core, thereby dismantling their structure.

That sounds like a lot of prep work just to get Uke completely willowed, but actually, it happens in an instant and is not as hard as it may sound.

Let us deepen our awareness of how all that work is accomplished by taking a step by step look at how to perform Aiki joint locks within the yin and yang of wrist locking: Kotegaeshi and Kotehineri.

Secret to Aiki Joint Locking

Wrists and Elbows First

For those with insights from 'Six Precepts', the explanation is simple: Kotehineri is an index pivot; Kotegaeshi is a pinky pivot, but this time, the pivot is in Uke.

Enough said? No? Ok, I guess we can get a bit more detailed than that.

Kotehineri (Inside Wrist Turn)

Kotehineri first, and only to the point of misaligning Uke's wrist and tensioning their elbow.

We will move to locking the shoulder in just a bit, but not surprisingly, locking the wrist and elbow is very important, for, within Kotehineri and Kotegaeshi, we cannot lock the shoulder effectively without first locking these 'more remote' joints (remote in relation to number of sequential joints; i.e., 'distance', from Uke's COB).

Here is how we do it... first, we need the correct Uke.

It will be best to find an Uke that has hands of similar size as your own, as for this discussion and example, we will need to be able to apply the Kotehineri technique with only our left hand performing the wrist locking action; we need a free hand to help guide us as we twist Uke's right wrist.

It also will be immensely helpful if Uke is not a closet contortionist.

My daughter can start by resting her palm on a table in a normal fashion and then turn her hand/forearm over so far that she can again be palm down on the table.

You do not want that type of Uke for this lesson. Find someone with what we will call a more 'common range of motion'.

The good news is that when we are done, even someone with forearms like my daughter will be susceptible to Aiki joint locks.

Found that perfect Uke? Great!

No lock right away.

Simply get into a position to apply Kotehineri on Uke's right hand using your left (as if you just made it to the wrist lock portion of technique thirteen of the Juu Nana Hon Kata: Tenkai Kotehineri (#13). (Figure 32)

Where Uke is Weak

For the purpose of clarity and for the non-Tomiki Aikido practitioners... here is the quick way to get into position. (This is not the way we do it in Kata)

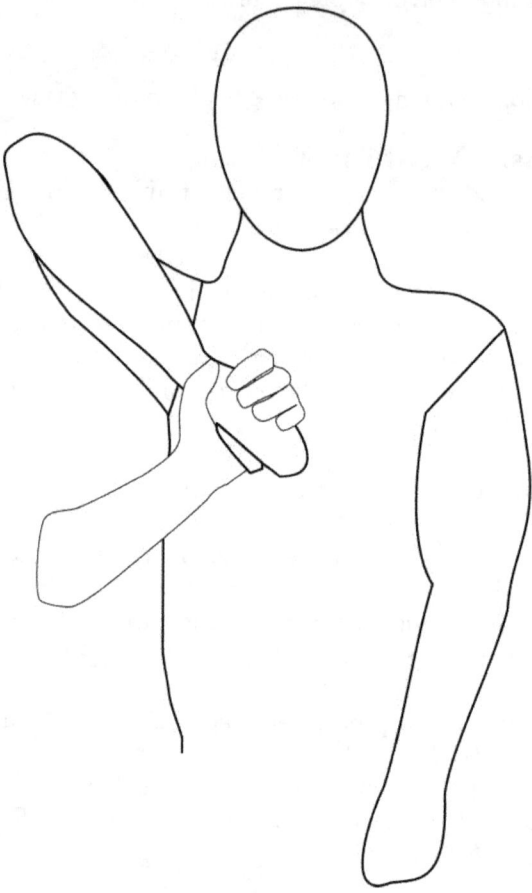

Figure 32: Us holding Uke's hand as we would in Tenkai Kotehineri (#13). Note: Uke's palm is facing us; Uke's thumb is pointed downward

Face Uke and have Uke give a 'thumbs down'; i.e., Uke raises their right elbow into a peak, turning the right-hand thumb downward.

Uke's right palm is toward you, position your left-hand palm toward yourself as well ('look at your phone') and place it on the back of Uke's right hand. (it might help to pretend Uke's right hand is your cell-phone)

Your fingers should curl around the pinky side of Uke's hand and your thumb is on the thumb/palm side of Uke's hand.

Secret to Aiki Joint Locking

The only additional guidance here is to be sure that Uke's wrist is not bent; i.e., keep Uke's elbow high and fingers pointing toward the floor as much as you can, keep your left hand as vertical as possible, in that you could point your left-hand thumb up toward the ceiling (and toward Uke's elbow).

We are not here to bend Uke's wrist, we are here to twist it (and soon you will see that we twist Uke's entire arm).

Once the learning is complete, you can go back to bending Uke's wrist if you so desire. Bending Uke's wrist is unnecessary, but can help with Aiki-Kotehineri once you know what you are doing.

Let us start by applying Kotehineri the 'less effective way'; in the sense of least effective in creating Aiki.

Uke get ready to 'tap out' anyway.

Have Uke extend their right-hand pinky (curl the other fingers slightly). Put your right hand outward like you were holding a plate, and place the tip of Uke's pinky on the palm of your open right hand. (See Figure 33)

Leave that conjunction point between Uke's pinky and your palm in place and rotate Uke's wrist; (i.e., apply a pinky pivot to Uke's hand.)

It may help to imagine the pinky was the tip of a pencil and your right palm a pencil sharpener, your right-hand palm stays stationary, the 'pencil point' of Uke's pinky spins in only one direction (If your right palm were the face of a clock, Uke's pinky spins counter-clockwise). Uke's thumb comes toward you.

You definitely can generate some pain this way, but you are not going to have a great deal of effect on Uke.

Take note of just how far you can twist Uke's wrist. You will see in just a moment that, in this pinky pivot, Uke's wrist can move a great distance further compared with what we do next.

Now it is time for a contrast in effectiveness.

Uke be ready to 'tap out'!

Use the same setup with one exception.

Where Uke is Weak

This time have Uke extend the right index finger (curl the others a bit) and place Uke's index finger on the palm of your open right hand. (ala Figure 33, but instead using Uke's right-hand index finger rather than as illustrated with Uke's pinky.)

Again, leave that conjunction point between Uke's index finger tip and your palm in place, and rotate Uke's wrist; (i.e., pencil point and sharpener; apply an index pivot to Uke's hand; same counter-clockwise turn, this time Uke's index finger stays in place as Uke's pinky moves away from you).

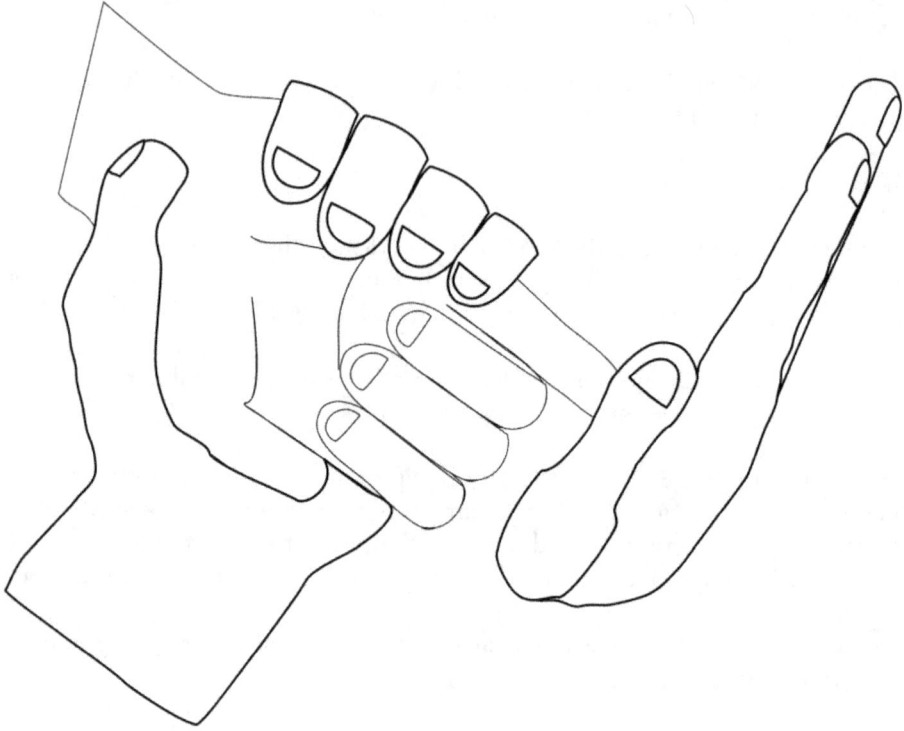

Figure 33: Detail of using both hands to 'sharpen the pencil' with Uke's pinky finger. (not nearly as effective as using Uke's index finger)

You should have discovered that the distance required to begin influencing Uke was incredibly reduced.

If you so desired, you could even create a good bit of pain, but what should also have happened this time is that Uke had to adjust the entire frame of their body in order to relieve the pressure. Done correctly, Uke will rotate leftward and try to face away from the hand.

Secret to Aiki Joint Locking

If you did not have this experience, I suggest that, as you twist Uke's forearm, you remember to move Uke's right 'hand blade' (the edge of Uke's hand between wrist and pinky) toward Uke's COB.

Do this with Uke's hand at, or about, the height of Uke's chest. (I have heard it suggested that we bring Uke's hand toward the chest bone, but it is much more accurate if you bring that hand to Uke's sagittal plane, above the COB, at the height of the chest bone (sternum)).

We have only just begun with our explanation, and will be adding more detail to ensure success, but truthfully, I would focus on this part of the lesson until you get it. Here is why.

Outside of Pain Compliance or control, the singular point of the Kotehineri lock at this stage is the ability to protect yourself from Uke's elbow.

Our Kotehineri should be keeping Uke's wrist from bending, while at the same time, stiffening Uke's elbow enough that it too cannot bend.

Uke should find it extremely difficult to attack you with their elbow. Protecting yourself from Uke's elbow in this way is quite an accomplishment, one that not all can claim they have mastered.

Test your skill by putting your right hand an inch from Uke's left elbow and as you twist Uke's wrist/forearm with your left hand, ask Uke to hit your right hand with their elbow. (See Figure 34)

If Uke can bring their elbow to your right hand, you are still not getting it. (Caveat: if Uke had to raise the elbow even further to try to strike you, that is good news, as we want that reaction.

You will see why when we lock Uke's shoulder. Additionally, do not fault yourself if Uke is super flexible. Move on, keep practicing, and you will develop the skill to stop them too.)

If you do get it, be sure to do your part as Uke and let your partner play the role of Tori for a bit. It is very important that you feel the contrast of index and pinky pivots within Kotehineri as both Uke and Tori. There is no replacement for this experience as this is the only way to appreciate a properly executed Kotehineri that stops with the tension in the elbow.

There is something else to take special notice of: the essence of the Kotehineri lock is the crossing of the forearm bones.

Where Uke is Weak

The tension created by the tip of Uke's pinky side forearm bone (Ulna) trying to cross over the top of the tip of Uke's stationary index-finger-side forearm bone (Radius); (i.e., the styloid process of Uke's Ulna trying to move around the styloid process of their Radius bone.) stiffens the wrist and elbow such that they do not function well.

Yes, that is correct. This 'wrist' lock has very little to do with the wrist and everything to do with the forearm bones. Why is that important?

It explains how and why we can perform the elbow lock when we have grabbed high on Uke's forearm. In other words, when we perform Kotehineri without Uke's actual wrist!

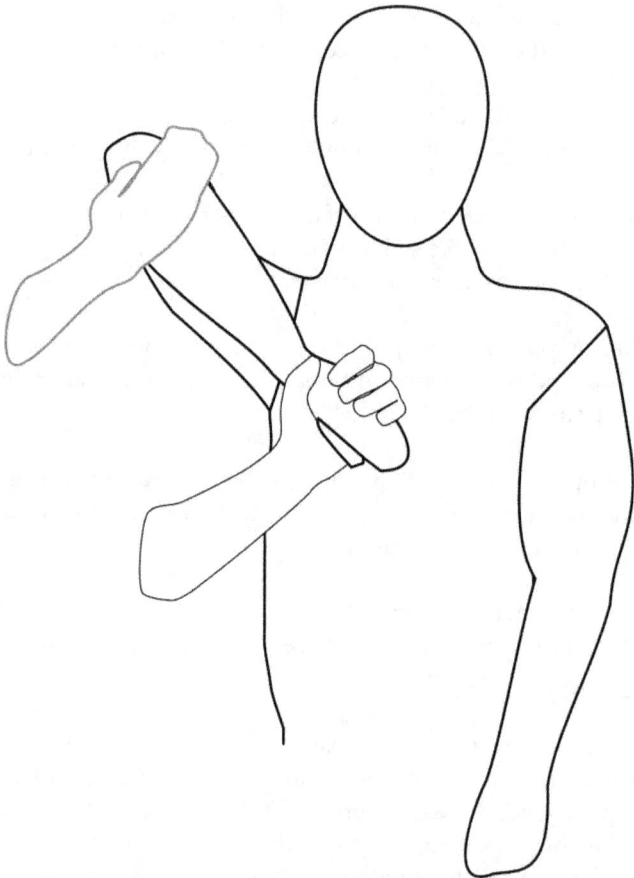

Figure 34: Use your right hand to test if you have correctly controlled Uke in Kotehineri (Uke cannot easily strike your right hand with their right elbow)

So, what did we learn? Kotehineri is an 'index pivot' and it collects two joints at the same time: Uke's wrist and elbow. We have a fool proof way to sense the difference between what works and what does not (sharpen the finger/pencil). We even included a test to see if we are really applying Kotehineri correctly; (i.e., can Uke hit you with their elbow?)

We are progressing nicely, but we are not even close to creating Aiki. We have to go through a few more joints before we can influence Uke's COB, but before we add the shoulder lock, let us first take a look at Kotegaeshi.

Kotegaeshi (Outside Wrist Turn)

From a bigger perspective, Kotegaeshi is very much the mirror image of Kotehineri.

That should lead you to guess that if Kotehineri is an index pivot, well then certainly, Kotegaeshi must be a 'pinky pivot'...

and you would be correct!

You might even conclude that Kotegaeshi is not all about the wrist, that its magic also lies in how the forearm bones cross over each other.

Again, you would be correct, but Kotegaeshi is a bit more complicated than Kotehineri in its application.

Very important nuance arises as we bend and/or straighten Uke's wrist in Kotegaeshi.

Let me go slowly and explain first how to perform the drill with Uke's wrist straight; Hang in there to discover what happens when the wrist is 'shifted'.

In the end, you will be able to confidently avoid two major points of confusion in how Kotegaeshi works. Points that take the pain out.

On with the drill...

Keeping Uke's wrist straight as we apply a Kotegaeshi is often awkward. (Most Uke's will actually try to bend their wrist a bit when we are applying the lock.)

The good news is that I have a sure-fire way to keep you from bending Uke's wrist while we perform the drill, we are going to use a wall!

Where Uke is Weak

Here is how we do it.

Have Uke stand facing a wall with the right-hand palm upward, much as though they were holding a plate at waist height; (i.e., put a ninety-degree bend in Uke's elbow, arm to the side or just in front of the hip, with the palm facing the ceiling.) Uke should extend the index finger and put the tip of this finger onto the wall. (Have Uke lightly curl the other fingers.) (See Figure 35)

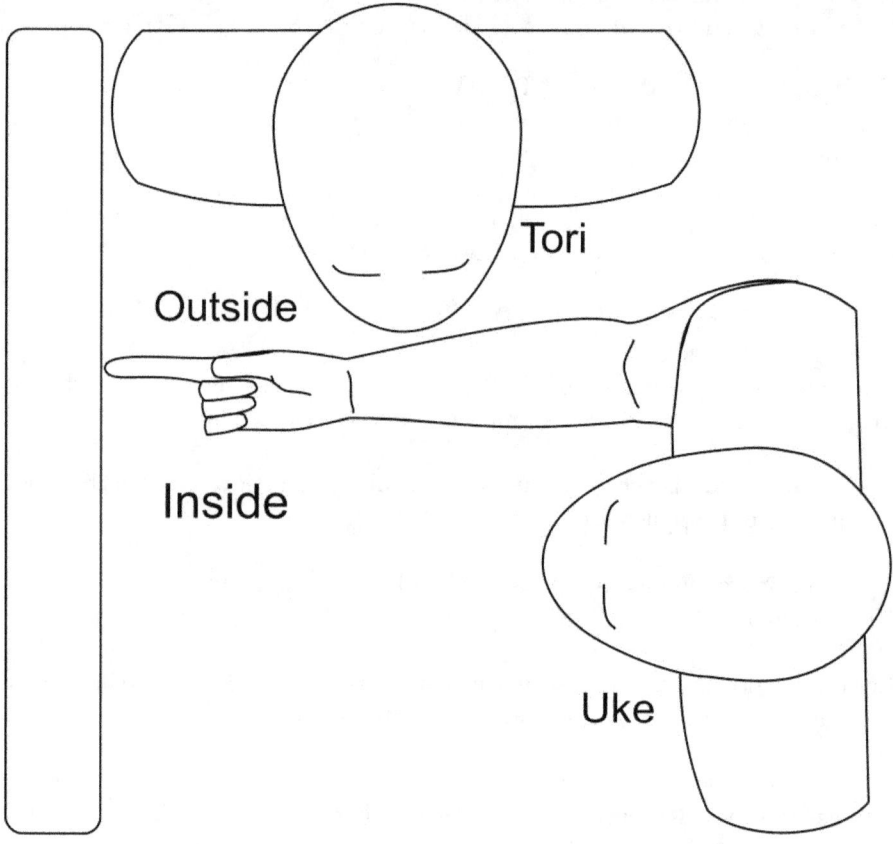

Figure 35: Uke positioned with outstretched index finger against the wall; prepared for (inefficient) Kotegaeshi drill; Tori positioned on Uke's outside

Depending upon the length of Uke's arm, or if you line up near a door (where you are inside the door frame and Uke is just to the side of the door) you might try to come at Uke's hand from the 'inside', or better said, standing directly in front of Uke, but since it will most likely be easier to approach Uke's hand from the 'outside' we will describe the drill in that way.

Secret to Aiki Joint Locking

I encourage practicing from both sides once you understand the drill.

If you find it impossible (due to lack of space) to use the wall, have a third person to hold their hand out as if it was a wall.

Having a third person act as the wall should also make it easier to approach Uke from either side or the arm.

Assuming you and Uke are the only two persons in the dojo today...

Let us first do this the inefficient way (less effect on Uke)

The task is to perform an index pivot in Uke from the outside position with Uke's index finger touching the wall (as in Figure 35).

Go slowly and be compassionate toward Uke. Uke, be ready to 'tap out' if necessary, but it is extremely unlikely that it will need to tap out this time.

Just as before, Uke's finger will resemble a pencil in a sharpener.

When coming at Uke from the 'outside', put both of your hands around Uke's. I suggest putting your left hand below Uke's (your left-hand palm up and on the back of Uke's right hand) and your right hand above; i.e., 'palm to palm with Uke's right hand.

With Uke's hand in yours, keep Uke's index finger in place and begin lifting Uke's pinky up and around their index.

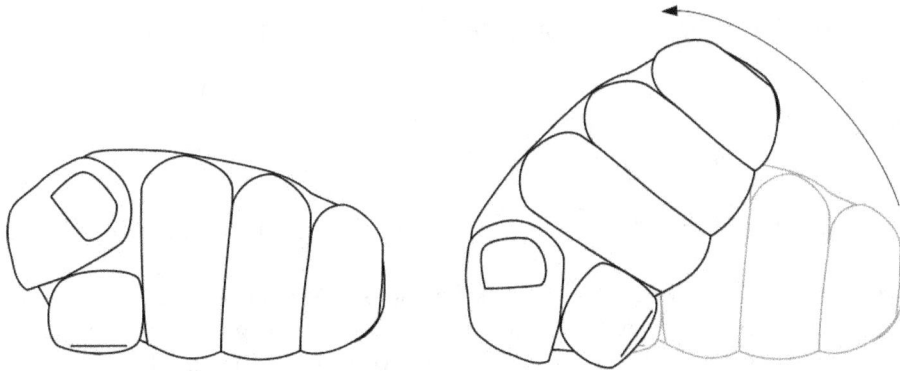

Figure 36: Index pivot applied to Uke's right hand; finger against the wall; as viewed by the wall during (inefficient) Kotegaeshi drill

Where Uke is Weak

The effect should be that Uke's index finger stays exactly where it is on the wall, and Uke's palm begins to move from facing the ceiling to facing toward Uke's right; (i.e., facing you, ala Figure 36).

I am going to generalize a bit, but most persons with moderate flexibility will be able to move about ninety degrees i.e., their palm is facing outward to the right, facing you, their pinky almost directly above their index finger. (That is where I can go without much pause, and being somewhat 'inflexible', if I can get there, likely most others can too).

Some people can go further, some less far. Again, go slowly and be compassionate toward Uke.

What I want you to notice is that, outside of being somewhat awkward and maybe a bit painful, this action had little to almost no effect on Uke's upper arm, and thusly, no effect on Uke's shoulder/torso. There was no reason for Uke to shift their torso.

So much for the uselessness, let us get it right this time. Reset and try again, but this time, with Uke's pinky extended to the wall.

Uke should be ready to tap out!

Same grip as before (your left hand under Uke's, your right hand above) keep Uke's pinky against the wall, and slowly (and I mean slowly!) rotate Uke's thumb downward in an arc, moving it such that the thumb would end up directly under the pinky; if we could get that far.

Feel for the 'motorcycle throttle' feeling in your right hand; or if that is awkward…

hold Uke's pinky in place with your right hand as you grab Uke's thumb with your left hand, then slowly swing your left hand under your right.

In either case, Go Slowly! as you push downward on Uke's thumb.

I will again admit, I am not the most flexible of people, but I cannot go more than about 45 degrees without feeling my entire arm beginning to rotate. If we go further, I am compelled to move my torso too. You should be sensing this same effect in Uke and concluding that this 'pinky pivot' has an immensely greater effect on Uke's torso than the index finger version.

In my assessment, the index pivot simply does not work at all in Kotegaeshi.

Additionally, besides the major flashback I expect you had as you 'Tilted the table' or 'Dropped the salt shaker' as in 'Six Precepts', I hope you also noticed that you were in an excellent position to easily have placed a lot of weight on Uke's thumb to make it turn.

That is going to pay major dividends when you are dealing with the Uke that has strong wrists. I have not yet met the person that can put a great deal of weight on just their thumb in this position and stand unaffected. (If you do meet that person, apologize right away and throw your wallet at that person as you run!)

The point is that Kotegaeshi is a 'pinky pivot' in Uke.

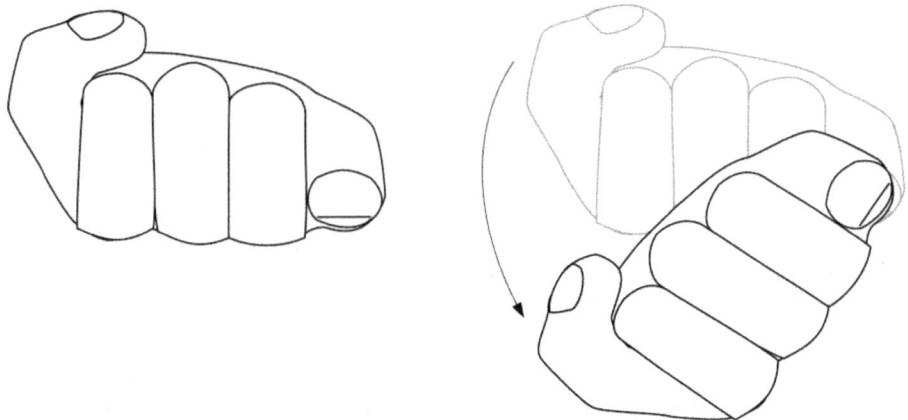

Figure 37: Pinky pivot applied to Uke's right hand (finger against the wall; as viewed by the wall during (correct) Kotegaeshi drill)

Now, some of you should be perplexed, as a good many of us have been told that Kotegaeshi is performed by pushing on that space on the back of Uke's right hand just south of Uke's ring finger base knuckle. (Triple Warmer 3 [TW3] for the acupuncture savvy).

It should not be too hard to imagine that, if Uke were to lift their right hand away from the wall and position that hand as though Uke were holding a book in front of them, with Uke's straight wrist, our pushing on the back of Uke's ring finger will not rotate Uke's forearm correctly.

At best [worst!], it creates an index pivot! No!!!

That might lead you to believe that 'pushing on the back of the ring finger' was bad advice. Maybe. It all depends. "Depends upon what?", you ask?

It all depends upon making that push against the back of Uke's ring finger the motivation for Uke's pinky pivot. That is what!

Would you like to know what they did not or simply forgot to tell you?

(Shhh, another secret! But I will tell you anyway. Share it with others! None of what is being shared should be a secret!)

A secret in Kotegaeshi; Bent wrist focus
Is there a way to position Uke's wrist so that pushing on the back of Uke's ring finger will assist us performing a pinky pivot; i.e., Kotegaeshi on Uke?

Yes! Of course there is!

When we say 'we have to bend Uke's wrist correctly', we have to first be in synch on which type of bend we mean.

Some might jump to the conclusion that we mean 'bringing the palm closer to or away from the bicep', but that would be a 'flexion' or an 'extension' respectively. (technically, 'Palmar' or 'Dorsi' flexion, look it up)

We do not want that. What we really want is what your Physical Therapist calls a 'deviation'. An 'Ulnar Deviation' to be specific. (See Figure 38)

For sake of continuity within these drills, I drew Figure 38 with the palm upward.

Unfortunately, you cannot see the TW3 pressure point on the back of the hand (in that Figure 38), but we can see the Heart 8 Shiatsu point on Uke's right palm. The Triple Warmer 3 (Triple Heater 3) point is in almost the same spot, but on the opposite side of Uke's hand.

I had to put focus on the positional relationship between the TW3 and Heart8 because we will be approaching this drill with Uke's right hand in position to hold a plate, so the Heart 8 will be all you can see...

but an issue arises with Uke's Ulnar deviation: Uke's hand is not positioned well to have a convenient 'pencil sharpener' action when it is deviated as in right side of Figure 38.

Thus, we cannot use the original drill setup.

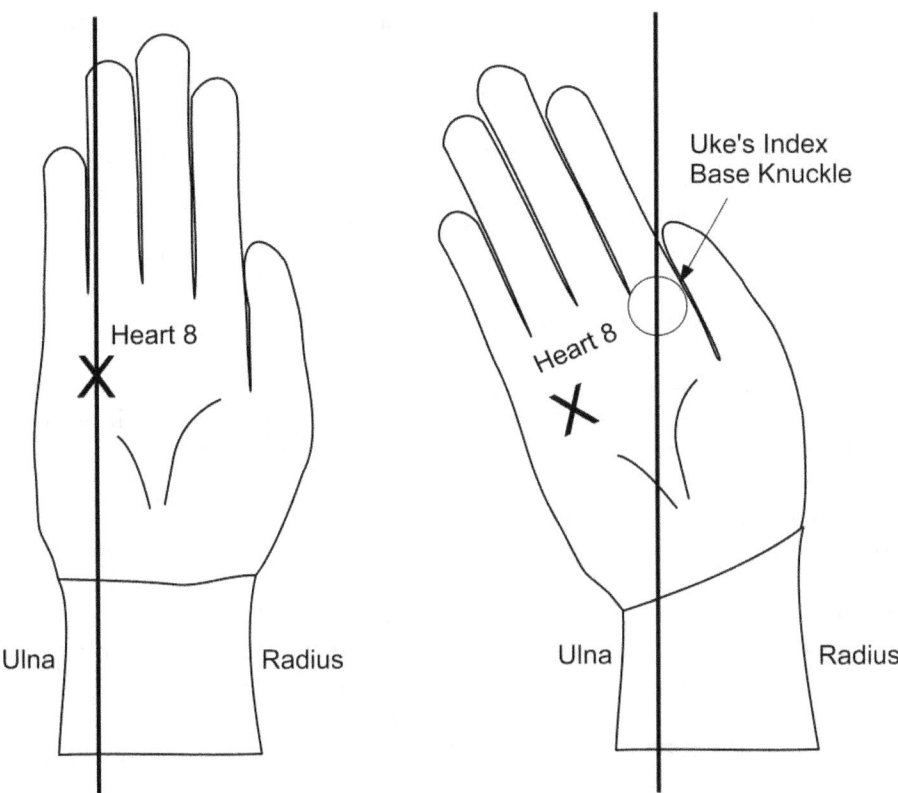

Figure 38: on the left, with the wrist in the straight position, the area behind the ring finger is directly on the 'pinky pivot line', thus pushing on TW3 will not spin the ulna; the ulnar deviation on the right shifts the TW3 spot into position to allow upward pressure on TW3/H8 to add to the downward pressure on Uke's thumb (thereby spinning the ulna in a 'pinky pivot')

So how do we drill this Ulnar Deviation?

Your best bet is to just be aware of keeping Uke's base index finger knuckle in its position in space (or ask someone to come over and use two fingers to kind of pinch it and hold it steady for you).

So, if you think you can hold things steady…

Uke stands with their hand out as before (holding the plate at hip height, palm up).

Where Uke is Weak

From the outside, reach over and grab Uke's right hand pinky with your right hand and under with your left to grab Uke's right thumb. (as if you grabbed the outsides of a plate, the far side with your right [pinky grab], and near side with your left. [thumb grab])

Shift Uke's wrist as in the right side of Figure 38 and slowly(!!!) lift up on the pinky side of Uke's hand with your right as you lever down on Uke's thumb with your left.

Yes, both of your hands move. Your left hand rotates Uke's thumb down as your right hand lifts Uke's pinky. Be sure the 'pivot' between your hands is the line of Uke's ulnar bone through their base index knuckle (Figure 38).

If you kept Uke's base index knuckle resolute in space; (i.e., straight out from the tip of the ulna), you should have experienced Uke performing the same as when we performed the pinky pivot with Uke's straight wrist.

Strike that...actually, I think this deviated wrist works much better than when we keep Uke's wrist straight.

The Ulnar Deviation is advantageous compared to a straight wrist. Sure, we lose a bit of landscape on Uke's thumb to push downward with, but we gain the ability to lift upward on the back of Uke's ring finger.

Both actions will support rotating the ulnar bone.

Uke's hand acts much like a wing nut; with the ulnar pivot line in the middle and both sides of Uke's hand acting as the wings.

It is a little trickier because we have to focus on keeping the pivot in just the right spot, but it provides a lot more leverage in a tighter circle (less motion).

Now you know yet another nuance to look for when you are having a bad day trying to perform a Shihonage (#14) or want to add the Kotegaeshi outward wrist twist to a Maeotoshi (#15)

The reality for all Aiki wrist locks is that Uke's pinky and/or index finger have little to do with the effectiveness of the technique, it is actually the forearm bones that do the work (the Radius and Ulna).

This is why Moe, Sensei Carlisle, and others like to forego Uke's hand altogether and simply use the tips of the forearm bones directly (also known as the styloid processes of the Ulna and Radius bones).

Secret to Aiki Joint Locking

Having large, strong hands helps, but you will see later that the aforementioned Sensei's do not use strength, the power is in 'fascial grasping'.

Take note that you very literally could have performed all the previous drills with Uke's wrist 'deviated' to any desired degree, and instead, used the tip of the radius bone as the 'tip of the pencil' to effect the Kotehineri (index side) and the tip of the ulna bone to perform Kotegaeshi (pinky side).

Pretty cool huh? I think so, as it explains the methodology for doing other immobilizations found in Aikido/Aiki Jujutsu/Hap Ki Do/Chin Na. etc.

Oh! So, you want to be able to use wrist flexion instead of deviation too?

For those that were taught to fully flex Uke's wrist when applying Kotegaeshi (bringing Uke's palm toward Uke's bicep, fingers pointing to the ceiling), the 'spinning the Ulna' idea is the same, but it is extremely awkward to perform this drill from the outside on a fully flexed wrist.

Switch to the inside of Uke's stance as you would in a traditional performance of Kotegaeshi (#12) kata; i.e., Stand perpendicular to Uke with your right side toward Uke; Uke 'holds a plate' with their right forearm at, or about, the height of your waist.

Have Uke bend their right hand so that their finger tips are pointed at the ceiling, Uke's palm facing Uke's bicep. (See Figure 39)

Put your left hand on the back of Uke's right; curl your left-hand fingers to grab Uke's right thumb. Put your right hand perpendicular to Uke's; i.e., lay the base of your right palm at Uke's right wrist/pisiform and align Uke's hand blade/pinky along the length of your palm and middle finger.

Use your left hand to keep Uke's hand tight against your right hand and slowly(!!!) extend your right hand at your wrist; i.e., keep your right wrist in place as you straighten it.

It resembles a tree falling. The base does not move, the upper portion topples forward; your fingers move away from you.

Go easy on Uke, this method has a lot of opportunity to become painful.

From a third person perspective, standing in front of Uke and looking straight into Uke's hand, when Kotegaeshi is applied to Uke's fully flexed wrist, it would look as on the right side in Figure 40.

Where Uke is Weak

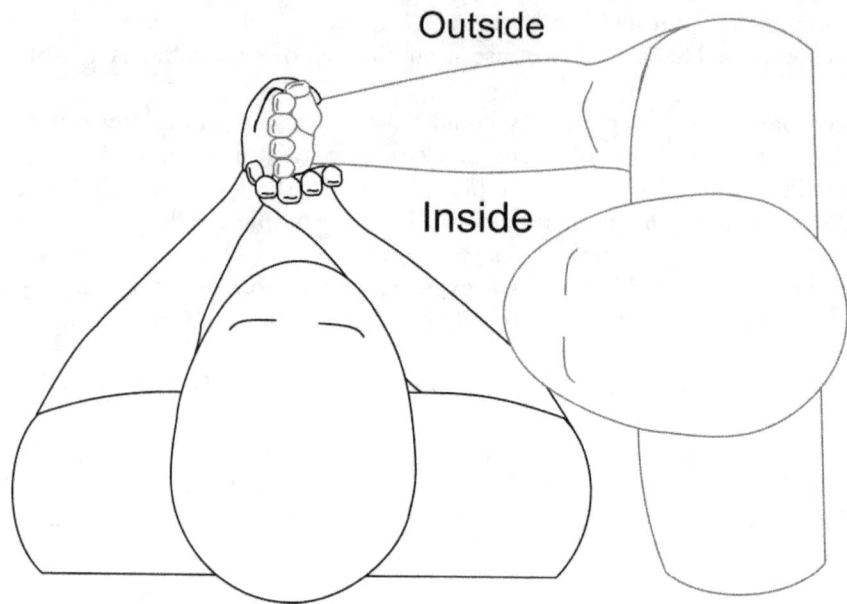

Figure 39: Approach Uke from their 'inside', flex Uke's wrist, and prepare to apply Kotehineri by 'Shooting a basketball' with your right hand

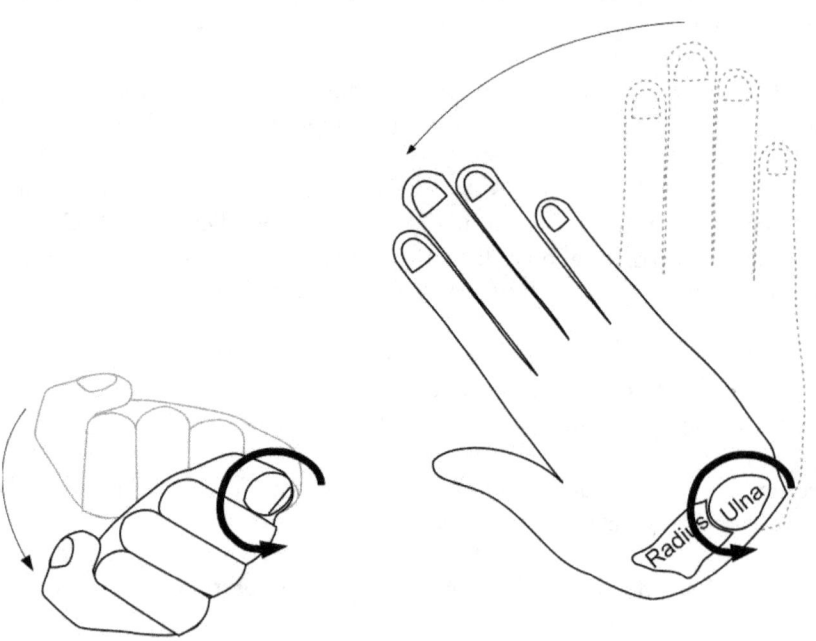

Figure 40: Applying a Pinky Pivot to Uke's in the earlier drill (on left) pivots on the ulna, same concept as pushing sideways on Uke's fully flexed wrist.

If we put this in terms of a 'wall' drill, this would be like placing the entirety of the back of Uke's hand on the wall and 'pencil sharpening' the styloid process (tip) of Uke's ulna.

From the perspective of Uke's ulna, both sides of Figure 40 are the same.

The Aiki magic is in the forearm bones, always. (never in the wrist)

We have done well so far, but there is an opportunity to go completely wrong when we approach Kotegaeshi as in Figure 39 (Flexed wrist)

I call that motion the 'Goose Neck', for Uke's hand/forearm is in a 'palmar flexion'; i.e., the shape one would use to cast the shadow silhouette of a goose on a wall.

The Goose Neck wrist lock is performed by putting Uke's hand in this flexed, fingers ninety degrees in relation to the forearm position, and then 'twisting' Uke's hand/wrist, instead of rotating Uke's Ulna.

The most common expression of this action is to put Uke's hand as in Figure 39 and then 'push against the back of Uke's ring finger'.

Pushing on Uke's ring finger when Uke is in a 'palmar flexion' serves no other purpose than to tear at the wrist joint, it does not rotate Uke's ulna. All it does is hurt!

In many ways, it is a lot like trying to twist the head off of a goose. (I do not really know what it is like to twist an animal's head off, and neither should you, but I hope I made my point: this action is gruesome.)

The 'goose neck' wrist lock is often used as a 'come along' maneuver by security staff or law enforcement. The 'goose neck' focuses on Pain Compliance and is an opening to Joint Destruction.

It relies heavily on a braced elbow; thus, it has little ability to collect other joints, therefore, not much for becoming an Aiki lock.

Even though the goose neck is not Aiki, I still whole-heartedly recommend that you learn how to use Pain Compliance and master the 'goose neck'.

There is a reason and an appropriate time to do almost anything, but the 'goose neck' 'come along' is not Kotegaeshi (not in my book anyway. Ha! Pun intended!)

Where Uke is Weak

Let us get back to Aiki as our focus…

Take these Kotegaeshi drills directly to your kata practice…

The more 'traditional' Aikido students (e.g., 'Aikikai') will likely find approaching Uke from the outside feels like the start of Shihonage (#14).

For us Tomiki Aikido practitioners, it is nearly the same, but we start the Juu Nana Hon Kata version of Shihonage (#14) from the inside.

In the end, it is all the same, just be sure that whichever way you practice Shihonage (maybe both ways!) that you are experimenting with Uke's hand/wrist in the straight and deviated positions.

When experimenting with Maeotoshi (#15), you will find a lot more fun when you add a good pinky/ulnar pivot, outside wrist turn, to Uke!

The drills that approached Uke from the inside should, of course, feel like a 'traditional' Kotegaeshi (#12), but…

Again, you should practice Kotegaeshi (#12) with Uke's wrist in all three positions (straight out, fully flexed [fingers up], and deviated), but remember not to hurt any 'geese'!

Another glaringly obvious 'outside wrist turn' technique is Hikiotoshi (#17). Using a straight wrist seems most comfortable, but again, deviated is an option.

I am sure if you think about it, there are more, just be sure to explore the advantages and disadvantages of having Uke's hand/wrist straight or in an ulnar deviation.

If all of that gets too easy, skip the wrist and start twisting Uke's forearm styloid processes!

Twisting Uke's forearm bones requires a bit of grip strength if you are not grabbing in just the right way; (i.e., using fascia instead of muscle), but since we cannot always count on having the perfect grip on Uke's hand, we should practice using Uke's forearm to be prepared for imperfect situations.

Most importantly, remember to go easy.

For many, this detailed awareness of Kotehineri and Kotegaeshi is new.

Secret to Aiki Joint Locking

To be blunt, you are a lot more powerful when you can focus your intent fully toward performing these joint locks effectively.

These techniques put a lot of torque into Uke's forearm, wrist, and elbow when done efficiently, but you do not need a lot of torque.

The torque is, of course, the point, but we can be subtle in way that these motions do not telegraph our intent to knock Uke down.

In both techniques, we only need to misalign Uke's forearm bones just enough to hinder Uke's ability to flex their elbow. Nothing more.

On that note, and before we move to the shoulder, let us finish this section with a short talk a bit about the elbow 'lock'.

Twisted Elbow Lock
Some of you are thinking, 'Wait a minute!!! We did not lock the elbow!!!'

The reality is, you did lock the elbow. Forcing Uke's forearm bones to cross over each other creates a great deal of tension in their elbow (where the forearm bones meet).

That 'one bone twisting over the other' condition is ever so subtly misaligning the elbow side tip of Uke's ulna called the 'olecranon'.

You are keeping the olecranon from smoothly entering a perfectly crafted depression on their upper arm bone (Humerus) called the 'olecranon fossa'.

Look it up on the web, but additionally, feel for it.

Bend your right elbow, right hand upward toward the ceiling, and with your left-hand finger tips, touch back of your right hand just below your pinky.

Slide your left-hand fingers downward toward your right elbow. (Figure 41)

Once you pass the wrist, you are sliding along the Ulna. You should be able to feel the bone sitting between the muscles the entire length of your forearm.

When you get to the elbow end, you are touching the olecranon.

Go around the tip and you should feel a depression. Most will say that is where the 'funny bone' lies.

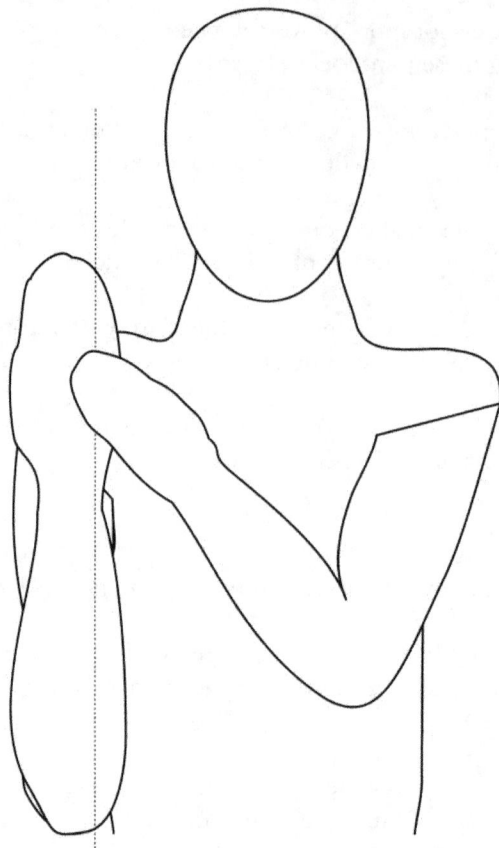

Figure 41: Putting your hands in position to feel for your right Ulna bone; progressing toward the elbow to wrap around and find the 'Olecranon Fossa'

Keep your finger on that depression as you straighten your arm. You should feel the tip of the Ulna (the olecranon) fill in the space.

The 'Aiki wrist lock' tension impedes that tight fit. The elbow 'grinds'.

(Secret tip:

Pay close attention to where the tip of the olecranon sits when your arm is fully extended.

Just above/beyond the tip of the olecranon is exactly where you want to push on when you want to separate (hyperextend) or even fracture an elbow (i.e., break off the olecranon) so be careful, but also know that it is exactly the spot we use to pin Uke to the floor.

Pinning Uke to the floor is an excellent topic to dive into, it is a lot more complex than it might seem, but this is a book about Aiki and not about the beauty of non-Aiki joint locks.

Hope that made sense, and that you use that knowledge wisely.)

This elbow tension is a lock, maybe not as extreme as the elbow lock produced when trying to hyper-extend the elbow; (i.e., fully extending Uke's arm, and then going further), but it is a lock and this is how we progress into control of the shoulder; i.e., Uke's elbow itself is just another point along the way to the COB, all it needs is a bit of tension/mis-alignment to keep it from flexing.

When acting as Uke, feel for the 'stiffness', almost grinding, in your elbow as a 'wrist lock' is applied.

You will get extra training benefit from the basic wrist warm ups most of us perform at the start of class if you occasionally focus on creating tension in your own elbow.

Now, onward and upward…

Locking the Shoulder (The Coat Hanger(s))
Quick review,

Index and pinky pivots in Uke are expressed as Kotehineri and Kotegaeshi (respectively). These two actions set up contortions of the forearm. When applied correctly, these contortions create a rotational tension (control) in Uke's elbow, but that is only the start! Two joints controlled, let us collect the next joint in the line: the shoulder.

The shoulder joint is an extremely mobile joint. In fact, it is so elusive to locking that many people do not even know how to lock it. (Really, stop right now and ask yourself and/or those around you how to lock a shoulder. If there was an answer, how precise was that explanation? Do not sweat it. You are going to master this joint in just a bit.)

How do so many not know how to lock a shoulder and yet there be so many excellent Aikido practitioners? Well…they either know how to avoid the need for upper body locks by using a ground-up lock, or they actually perform the shoulder lock but do not have the awareness required to fully explain their actions. Neither is bad, it just does not help us learn how to lock the shoulder (except maybe via a 'do as I do approach'. Blech!)

Where Uke is Weak

A vivid understanding of 'collecting the shoulder joint' relies upon three realizations…

First, we find the similarities between your shoulder and a joint you already know how to lock: the hip joint (learned in detail in 'Six Precepts'). Hips and shoulders both contain ball joints, and they lock in a very similar way.

Second, I said 'similarities', so we have to also understand where the differences lie. The differences are primarily due to the flexibility in the shoulder that we do not have in our hips. To control that 'slack', we apply a framework that gives us control of Uke's 'shoulder kua'. It is called 'the coat hanger' and understanding 'The Hypotenuse' within that 'coat hanger' triangle will make us truly efficient at creating the desired effect.

Third, we complete the shoulder control by contending with the shoulder as two joints, not one. Yes, you read that correctly, your shoulder has two areas, actually two different joints, that we need to control.

Curious yet?

Let us start with a word to wise on where to review…

We briefly described control of the hip kua in the earlier 'Three-Legged Dog Spots' section; for more review and investigation, I suggest a re-read of 'Tier 1 Advanced' section of 'Six Precepts' where we highlight the ball joint in your hips and the control of the 'hip kua'. Read that section with intent and then let us mull the control of the shoulder…

The key take-away in the comparison between hip and shoulder is that they are both ball joints. As such, they move, let us say 'rotate', in a very wide range of motion, but have one dimension of mobility that is rather limited in its range of motion.

In the hips it is expressed as 'toes in' or 'toes out' also known as 'opening' or 'closing' the 'hip kua'. It is primarily the same in the shoulder.

See Figure 42 to find the shoulder's parallel to 'pigeon toes' and 'duck feet' in the hands.

The ball joint part of shoulder finds its limit of motion identically in every way to the hips. (just the ball joint portion)

Look even closer at the effects on the shoulder.

Secret to Aiki Joint Locking

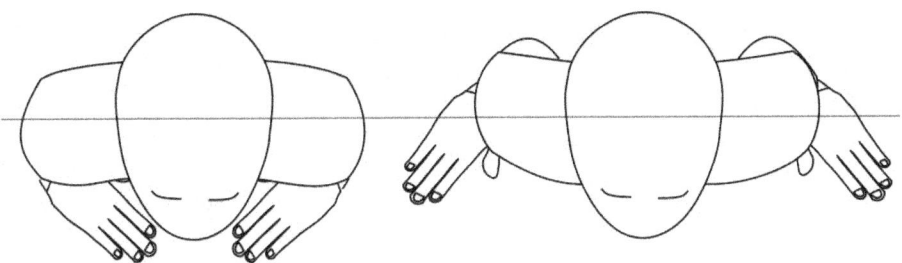

Figure 42: with arms straightened and at the sides, wrists in full extension, fingers forward, (mimicking our feet), the above shows the limit of motion of the shoulders as we spread our shoulder blades (on the left: sunken chest; Kotehineri) and as we pinch our shoulder blades together (on the right: puffed chest; Kotegaeshi). In this way, shoulders work the same as our hips

Yes, the shoulder blades are shifting closer and away from each other, but that is only part of the equation.

The real magic at this point is in the upper arm bone and its position in the 'shoulder socket' of your shoulder blade.

On the left, the 'sunken chest' pose includes an upper arm bone (Humerus) that has reached its extreme range of motion when trying to bring the outer tip of the Humerus forward toward the chest.

On the right, the Humerus is instead arced toward the rear and has again found its limit.

We describe these actions as 'closing' and 'opening' the 'shoulder kua'. ('shoulder kua' = the line between the chest and front shoulder muscle (deltoid), think 'inguinal crease', but between your chest and arm instead of between pelvis and thigh)

Kotehineri will close the shoulder kua, and Kotegaeshi will open it.

To get specific, the 'tip' of the Humerus that I am describing is called the Greater Tubercle. It is a 'bump' on the upper part of your upper arm bone.

Do not get confused by the tip of the shoulder blade called the Acromion.

The acromion is an easily visually notable bony projection in your shoulder blade that puts a roof over the Greater Tubercle and makes it hard to see the motion of the upper arm bone. (Look it up or browse ahead at Figure 54)

Fear not, as it is fairly easy to sense the motion in the upper arm bone via touch.

You can almost certainly feel the upper arm bone moving if you simply cup your left hand over your right shoulder muscles [deltoid muscles], keep your right arm at your side, and move between the fingers in/out poses in Figure 42.

As you twist, feel for the motion.

The acromion will not be moving, but just below that stationary 'shelf', all of that is the greater tubercle moving 'toward the front' and 'toward the rear'.

Your muscles may still be hiding the greater tubercle moving (lucky you!), but you should get the gist.

Those two motions are pretty much all your hip can do in this dimension, e.g., 'close' the kua as the Femur (upper leg bone) moves forward (toes in), and 'open' the kua as the Femur rotates rearward (toes out). Keeping your arms down, palms to the floor (fingers out forward) position is an easy top-down comparison between hips and shoulders, but...

the reality is that your arms do not stay downward much. Most times we are pointing, punching, grabbing, poking, etc. at something. These actions rotate the upper arm forward and thus complicate the ability to pin the greater tubercle to its extreme range of motion.

Let us see if you can again find the greater tubercle when twisting an arm that is straight out (e.g., pointing at something).

In an 'arm outward' position, (i.e., extend your arm horizontally and point at something) twist your entire arm (not just the forearm) to move your index finger as if trying to 'Sharpen the pencil'. It will require you to move your elbow so that it toggles between facing downward and facing outward.

As you move your elbow from downward to outward, the greater tubercle will follow and will be moving upward and downward.

You should note that the 'shoulder kua' has somewhat moved as well.

The 'opening' and 'closing' of the 'shoulder kua', thusly the opening and closing of the shoulder joint, is still the line between the chest and shoulder, but the shoulder kua 'kind of faces upward'.

Secret to Aiki Joint Locking

The change in the shoulder kua is important as it changes how we control the shoulder ball joint. No worries, we are going to make this simple to deal with, the re-positioning of the shoulder kua is just something to note. Let us begin.

Our journey toward understanding how to lock the shoulder starts with Uke's arm at rest along the side of Uke's body as we apply a Kotegaeshi pinky pivot.

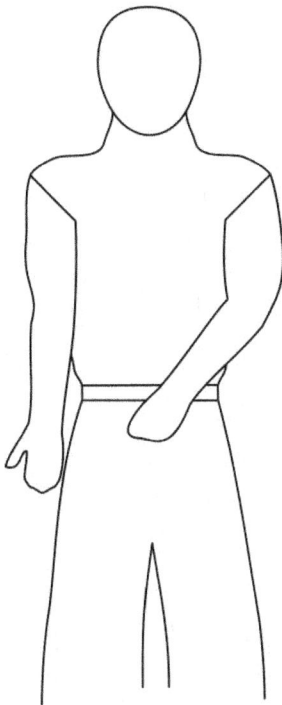

Figure 43: Uke standing prepared for the initial shoulder lock drill; Uke with arm to the side, elbow straight, pinky side of the hand on the hip

Have Uke stand with the right arm downward at the side, elbow straight, hand with pinky side touching Uke's hip, and their thumb as far from their body as possible.

Let us call this pose 'palm forward' with the arm dropped. (pretty close to the standard pose used in most anatomy lessons)

Take a moment to breathe, because, as we typically do, we are going to sense failure before we sense success.

You are about to move Uke's arm as if performing a pinky pivot.

Where Uke is Weak

It is exactly the correct thing to do, the best we can do, but it is not going to have a lot of effect on Uke. It is by (divine) design that it does not work, so hang in for a process that includes a couple of improvements as we progress.

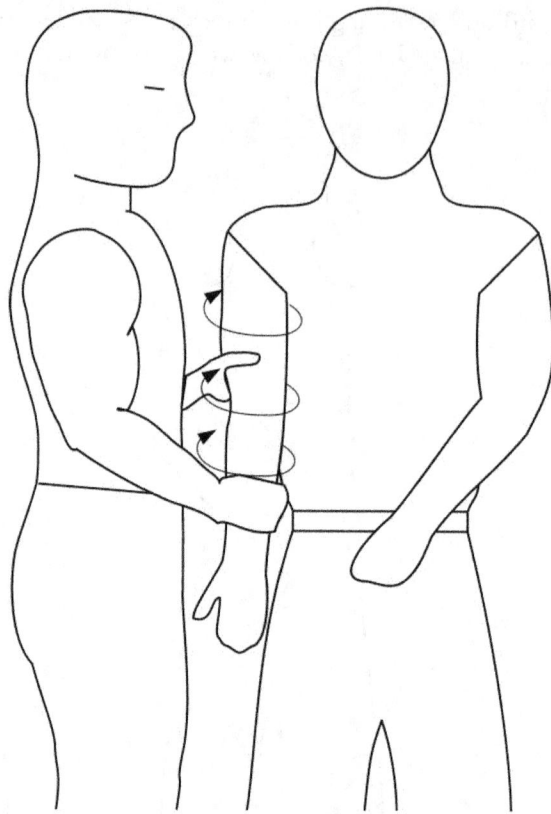

Figure 44: Uke standing prepared for the initial shoulder lock drill; Uke with arm to the side, elbow straight, pinky side of the hand on the hip

Attach to Uke as above in Figure 44 and do your best to lock Uke's shoulder by rotating Uke's entire arm, thumb tracing an arc around the pinky and toward the back of Uke's body; i.e., open Uke's shoulder kua by keeping Uke's right pinky in place and rotating Uke's right arm so that Uke's right palm faces outward, away from Uke's body and toward you.

If you cannot make sense of the motion, simply have Uke rotate their thumb back as far as they can go. That is what you want to make Uke do.

It may help to use Uke's body as the pivot by keeping Uke's pinky, ulna, and triceps against Uke's body as you 'roll' Uke's arm toward Uke's backside.

Secret to Aiki Joint Locking

Did you find it harder to affect Uke in this drill than it was in the earlier Kotegaeshi wrist/forearm drill with Uke's pinky in the wall?

You should have! In fact, it is a lot harder!

It requires a lot of strength and/or a great deal of precision to apply a Kotegaeshi Aiki Joint Lock when Uke's arm is straight. (important: read that last sentence again.)

So how do we gain some leverage? We bend Uke's elbow!

To find the opposite extreme of shoulder ball joint locking (extreme efficiency), let us return Uke to the earlier palm up position with Uke's finger touching the wall.

Figure 45: Rotating ('arcing') Uke's arm to create the shoulder lock

This time, instead of keeping a pinky finger in 'the pencil sharpener', we slowly swing/arc/rotate Uke's hand to the side as if Uke was trying to thumb a ride; i.e., arc Uke's hand toward the right, keep Uke's elbow at Uke's side. (See Figure 45)

Where Uke is Weak

It should not take much of an arc before you see Uke puff their chest forward in response to your actions. Stop before it hurts.

Your Uke will have to be very flexible in their shoulder if they hope to reach a full ninety-degree arc. You should never expect to get them to ninety degrees.

Now try it yourself. You do not need anyone else to help you along, just set up as in Figure 45 and move your arm yourself.

Feel the tension in your shoulder and how far you have to go to find yourself projecting your chest outward and your spine starting to form a 'back teeter'. Sense your upper arm bone and the tension in your socket joint. You are pressing the back of the head of your Humerus against the rear edge of your 'shoulder socket' (aka the glenoid cavity).

Instant and undeniable., but still not an extremely likely action to have happen in the thick of a conflict.

In that I mean, although Uke is going to keep their upper arm dropped and near their side most of the time, what Uke is not likely to do is approach you with their elbow at a ninety-degree bend, as if trying to serve you something on a platter.

If Uke should assume the 'holding a platter' pose, with a bit of quick stepping and the proper hand placement, you can put this arc into Uke's arm and make even the biggest Uke regret giving you the opportunity, (remember when I said we can shift easily between Pain Compliance and Aiki?!) but again, it is not practical to expect Uke to assume this position.

Yet, we still want to lock Uke's shoulder. What do we do?

We get creative about effecting the desired twist in Uke's upper arm when Uke is in more 'traditional' arm positions.

Here is how we do it in 'The Mojo': we use the 'Coat Hanger'!

The Coat Hanger (The Hypotenuse)

We need a way to lock that ball joint portion of Uke's shoulder no matter how that joint is positioned.

The 'Coat Hanger' gives us that capability. Here is how:

Secret to Aiki Joint Locking

This time, have Uke set up with the right arm extended outward, hand in front at chest height, palm facing inward, with the elbow dipped downward (I sometimes liken this position as Uke holding a book at a distance due to needing bifocals. Am I getting old? See Figure 46)

The position of Uke's feet does not matter, but I will suggest a right foot forward stance for sake of making a decision.

The focus here is the 'almost triangle' formed by Uke's arm.

The key points of which are the wrist (tips of the forearm bones), the elbow, and the tip of the shoulder (the acromion: bone sticking out of the shoulder area).

I call this triangle the 'coat hanger', as, on my arm, the triangle is almost exactly the size of a typical plastic coat hanger. (Go ahead, go get one from your closet. I will wait. We keep a coat hanger in the dojo to facilitate explanation and demonstrating certain techniques. After today, I will bet you will too!)

The three sides of the triangle are the:
1) Upper Side: upper arm side,
2) Lower Side: the forearm and
3) the Long Side: nicknamed the Hypotenuse (in homage to the longest side of a right triangle).

Note: The Hypotenuse is not tangible; (i.e., there is no body part here). It is imaginary and always extends from the wrist to the tip of the shoulder.

Technically, depending upon the type of wrist lock you are performing, the Hypotenuse typically extends from the tip of the shoulder to the wrist at the tip of the Radius or Ulna (Kotehineri or Kotegaeshi respectively).

We use the different sides of the coat hanger triangle to add detail and focus to how we manipulate Uke's arm. For example, when we were rotating our arm away from the wall in the previous drill (Figure 45), we would describe it as 'spinning the upper side' of this coat hanger. Let us figure out why.

Take that coat hanger that you are likely holding and put it against the inside of your arm. (Adjust/lower your forearm a bit to match the hanger)

As you perform a 'thumbing a ride' action, take notice of how the 'upper side' of the coat hanger solely spins, the other two sides 'move'.

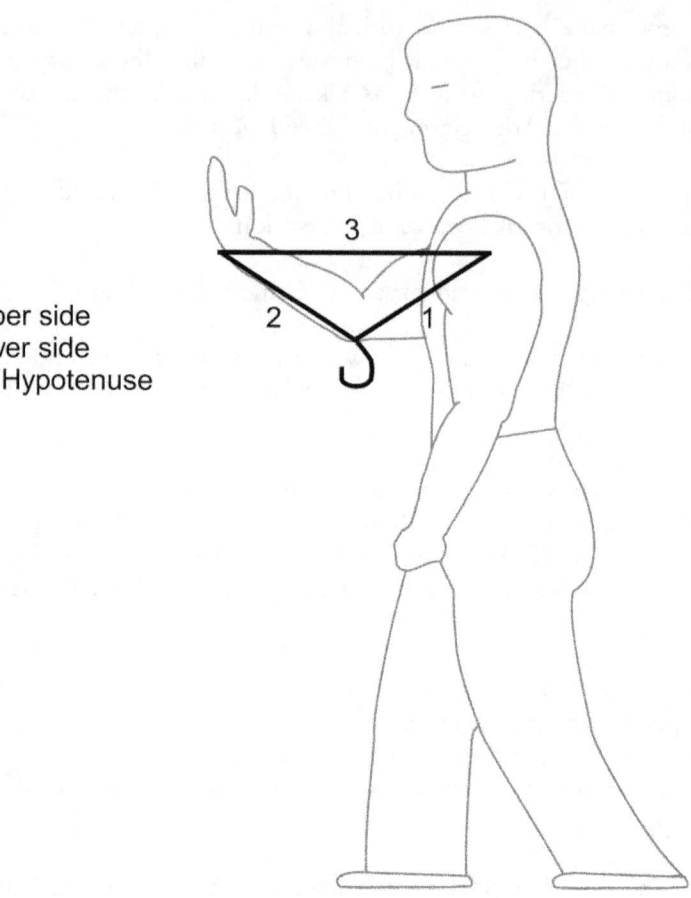

Figure 46: The 'Coat Hanger' in Uke's arm

As mentioned before when exploring the previous drill ('arcing' Uke's arm when Uke is holding a plate, Figure 45), even though spinning the upper side of the Coat Hanger is the most direct path to locking the shoulder, unfortunately, it is not often practical in the sense of opportunity.

This is why the Hypotenuse is almost always the most important focus of attention in class. With this in mind…

Let us introduce the two primary ways within a dropped-elbow coat-hanger (Kotegaeshi) to create the shoulder lock by 'turning the coat hanger'

The first focuses on the Coat Hanger 'hook' and we call it 'Over the Hook' or sometimes 'Set the Hook'. The other focuses on spinning the Hypotenuse. It is called 'Under the Hypotenuse' or 'Rotate the Hypotenuse'.

Over the Hook (Set the Hook)

The first action we perform to Uke's elbow-dropped coat-hanger is 'over the hook' aka 'set the hook'. (These names are purposefully descriptive.)

We are going to keep Uke's elbow (i.e., the 'hook' corner of the coat hanger) steady in space as we move the hypotenuse in an arc 'over' this elbow pivot point (note that I said 'move the hypotenuse' not 'arm wrestle with Uke' You will understand the significance/difference in just a bit.)

You can imagine this as though the coat-hanger triangle had fallen to the ground point/hook first, and is now toppling over, much like we have described 'falling fences' in previous sections.

Let us give it a try…

With Uke in the starting position (holding the book at a distance, hypotenuse parallel with the floor), approach Uke from the 'outside'.

Go palm-to-palm perpendicularly with your and Uke's right hand (as if performing the early portion of Shihonage (#14); i.e., the part just before we dip and draw Uke's arm over our heads).

Use your left hand to cup Uke's elbow. For your left thumb's safety, I strongly suggest keeping your left thumb along the side of your other fingers; not sticking upward and in the crease of Uke's elbow. (see Figure 47)

Tension Uke's elbow by applying some 'pinky pivot' to Uke's right hand. (ala the earlier Kotegaeshi drill) I suggest thinking 'push down on Uke's thumb with your right palm heel'. ('Rev your motor cycle').

You will need some, but not a lot of tension in Uke's elbow, so do not focus much on this twist.

Instead, be sure you have established 'extension connection' by moving the entirety of Uke's Coat Hanger away from Uke along the line of the Hypotenuse.

Do not cause Uke to step, simply set the twist in Uke's wrist, which, in turn (pun!), tensions Uke's elbow, and then establish the extension connection by moving Uke's 'coat hanger' away from their body.

The goal now is to move Uke's entire hypotenuse such that it falls in an arc toward you. (See Figure 48)

Where Uke is Weak

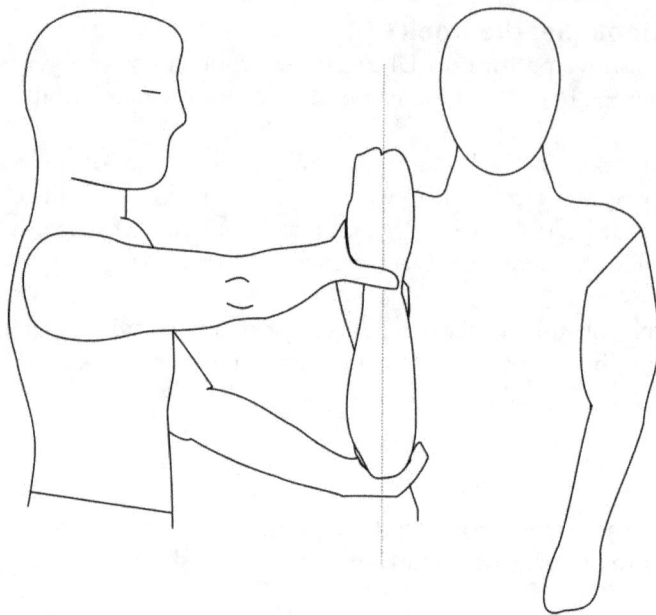

Figure 47: Preparing to turn the 'Coat Hanger' in Uke's arm

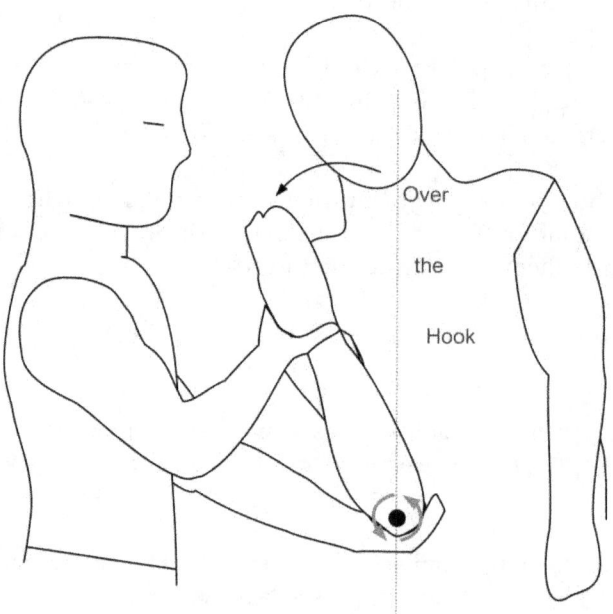

Figure 48: Uke's Hypotenuse has fallen toward us; Uke's elbow stayed in place; successful 'over the hook'/'set the hook' shoulder lock

Secret to Aiki Joint Locking

This means that Uke's elbow should roll inside your stationary left palm and/as Uke's shoulder and hand begin to fall/rotate toward you.

Both Uke's shoulder and hand must move or you are not performing this drill correctly. (See Figure 48 again for a look at success.) Uke will sense their shoulder binding up significantly.

Some find this tough to do, and end up moving just Uke's hand/forearm, leaving Uke's shoulder, upper arm, and elbow in place.

When we are only moving Uke's forearm, it is called 'arm-wrestling', because on most arcade arm-wrestling machines, the fake shoulder, upper arm, and elbow stay in place, only the hand/forearm move.

The trick is to use the tension in Uke's elbow to make the shoulder drop in unison with the hand; (i.e., The 'elbow lock' is the secret to forming the coat hanger)

If you are experiencing difficulty getting the desired result, here is another focus of your attention (intention) that has helped.

Have Uke assume the 'book at a distance' position again.

Engage Uke's arm as before, but this time, when attempting to 'set the hook' and letting Uke's Hypotenuse arc around Uke's elbow, focus your attention toward rotating Uke's elbow in your left hand so that the inside area of Uke's elbow (the part where the funny bone passes through) becomes visible to you.

You will not be able to see that crease at the start of the move, and that is the point. If you arc the hypotenuse with the elbow as the pivot, Uke's elbow should roll inside your left palm to the point that you will see Uke's funny bone crease.

Approaching this 'setting the hook' Kotegaeshi action by way of visually exposing the inside of the elbow has had a high likelihood of success.

My guess is that it is because you tend to not move the object you are trying to view, so it reinforces the intent of keeping Uke's elbow in place as you move the rest of Uke's body to accommodate your view.

When you think you have it, play a little bit and try this drill from a traditional Kotegaeshi approach. (from the 'inside' of Uke's stance) i.e., …

Where Uke is Weak

Put Uke in same 'read the book position' and this time approach Uke from the 'inside'.

Use just your left hand to engage Uke's right with traditional Kotegaeshi grip; put your right hand under Uke's elbow. Use the secrets you just learned regarding the pinky/ulnar pivot to tension Uke's elbow and create some extension connection by moving the entire frame of Uke's Coat Hanger away from Uke; along the line of the Hypotenuse.

Now, slowly and carefully, tip Uke's Hypotenuse away from you.

Again, I suggest you could think of it as trying to roll Uke's elbow in your right hand to better expose Uke's funny bone channel to you.

Inside or outside, 'setting the hook' works the same.

Let us have some different fun. Let us not rely on having Uke's elbow in our hand and let us also learn an important aspect of Kotegaeshi.

Have Uke assume the right-handed 'look at the book' pose and, this time, grasp Uke's right hand with two hands as you would in a typical Kotegaeshi technique and…

Stretch/Lengthen/Extend Uke's Hypotenuse
Willow Uke a bit by applying that subtle wrist turn to stiffen Uke's elbow and then move Uke's entire Coat Hanger outward from Uke's body along the line of the Hypotenuse; (i.e., create the extension connection)

Once you get Uke willowed, please go slowly (very slowly!) and, with Uke ready to take a fall or tap out quickly, begin extending the length of Uke's Hypotenuse by moving Uke's wrist away from the tip of their shoulder; i.e., let Uke's elbow rise and straighten some as you stretch Uke's Hypotenuse.

Extending Uke's Hypotenuse in this way will add torque to Uke's shoulder, but they should feel little to no additional stress in neither their wrist nor their elbow. (If they do sense additional pain, that is you adding twist and not a function of extending Uke's hypotenuse. Soften the pinky pivot.)

The danger is that you might dump them on their head, be careful not to.

The useful point here is that extending Uke's Hypotenuse in Kotegaeshi adds to the control. (Kotegaeshi works best with extension connection! Grasping Uke's wrist and walking away from them maintains the lock.)

In this latest example, we kept Uke's Hypotenuse parallel to the floor as we lengthened it.

Additionally (alternatively?), you will be almost sure to cause Uke to take an aerial fall if you perform the 'snow shovel' with Uke's Hypotenuse as the substitute for the Jo staff.

By using the snow shovel, I mean, keep Uke's wrist 'resolute' as you drop it directly downward; i.e., solely move Uke's wrist downward on the 'Z axis'; (e.g., Put your foot directly below Uke's wrist and then guide Uke's wrist to your foot)

Uke's Hypotenuse takes on the role of the jo staff in our snow shovel drill and the action of 'keeping the tip of Uke's Hypotenuse against the (imaginary) wall' keeps the tension in Uke's Hypotenuse (maintains the Coat Hanger) as it uproots them from their stance.

Go slowly and be very careful with Uke!

It takes some practice, but you should find that you can throw Uke without hurting the wrist. Most importantly, YOU will have thrown Uke, Uke does not get to influence or object to the situation.

Under the Hypotenuse (Rotate Uke's Hypotenuse)
Are you comfortable with 'setting the hook'? Let us look at the second coat-hanger twist, 'under the Hypotenuse', aka 'rotate Uke's Hypotenuse'.

Uke will again assume the right-handed 'book at a distance' starting position and we again approach from the outside. (Review Figure 47 again) Be certain that Uke's hypotenuse is parallel to the floor.

Are you joined to Uke with your right hand perpendicular and palm-to-palm with Uke's right hand? Is your left hand under Uke's elbow? Great!

Tension Uke's right wrist and elbow with a pinky pivot, again create some extension connection (by moving Uke's entire Coat Hanger away from their body along the line of the Hypotenuse), and then, this time, keep Uke's shoulder and hand in place as you use your left hand to 'swing' Uke's elbow away and slightly upward from you; i.e., keep the Hypotenuse stationary as you spin it by arcing Uke's elbow away and up. (See Figure 49)

You can use the intent of 'showing Uke's funny bone to someone standing on the other side of Uke'.

Where Uke is Weak

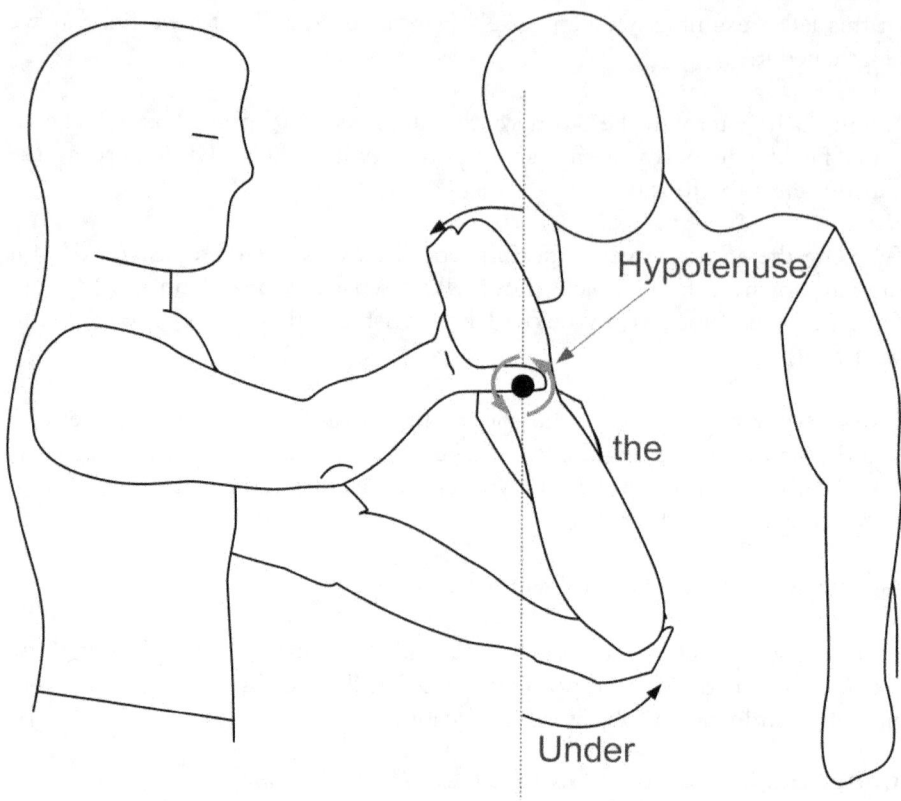

Figure 49: Uke's Hypotenuse stays in place and rotates. The elbow (Coat Hanger 'hook') has moved away from us and slightly upward.

Quick observation!

Take a close look at the results of the 'Under the Hypotenuse' and the previous 'Set the Hook' versions of this shoulder lock.

Both bring Uke to a 'Box' condition, but this 'Under the Hypotenuse' action is more likely to move Uke's COB across to Uke's far side (left side) and create a 'Stretch Box'; whereas our 'Over the Hook' action is likely to keep Uke weighted on their right and create a right side 'Crush Box'.

You ultimately decide which foot Uke will be weighted upon, but it helps to know what the natural tendencies are so you can scratch Uke's COB along their '20' to define the circumstances of the throw.

Starting to put things together yet?!?! Back to the action!

Secret to Aiki Joint Locking

This 'Hypotenuse spin' aka 'Under the Hypotenuse' is a very important skill to learn. It is a bit more direct than the 'Over the Hook' version in its ability to lock out Uke's shoulder.

In my opinion, spinning Uke's hypotenuse is easier to use, and will present itself as an opportunity more often than 'setting the hook', especially when grappling; (i.e., groundwork, not standing. Yes! These locks, and the principles we are exploring, still apply when we are performing traditional wrestling/Jujutsu moves!)

We will not belabor this discussion with detailed examples of how to affix ourselves to Uke and take advantage of 'Rotating Uke's Hypotenuse'.

Even novice Aikido practitioners, should be able to apply these precepts to Udegaeshi (#7), Shihonage (#14), Maeotoshi (#15), and of course Kotegaeshi (#12). (Personally, I would not apply 'Under the Hypotenuse' to Hikiotoshi (#17). It can be done, but it does not seem natural to me.)

What we should do in detail, is take note of what can go wrong.

With shoulder locks, success relies upon one important, but subtle nuance: the amount of slack in Uke's shoulder. (This is a shoulder lock after all!)

Even though Kotegaeshi can be done with either a raised or dropped shoulder (Uke's collar bone either raised or parallel to the floor), Kotegaeshi works best when Uke's shoulder is dropped. (Having Uke's shoulder dropped is typically a bit safer for Uke's shoulder joint.)

Most Uke will react to the pinky pivot by dropping their shoulder as a way to protect their shoulder joint, but some might not be so wary.

If Uke's shoulder is not already down (i.e., Linked! Basic Terms), we can help Uke put it down by adding a bit more pinky pivot or by lowering the entire coat hanger assembly for them.

Outside of helping Uke stay safe (a primary goal!) I would not spend too much time focusing on dropping Uke's shoulder.

The most efficient way to take the slack out of Uke's shoulder joint is still moving Uke's arm away from or toward their body along the path of the Hypotenuse (thus creating extension or compression connection respectively), but it is helpful to be conscious of Uke's collar bone as well.

Where Uke is Weak

With Under the Hypotenuse, there is really only one true mistake to be made: do not stick Uke's elbow directly into their body!

You will find it extremely difficult to move Uke's elbow across the body if their body is in the way. So be sure to position Uke's elbow at least slightly in front of their body. (again, another benefit offered to us by creating extension connection by moving the entirety of Uke's Coat Hanger away from Uke's body along the line of the Hypotenuse)

Scratch Uke's elbow along their belly, or go around Uke's belly if you have to.

After you have gained some success with the aforementioned drills, you can mix things up a bit by having Uke start at various degrees of angle in Uke's shoulder; (i.e., with Uke's Hypotenuse parallel to the floor, dropped almost straight downward, and any angle in between).

Additionally, be certain to change the starting length of Uke's hypotenuse (various distances between Uke's wrist and shoulder.)

Why?

Well, yes, because we can all use some practice, but primarily so that you begin to see the Hypotenuse in various dimensions and sense how it locks Uke's shoulder. Shoulder locking applies in more places than you might have first thought.

Sure Udegaeshi (#7), Kotegaeshi (#12), Shihonage (#14), Maeotoshi (#15), and Hikiotoshi (#17) are obvious beneficiaries of this type of Kotegaeshi shoulder lock, but additionally, take the time to try to rotate Uke's Hypotenuse as you perform the initial blocks we apply to Uke's arm as we enter into techniques such as Aigamaeate (#2), and Ushiroate (#5).

Yes, you read that correctly, your blocks should also collect the joints, not just the joint locks themselves.

You will find convenience when you apply this 'rotate the Hypotenuse'/'Under the Hypotenuse' Kotegaeshi shoulder lock in any technique that requires putting Uke into a 'back teeter'.

One last highlight before we close this dropped-elbow (Kotegaeshi) coat-hanger discussion.

You have seen the effects of lengthening Uke's Hypotenuse in this Kotegaeshi drill; (i.e., when we move Uke's wrist away from their shoulder along the line of the Hypotenuse). Done correctly, the action will create additional locking tension in Uke's shoulder, not pain, but…

what happens when we shorten Uke's Hypotenuse in the Kotegaeshi drill?

Shortening Uke's Hypotenuse in Kotegaeshi has the tendency to release the lock in Uke's shoulder and gives Uke some freedom to move. (not typically what we want to have happen, but, at times, it could be.)

When applying 'outside wrist turn' (Kotegaeshi) on Uke, if your technique requires allowing Uke's hand to move closer to their shoulder; (i.e., shortening the Hypotenuse and adding slack to the shoulder joint lock), simply compensate by adding rotation to 'the hook' in 'Over the Hook' (first drill) or add spin to Uke's Hypotenuse in 'Under the Hypotenuse'.

Raised elbow (back to Kotehineri)

There is one more joint in Uke's shoulder that we need to control, but since that joint is common in both outside and inside wrist turn (Kotegaeshi and Kotehineri), let us first explore how the coat hanger is used in Kotehineri.

In a perfect world…, we would walk ourselves through the exact same sequence of events we used in dropped-elbow coat-hanger (Kotegaeshi) and we would find a simple complement of each situation.

Well, we do not live in a perfect world, but the good news is that we are close.

We will use a similar sequence to examine the nuances that make Kotehineri unique; so, let us start by getting into a 'Kotehineri state of mind'; elbows up!

In Kotehineri, Uke's coat hanger is positioned as a coat-hanger normally would be when hanging clothes in a closet; i.e., The 'hook' (elbow) is at the peak and the Hypotenuse is below. (See Figure 50)

As we learned in the 'wrists and elbows' section, Kotehineri is an 'index pivot' in Uke.

Technically, it is sourced by the radius bone fully crossing over the ulna and then trying to progress Uke's ulna around the radius.

For a quick sensory review, put your right-hand palm down on a table [take note that your radius bone is crossed over your ulna].

Where Uke is Weak

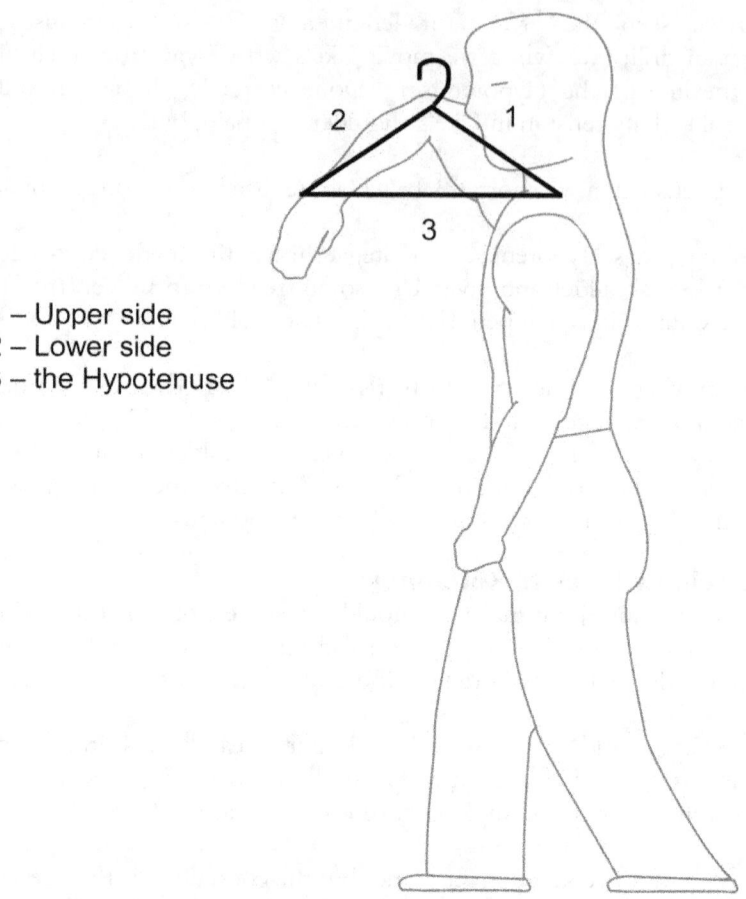

1 – Upper side
2 – Lower side
3 – the Hypotenuse

Figure 50: The position of Uke's 'Coat Hanger' when their elbow is raised

Tuck your right thumb under your right index finger, keep it in contact with the table as you use your other hand to lift/rotate the pinky side of your tabled hand over your index finger [i.e., move the ulna around the radius])

I expect you are having flashbacks to the Kotehineri drill way back in the early part of this section (near Figure 32).

We are now ready to get working with Kotehineri again.

Let us start collecting some joints!

Raised Elbow Coat-Hanger (Kotehineri)
This time let us change our grip on Uke a bit and approach Uke with a more traditional Kotehineri (#11) grasp.

Secret to Aiki Joint Locking

Place your right hand on Uke's right, palm-to-palm, with your left hand cupping Uke's right elbow; i.e., the same as you use to approach Uke when performing Oshitaoshi (#6) aka a traditional Ikkyo [See Figure 51]

Our right hand controlling Uke's right wrist, and our left controlling Uke's elbow, we begin by using our right hand to create an index pivot in Uke's right hand; twisting just enough to put tension into Uke's elbow.

Hooray! We have again collected the wrist and elbow, but there is a potential pitfall coming!

You probably went right ahead and removed some slack from Uke's shoulder by moving the entire frame of Uke's Coat Hanger away from Uke's body along the line of Uke's Hypotenuse. (Extension connection)

If so, great!

That is a proper way to remove slack from Uke's shoulder ball joint, but we must take care to move Uke's entire coat hanger, not just move their hand.

The risk here is that moving solely Uke's hand along the line of the Hypotenuse will lengthen Uke's Hypotenuse and release tension in the shoulder lock. (Opposite effect of Kotegaeshi! Kotehineri and Kotegaeshi are truly complements of each other.)

If you have not already, go ahead and make the mistake of solely moving Uke's hand.

First, 'Sharpen the pencil' to gain control of Uke's wrist and elbow, then try solely moving Uke's hand away from their shoulder; take notice of how the Kotehineri control starts to wane.

In contrast, bring Uke's hand closer to their shoulder (i.e., shorten Uke's Hypotenuse) and sense the tension in Uke's shoulder increase.

This important distinction, that Kotehineri gets tighter as Uke's Hypotenuse shortens, and that the opposite is true for Kotegaeshi, can be a source of frustration, but also a source of improved control.

You now know how to control joint tension (and potentially pain) in Uke's shoulder. Remember, 'just right, is just right'; Aiki is control, not pain.

Let us get back to doing things correctly.

Where Uke is Weak

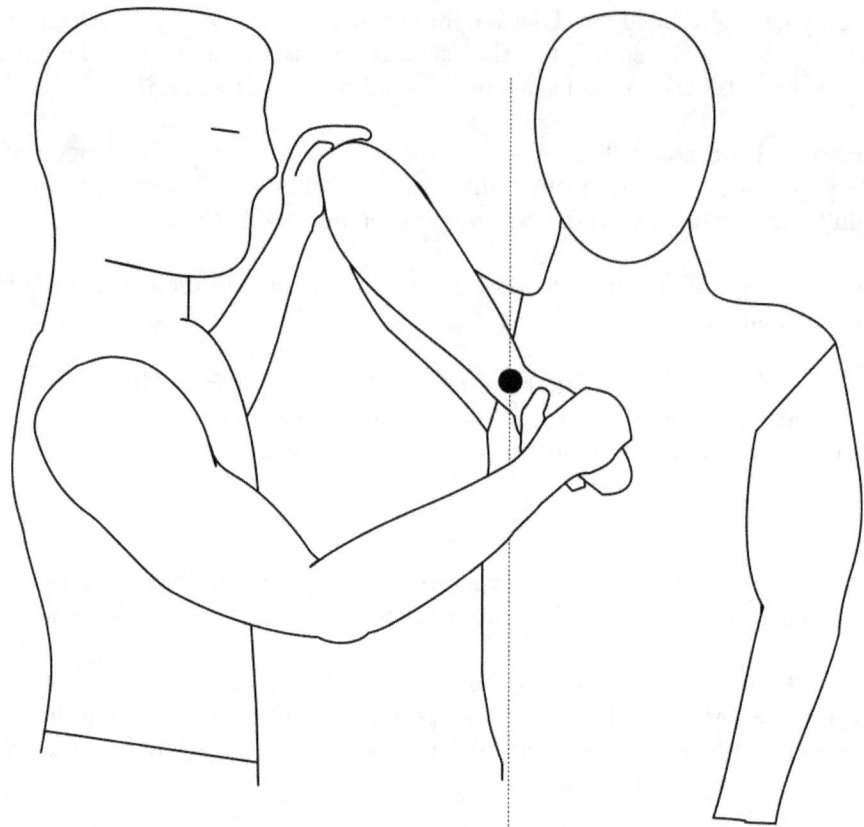

Figure 51: More Kotehineri practice! This time with our right hand controlling Uke's right wrist, our left hand controlling Uke's elbow

Place Uke once again into the Figure 51 pose.

Be sure you control the tension in Uke's elbow as a safeguard against changing the length of Uke's Hypotenuse (Uke's Hypotenuse cannot change in length if Uke's elbow is stiff and cannot bend or straighten!), and then move the entire frame away from Uke's shoulder along the line of the hypotenuse.

With the base slack removed from Uke's shoulder, and extension connection established…

we have two ways to collect Uke's shoulder when using extension connection in Kotehineri.

Not surprisingly, they are named similarly to before.

Over the Hypotenuse

The first action is 'Over the Hypotenuse'.

Keep Uke's Hypotenuse in place and rotate (pivot) it! (classic Kotehineri (#11), or when we do not rely much on the wrist twisting, it becomes an Ikkyo/Oshitaoshi (#6)) [Figure 52]

In other words, 'Rotate Uke's Hypotenuse' by tipping Uke's elbow 'over' (to Uke's front) and orbiting the stationary but rotating Hypotenuse.

It may help to think about 'letting the Coat-Hanger topple over', as if you had set a real coat-hanger on the floor, hook upward, long-side (Hypotenuse) on the ground and let go.

I am sure you have done this technique a million times, but pay close attention to Uke's shoulder.

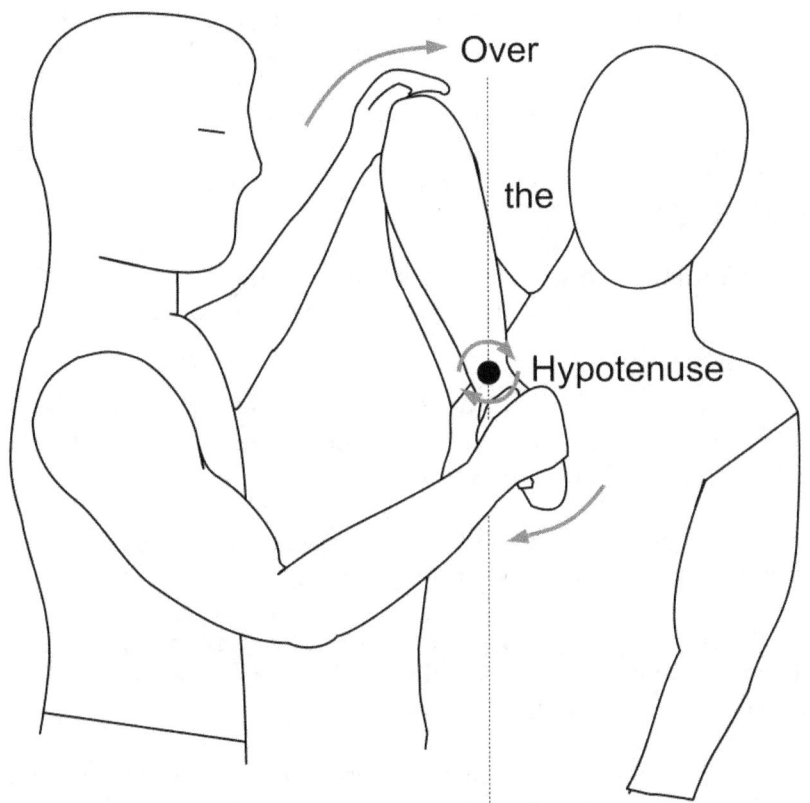

Figure 52: 'Over the Hypotenuse' aka 'Rotate Uke's Hypotenuse' in Kotehineri

Can you feel the shoulder lock?

Done correctly, you have a much more solid contact with Uke's upper torso, and thus, a much better ability to 'tip' Uke into 'Crush Box' or 'Stretch Box'.

The choice is yours, either Box works.

Under the Hook
The second method is to 'Set the Hook' by keeping Uke's elbow stationary and swinging the hypotenuse 'under' like a swing a small doll might sit on (if the Hypotenuse was actually tangible). (See Figure 53)

'Setting the Hook' in Kotehineri is a bit tricky to perform if we do not remember to create and maintain tension in Uke elbow, lest we only move Uke's hand and leave Uke's shoulder in place.

It may also help to (re)use the intention of revealing initially obscured parts of Uke's elbow to yourself or someone on the other side of Uke.

When performing 'Under-the-Hook' well, as you swing Uke's Hypotenuse, Uke will fold, so be sure to drop Uke's elbow directly downward toward the ground as Uke's right elbow rolls inside your left hand. [Figure 53 look for the line leading downward.]

Up on 2; Shorten/Compress Uke's Hypotenuse
For extra fun, and to add in 'Basic Terms': extension connection Kotehineri combines well with 'up on 2'.

Try moving the peak of Uke's Coat Hanger (aka Uke's elbow) over 'the 2' in their right-foot 'Plank' (Basic Terms), from there, use either Kotehineri shoulder locking method as you see fit.

We have explored and gained insight to Kotehineri using extension connection, but what about compression connection?

When it comes to performing Kotehineri in compression connection, both the 'Rotating Hypotenuse' and 'Set the Hook' work (although 'Set the Hook' tends to be a bit more painful for Uke.)

There is yet a third option to put turn directly into Uke's shoulder: shorten/compress Uke's Hypotenuse by moving Uke's hand closer to their shoulder via the Hypotenuse path (as we learned at the start of this section.)

Secret to Aiki Joint Locking

Set up again as in Figure 51, set the tension in Uke's wrist, stiffen Uke's elbow, and move Uke's entire Coat Hanger along the line of the Hypotenuse and into Uke's torso (shoulder socket). This will remove most of the slack in Uke's shoulder and should be providing us compression connection.

The first thing we need to do is to release just enough tension in Uke's elbow to allow us to move Uke's wrist closer to Uke's shoulder/armpit by way of the hypotenuse; (i.e., shorten Uke's Hypotenuse).

This will cause additional tension in Uke's shoulder and you will likely notice a bit of a 'spinning' in Uke's Hypotenuse; (i.e., the wrist and shoulder stay in place as the elbow 'hook' orbits over/around the Hypotenuse and toward the front of Uke.)

It happens almost organically and looks almost exactly like 'Over the Hypotenuse'.

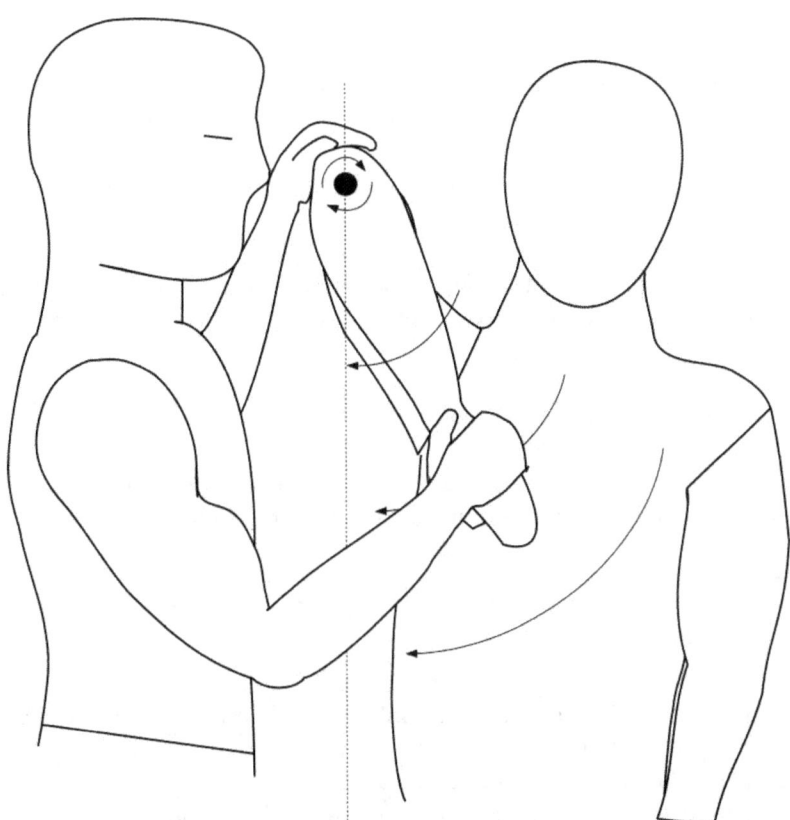

Figure 53: 'Under the Hook' aka 'Set the Hook' in Kotehineri

The trick to this 'Hypotenuse shortening' method of Kotehineri is to keep just enough twist to keep Uke's Index Finger/Ulna spinning, but allowing enough flex in Uke's elbow that you can move their hand closer to their shoulder… all the while rotating Uke's hypotenuse as they head toward the ground.

Sounds hard, but it is not as bad as it reads. Just be conscious of how much twist you are placing in Uke's Index Finger/Ulna and it should be manageable (although it might take practice to make it painless for Uke).

The Sternoclavicular Joint (SC Joint)
We have come a long, long way to get here, and so far, so good! But..!

Most of us likely found that the Kotegaeshi, outside wrist turn drills worked much more consistently than the Kotehineri drills…

but, why?

In Kotehineri, we applied an index pivot to Uke's right hand in order to tension Uke's elbow to create a coat-hanger in Uke's right arm.

We attempted to lock Uke's shoulder with 'Over-the-Hypotenuse'/ 'rotating Uke's Hypotenuse'. (ala Figure 52)

Everything we have done has given us a perfect shoulder lock and Uke folds as we guide their elbow across their body in a perfect performance of Oshitaoshi (#6) (Ikkyo). Right?

Well, maybe, but many, many times, it does not.

In fact, if we continue rotating Uke's Hypotenuse, Uke is almost surely going to be able to slip out of the technique. Something is still missing.

Remember when I said there are two joints to control in Uke's shoulder?

Well, two that matter most.

Technically, there are four joints, but only the GH and SC matter, the AC and ST are irrelevant for this explanation. [Is the ST really a joint? Not in my opinion.] I digress. If you understand those acronyms, you are likely medically trained and should relax as I focus this discussion for everyone else.

In this discussion, two joints!

Secret to Aiki Joint Locking

Rotating the Hypotenuse will certainly control the 'ball joint' where the upper arm bone (Humerus) meets the shoulder socket (Glenoid) but the source of trouble is not in that GH (Glenohumeral) joint, the issue is lack of control of Uke's Sternoclavicular (SC) joint.

This tiny joint is responsible for more failed Aiki throws than any other joint in the entire body, bar none!

Let us detail what this joint does and see why.

Unless you paid extremely close attention in biology class, you most likely did not notice that the only 'bone to bone' joint that joins your shoulders and arms to the rest of your body is this tiny little joint on the end of your collar bone called the Sternoclavicular (SC) joint. (again, ignore the ST 'joint' on your back, it is a group of muscles, I do not consider it a joint)

The Sternoclavicular (SC) joint is named this way as it joins your Sternum (breast bone) to your Clavicle (Collar bone). Find it on yourself.

Look at your 'throat notch' at the top of your breast bone (Suprasternal notch). The bony projections on either side are where the ends of your collar bones meet your torso. Those are joints! (Sneaky joints! Joints that love to play tricks on us!)

So why is this Sternoclavicular (SC) joint so important in Kotehineri (and in Oshitaoshi (#6) aka Ikkyo? and why was it not an issue in Kotegaeshi? Great questions! (You always seem to know what to ask! <Wink>!)

It all starts with deeper understanding the shoulder joints.

Take a look at a top down view of your shoulder assembly (Figure 54) to notice that 'less than' symbol (<) created by the 'V' of your shoulder blade and collar bone?

Well, that 'V' is rigidly connected to the front of your body (i.e., bone to bone jointed), but the shoulder blade 'floats'; connected to your back by only by muscle and tendons (aka the Scapulothoracic ST joint. Is it really a 'joint' if it is muscles and tendons?)

The 'floating shoulder blade' is kept stable by the collar bone and the Sternoclavicular [SC] joint.

Break your collar bone, and you have effectively disabled the associated arm.

Where Uke is Weak

(I have witnessed this injury twice and both times, that person had a hard time sitting or standing, they wanted to lie down to take the pressure off of their hanging shoulder. Not pretty, looked very painful. The rehab/traction that followed did not look like fun either!).

When you consider the frailty of this set up, it is no wonder that the collar bone is such a desirable target for martial arts that focus on bone breaks!

We are not going to break Uke's Collar Bone, but we are going to have some fun with their SC joint.

Here is how...

The SC joint allows us the freedom to do two things with our shoulders.

On the horizontal plane, the SC joint allows us to puff our chest out by pinching our shoulder blades together and conversely, to sink our chest by spreading our shoulder blades apart.

On the vertical plane, this joint allows us to 'shrug' (lift) and 'drop' our collar bones. ('Drop' and 'link' our shoulders to our torso. Basic terms!)

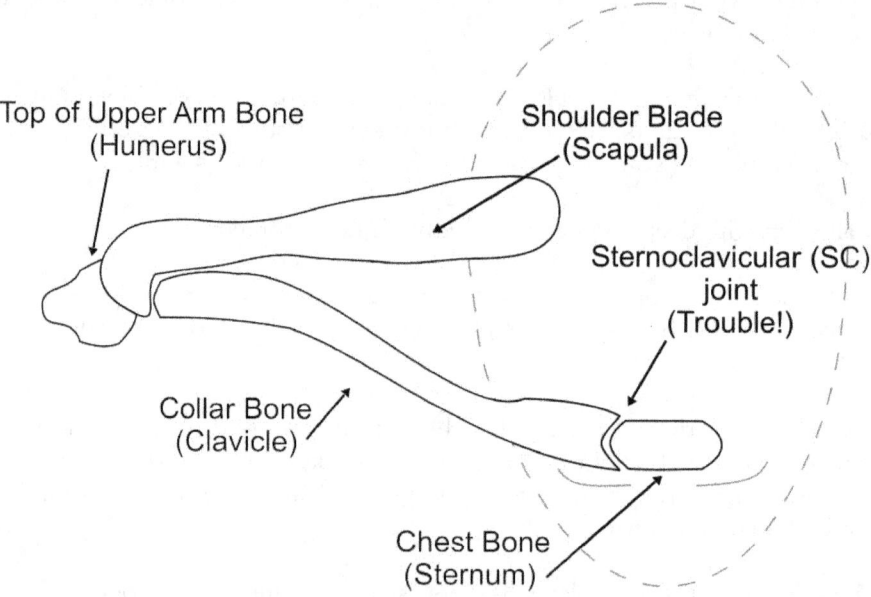

Figure 54: Top-down view of your shoulder assembly. The bone to bone joint that connects your shoulder to your body is between the collar bone and chest bone (Sternoclavicular). Note that the shoulder blade and collar bone form a 'v' in the sense of a 'less than' symbol: <

Secret to Aiki Joint Locking

Depending upon how you slice that, maybe that is four things, but I think you get the point. Forward/back (Sink/Puff the chest) and up/down (Shrug/link the shoulders) all create action in the SC (sternoclavicular joint).

As Uke, we use both the vertical and horizontal mobility of our shoulder to squirm out of a lot of throws, most commonly, to get out of being thrown in Oshitaoshi (#6)/Ikkyo; i.e., we can shift/move our shoulder blade to relax the connection Tori had with our torso.

The loss of connection means loss of control of our COB, and no Aiki!

So how do we, as Tori, control Uke's SC (Sternoclavicular) joint?

Lay your left index finger against your right collar bone as in Figure 55 (base index finger knuckle touching your SC joint, palm down, pinky away from the body, thumb tucked).

Figure 55: Lay your left index finger against your right collar bone; base index finger knuckle on the SC joint

Where Uke is Weak

The objective here is to keep that finger against your collar bone as you shrug and drop your shoulder (collar bone). (Go ahead, shrug and drop)

Next, with collar bone dropped, spread and close the distance between your shoulder blades; (i.e., puff and collapse your chest). Keep your finger attached to your collar bone.

Did you notice how you can easily articulate the finger knuckle joint to match the rise and fall of your collar bone as you shrugged/raised your shoulder?

It should have also been easy to shift the finger sideways to match the collar bone as you puffed and sank your chest. Your finger's range of motion in this dimension is small, but it keeps pace with your collar bone.

What did we learn? For practical purposes, the SC joint works similarly to the knuckle at the base of your fingers. Up and down very fluidly, and a decent range of motion forward to facilitate spreading your shoulder blades, and even backward to close your shoulder blades.

Very interesting, but in all things 'joint lock', the important question to ask is, 'In which way does the targeted joint not work?'

In the negative sense, the SC and base knuckle also share a commonality in how they do not work: they do not like to be twisted!

Grab one of your index fingers and give it a light motorcycle throttle twist toward your thumb. It should feel like an index pivot we use in Kotehineri.

Actually, it is a form of Kotehineri, and you can do Kotehineri on Uke's index finger in this way (and it does not have to hurt).

What I want you to notice is that once the knuckle gets twisted to the point of seizing, it begins to want to move everything else around it.

The very same thing happens in the SC joint. Once Uke's SC is twisted to its extreme (which is a very short distance) it will begin to have effect on the breast bone, which in turn becomes influence on the upper parts of Uke's rib cage/torso.

We care about that because the rib cage is directly linked to the most important bones we can hope to control when generating Aiki: the vertebrae of Uke's spine!

Secret to Aiki Joint Locking

So, from now on, just reach out, grab Uke's collar bone and give it a twist and you will have control of Uke's upper torso!

What?!?! You cannot grab Uke's collar bone? Look again. Are you sure?

Hmm… you are right, we cannot grab Uke's collar bone.

We have a what, but not the how. What to do? What to do…?

Did you figure it out already? Maybe you did! Let us compare ideas.

I suggest we have a really big bone on the far side of the collar bone that we can manipulate to act as a handle to spin Uke's collar bone and drive rotational tension into Uke's SC. It is called 'Uke's shoulder blade'.

That shoulder blade is connected to the collar bone at the Acromioclavicular [AC] joint, but that joint does not matter to us as a joint.

Instead, to us, it glues two bones together and if we learn to shift the shoulder blade correctly, it puts the desired twist into Uke's collar bone/SC joint!

(You had the same answer too! Impressive!)

What does rotational tension in an SC joint feel like?

Try it yourself first.

The quick description: with arms to your sides (start with palms on hips), rotate your thumbs inward and backward until the back of your hands touch your thighs; i.e., spread your shoulder blades some and then try to roll them up and over your shoulder, as if trying to slide it over your body and onto your chest. (Do your best to keep your collar bone parallel to the floor)

As you perform that action, you might notice that the bottom of your shoulder blade will be trying to lift away from the rib cage a slight bit.

Take even closer notice that your shoulder kua wants to close.

You should feel tension building in your SC joint.

Now do the reverse and simply puff out your chest as you rotate your thumbs outward until pointing backward; back of your hands on your hips if you can get that far. (Expect to stop at inner arms against your sides.)

Where Uke is Weak

Can you feel your shoulder blades dropping and coming together (toward your spine)? Did you notice your shoulder kua opening as your SC joint starts to twist?

Besides being an excellent way to get rid of tension in your shoulders (try it a few times and see just how much less tension you feel when you return to the neutral position), this action also tensions the SC joint, and via the ribs, transfers movement (control) into Uke's upper back (thoracic vertebrae). [Not sure where those thoracic vertebrae start? Tip your head forward and feel for the most prominent bone at the base of your neck. That is your first thoracic vertebrae and it is called 'Thoracic 1' or 'T1'. The topmost bone of your torso. Twelve bones down (about kidney level) is the other end of the thoracic portion of your spine. Your ribs connect to these thoracic vertebrae.]

The SC joint is our 'connection' to gain control of Uke's upper torso!

For all intents and purposes, the shoulder blade has the potential to act as the handle of a socket wrench (ratchet) and can convert the collar bone into the equivalent of a rather long socket arm that twists the SC joint.

I said the shoulder blade has 'potential' to be the handle that turns the collar bone, for, how do we grab a bone that is flat against Uke's body?

It sure would be convenient to have another handle to help us move the shoulder blade... figure it out yet?

I expect that you caught on quickly to the reason we spent so much time detailing how to bind Uke's shoulder ball joint (by using the Hypotenuse).

Locking Uke's shoulder ball joint is the prerequisite action to having what is as close as we can get to having a 'handle' with which to manipulate Uke's entire shoulder blade complex.

When Uke's ball joint is locked via twisting of their Hypotenuse (or by 'setting the hook') the upper arm bone (Humerus) becomes the handle we use to manipulate Uke's shoulder blade so that it can twist Uke's SC joint.

Sing it with me... the Humerus locks into the Glenoid, then leverage on the Humerus moves the scapula, which binds the AC (acromioclavicular) joint, the bind is so quick the leverage transfers across the clavicle to the SC, the SC binds up even quicker and moves the sternum and Uke's ribs, Uke's ribs affect T1 through T12...

Secret to Aiki Joint Locking

OK, so I am no song writer and you almost certainly sing better than I do, so let us look at it like this…

  with Uke's shoulder ball joint locked via hypotenuse or hook rotation, we use Uke's upper arm bone like a handle to manipulate Uke's shoulder blade to try to go 'up and over the shoulder, and onto Uke's chest' or 'downward toward Uke's kidneys'.

These actions roll Uke's collar bone forward and downward, or back and downward, respectively. As the collar bone twists, so too does Uke's SC.

Twisting the SC joint delivers direct access to the top of Uke's spine and rib cage. Uke's T1 will 'tip forward' or 'tip backward' per our manipulation of Uke's shoulder blade.

What results is a fold in Uke's torso at the level of their diaphragm.

When Uke's shoulder blade is moving toward Uke's chest, you should recognize Uke's chest collapsing at the diaphragm, and the closing of Uke's shoulder kua as the start of an excellently executed Oshitaoshi (#6) (Ikkyo) or Kotehineri (#11).

Alternatively, dropping Uke's shoulder blade toward the kidneys will 'open' Uke's shoulder kua and puff their chest outward, allowing Uke to form a 'back teeter'.

This will bode well for all of the techniques that seek that open shoulder kua and Uke's head moving backward; e.g., Kotegaeshi (#12), Udegaeshi (#7), and Shihonage (#14).

We are so close to Uke's COB that we can almost manipulate it, but there is one last nuance I feel I should mention before closing this shoulder locking section, the answer to 'Why Kotegaeshi was working more consistently?'

It is a hidden in plain sight difference between rolling Uke's collar bone backward and down (shoulder blade toward the kidneys; Kotegaeshi) versus the closed shoulder kua we produce in Uke when rolling their collar bone forward and down onto Uke's chest (in Kotehineri).

In the 'toward Uke's back', a.k.a. 'Shoulder blade to the kidneys' action, the action itself drives Uke's shoulder onto their body.

In other words, it naturally 'links' Uke's shoulder.

This fact simplifies the control of Uke's shoulder, in that it restricts mobility. Every bit of the effort is directed toward keeping Uke's shoulder in a static relationship with Uke's torso.

In the opposite motion, the Kotehineri type, where we (attempt to) bring Uke's shoulder blade onto Uke's chest, we are not so lucky.

If you are not careful, it is easy to have Uke's collar bone 'lift' and separate from their torso.

If you are not skilled, that lift can release the twisting tension in Uke's collar bone and destroy any connection, thus Aiki, you might have created.

The goal with the 'roll the collar bone forward and down' version is to be able to keep the twist in Uke's SC joint no matter the position of Uke's collar bone. (raised, 'linked', or anywhere in between).

Thus, the SC joint is the sneakiest joint of all to control.

It requires control of Uke's shoulder ball joint and the position of Uke's collar bone, and even then, a bit of finesse to properly control their SC joint.

At least now you know what is happening and can work intentionally toward a goal of twisting Uke's SC joint and gaining control of the top of Uke's spine.

Locking the Spine (Boxes, Teeters, and 'QL's')
We are so close to controlling Uke's COB you should almost taste it!

Stay with me. I gave away a lot in the title of this section, but there are some intriguing details you want to know before you run off and practice what we have covered thus far.

Let us finish the joint collection.

Where are we? Oh yeah!

We have collected Uke's wrist, elbow, and the two relevant joints in Uke's the shoulder (ball joint and SC). So, what is next?

Uke's spine is next, and technically, you already have half of Uke's spine under control!

Secret to Aiki Joint Locking

Our Aiki 'joint collection' thus far, in either technique (both Kotegaeshi and Kotehineri) has started to form control of Uke's rib cage and upper back (twelve thoracic vertebrae). Kotehineri tips Uke's upper spine forward and Kotegaeshi tips it backward.

The upper back feels like a 'wobbly stack of plates' ('Six Precepts') and we are exerting a good amount of indirect vertical pressure on Uke's COB.

That is a lot of control and it is very deep into Uke's frame, but the control stops just below Uke's ribs on both the front and in the back. I think of it as 'folding Uke forward at the diaphragm' on the front and 'Folding Uke backward at the kidneys' in the back.

'Folding' Uke further is not going to take control of Uke's abdomen and that is a bad situation, for, even with all of the success thus far, Uke can put up quite a bit of resistance if we have not attacked Uke's 'core'.

Now, there is a lot, and I mean a lot, going on in your core area. Much more than is needed to understand in order to do good Aiki. So, we are going to skip even trying to comprehend and explain it all. Let us focus on just enough detail to get our job done.

Uke's and our core (abdomen) has a lot of moving parts. They all work in concert to keep us moving. Let us pick on just one part of that complex, integrated machine, and have it throw off the entire operation.

Since we are moving our control through Uke's bones, specifically in this case: Uke's spine, let us attack Uke's lumbar region. Key word: region.

When we say we are going to attack the lumbar portion of Uke's spine, we are not actually going to directly lock these lumbar bones. In fact, neither are we going to tension these bones/joints.

We are going to stress their good friends and neighbors that often help keep these lumbar bones stay erect: the Quadratus Lumborum muscles. (Better known and easier said as 'the QL's'.)

Unfortunately, too many of us know these muscles intimately. Ever have lower back pain? The pain was very likely due to issues with these muscles.

If you have had QL issues, just how powerful were you at that time?

Answer: not very (if at all)

Where Uke is Weak

That is the point. The QL's themselves are not super strong but they account for an amazing amount of the reason why you can even stand up straight, let alone bear weight on your frame.

I cannot think of a better target to attack that is as exposed, vulnerable, and capable of completely 'de-structuring' Uke.

What do the QL's do?

Technically, they contract to bring your hip closer to your ribs. (Think sideways bend at your belt line crushing one of your 'love handles' as you lift a straight leg off of the floor.)

That and they often assist your true back-straightening muscles (erector spinae muscles) when they find themselves unable to get the job done.

Where are the QL's?

They live on both sides of your lower (lumbar) spine; (i.e., lower torso where you have no ribs). You can feel them contract if place your hands on your lower back and raise a hip.

Another great way to get familiar with the QL's is to put one hand on your lower back; the other hand outward to your side and against the wall. Now push against the wall and you should feel the QL's contracting.

You could try to sense them contracting when you go into a plank position. (I would say you could directly feel them with your hand, but you kind of need two hands on the floor to do a plank. Try a one-handed push up maybe?)

Now that you know where the QL's live, what does 'attacking the QL's' look like? Well, you already know!

The excellent news here is that we have already explored lower spine locking within this very book.

Locking the lower spine reveals itself in the various distortions in Uke's body that we described earlier as the 'Boxes' and 'Teeters'. (Basic terms! Not so basic huh? Skipped that section? You got some reading to do! This may be a good time to point out, that although most of the explanations to this point are detailed, there is a reason for it, every concept in this book integrates with another.)

Secret to Aiki Joint Locking

With Uke folding at the bottom of their ribs (i.e., upper torso control), it is time to complete the control by bringing Uke's hips closer to one side of the rib cage and further from the other.

Identifying which side has Uke's COB tells us if we have Uke in a 'stretch box' or 'crush box'. Simple, efficient, and oh so disruptive to Uke's ability to generate power.

So, what is the key here? What is the magic in the 'boxes'?

The actual point of the 'box' distortions is what they do to the QL's: one side's QL is longer and flexed (stressed) more compared to the other.

An imbalance in the QL's depletes much of Uke's structure. In some cases, it is all that is required to completely break down Uke's structure. Uke falls because they cannot keep their spine erect.

This is a very different approach to making Uke fall compared to 'Pin and Spin'. We call it 'de-structuring' Uke when we make Uke's QL's fail.

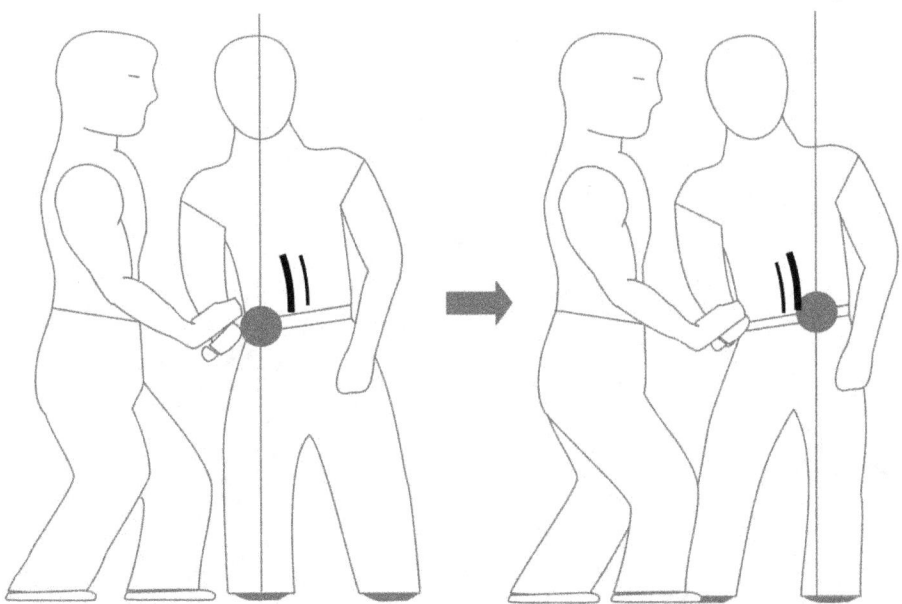

Figure 56: A view of Uke's QL's under stress in the 'Boxes. (left) Uke in right side 'Crush Box' leaves their right side QL more stressed than the left. (right) Uke's left QL under increased stress in left-side 'Stretch Box'

Where Uke is Weak

We can stress the QL's in the Teeters as well.

The back teeter pose hides (exposes?) the act of completely contracting both of Uke's QL's (left side of Figure 57). The QL's have nothing more to give in that pose.

In the side teeter (the 'sprinkler') the QL's are still of equal length, but one is more stressed than the other due to shape of your body. (right side of Figure 57) The side that is higher than the other is experiencing more stress as it tries to fight gravity doing its best to pull you into a side bend.

It is the imbalance in stress that takes the starch out of Uke's form. Letting gravity do this work for you is very efficient as you do not have to exert as much effort.

Because you asked, <wink> Yes, there are ways to stretch Uke's QL's to an extreme (e.g., bending Uke over in the forward direction), but I find it is extremely difficult to the point of nearly impractical. Maybe you will find better success then I.

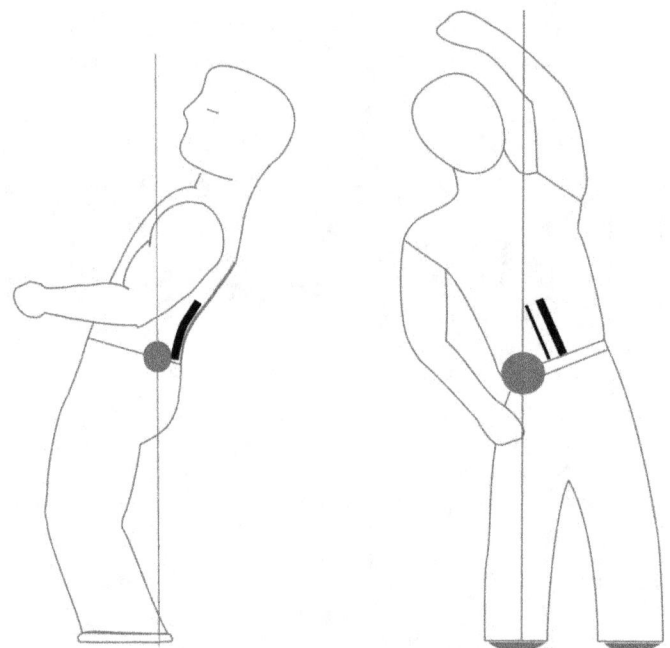

Figure 57: A view of Uke's QL's under stress in the 'Teeters' (left) in Back-Teeter, both QL's are contracted; (right) Side-Teeter the 'top most' QL bears the major stress in this pose

Secret to Aiki Joint Locking

Twisting Uke's spine can stress Uke's QL's to an extreme, but even though the twisted QL imbalance will certainly take the strength out of Uke, any pose is made robust when coupled with a Pin!

To tie this in with earlier lessons: de-stabilizing (willowing) Uke by bringing Uke's COB to the edge of their stance, coupled with QL instability, is complete (temporary) physical dominance (a 'Checked' condition!).

Make it your goal.

So, what do we do with an Uke that is under this complete control? (where we have locked all the joints from Uke's wrist, all the way to and through Uke's QL's?) Throw them, or continue to stress Uke's QL's till they give out.

What else can we do?

Can we go even further down Uke's chain?

Yes!

Why not take the collection all the way, all the way to the ground that is!

Now that we have control of Uke's chain from Conjunction Point (CP) to Uke's COB, let us control the joints in the lower half of Uke's body.

Instead of 'Ground-up', let us call it 'From COB to the ground'!

All we need is a small bit of horizontal rotation in Uke's COB to create that 'opening' or 'closing' in Uke's hip-kua; and if we have 'Pinned' Uke with Uke's COB sitting comfortably on that instep side of the blade of Uke's foot... the stress goes to Uke's knee, and eventually into their ankle.

Remember those 'Three-Legged Dog Spots'?

At this point, with Uke's entire 'chain' tensioned (every joint from Conjunction Point all the way to the ground), just nudge Uke toward Uke's 'dog spot(s)', and Uke might just take the prettiest aerial fall you have ever induced in an Uke. (The real measure of success: you did not hurt Uke! Well, that comment assumes Uke knew how to do Ukemi <smile>)

Incredible and complete control, but was the control of Uke's entire Chain necessary?

Where Uke is Weak

No!

We could have attacked Uke at any point along the 'collection of joints'.

You will begin to find many points along the path to controlling Uke's COB where you can easily knock Uke down.

Throwing Uke early along the path to the QL's (i.e., abdomen, where the COB lives) might take some additional force, or maybe even less if you use a 'Pin and Spin', but the point is, Uke has lots of vulnerability all along the path to directly controlling the COB.

You do not have to make it all the way to the QL's. The advantage of taking control of Uke's COB is the much more complete control and that it opens the gate to 'de-structuring' (over-burdening Uke's QL's).

Please allow me to close this section with a quick thought.

Everything you read (and have yet to read in this book) still supports (strongly) the 'Six Precepts', in fact, all of this is just extension of those root concepts.

Keep going!

but...

please remember that these lessons are simply parts of a singular approach...

the underlying protocol of (physical) Aiki...

the control of stability and balance (yours and Uke's!)

Where We are Strong

Knowing where Uke is weak is truly a good thing, but self-awareness is its complement. How could we ever hope to master the Tori/Uke relationship without a strong awareness of both sides of that equation?

Presence

Not everything in Aiki is physical.

No, we are not going to get all esoteric and wax philosophical about non-discernable forces flowing out of our fingers. Trust me, no one wants to be a Jedi [Sith?] more than me, but it is just not going to happen.

Instead let us talk about the reality and strength of having 'Presence'.

Presence is an "atmosphere of intent". It is the mindset best suited for performing Aiki. In the same way that it is difficult to perform a sit-up while breathing in, it is equally as unnatural to attempt to perform Aiki without a keen projection and sense of 'Presence'.

So, what exactly is Presence?

Presence is an unwavering sense of self, rooted in the commitment to make a stand to maintain, to the best of your abilities, your right to co-exist.

That word "Co-Exist" is a tricky one.

It means that you not only want "to be"; i.e., self-preservation, but it also includes the desire "to be together"; i.e., part of a group. Actually, to be even more precise, "to be part of a team"

Why a team?

When on a team, interacting with your team mates is fundamental. We are predisposed to engage with the other team members; zero hesitation, perhaps even having a predilection to interact.

Being on a team also means you have to have other team mates! We do not seek to eliminate persons from our team!

Oh, and the entire team is 'on the same side'. Our team is strongest when we all find our place and move toward a common goal together.

Have you figured out that everyone you meet is 'on the team' without exception? Membership is automatic and never rescinded.

Holding steadfast in this mindset can be a challenge when one of your 'teammates' (e.g., Uke) is attacking you, but wavering in your resolve to maintain yourself and the team creates serious issues.

That is why Presence is important.

It is truly aberrant behavior to perform violence upon another human. We are wired to propagate our species, not decimate it (if you feel otherwise, you have an issue to resolve, no joke, talk with a professional about it.)

Harm onto another is shocking and unnatural, even in the de-escalated, play sense that we use in training scenarios. Mock striking someone will influence our psyche. Being at fault for accidental harm unto others can have just as much impact as intentional actions. It is only through desensitization and conditioning that harm to another seems a norm. We may suffer even when the action is 'justified'; i.e., necessary to maintain our existence and/or that of our team.

This is why Presence has a few jobs to do all at once.

Presence arms us with a mental toughness to interact without hesitation to the ugly situation where someone would do harm to us or others around us.

Overthinking or overanalyzing will separate your body from your mind. There is no time to cower, or negotiate. To get past the paralysis, we must be pre-resolved to the reality that bad situations can arise and we will act instantly to the best of our abilities to maintain the team, especially our spot on that team (i.e., self-preservation); for, purposefully allowing harm to come to oneself is equally as distorted as acting violently toward another. Self-sacrifice is not an option, it is the 'last action available to protect the team' or your loss was not sacrifice/heroism, it was something else.

Presence additionally shields us from becoming that which we hope to resolve.

The act of Aiki is the neutralization of a potential threat and guiding that situation to a peaceful resolution. Peaceful resolution comes in the form of many outcomes, sometimes even to the extreme of lethality, but the inherent, integrated, and constant intent is a desire to exist within the context of co-existence; i.e., there is no violence in Presence.

Presence

Presence is neither aggression, nor predation.

It is desire to do the least amount of correction necessary to find peace within the team; to maintain the team. This is the mind of the peacekeeper. (the virtue within the intent to serve and protect.)

Presence is respectfully and honorably Alpha.

Alpha is not 'king of the hill'. It is an assured sense of place within a group. None above, none below, it seeks to help others find their place and assist their success as well. Open to engagement, it promotes the strength and order of the team, ready to channel chaos back to civility.

The benefits that you receive from Presence are lasting.

Presence provides mental strength to deal with the after-effects of sordid events. Restoring order through the intent of Presence keeps us unblemished ethically, which supports us in coping with extreme outcomes (e.g., lethality) where we may have had to have become the extension of the 'Tao' upon which our team mate breaks themselves upon.

Is it easy to have this mindset? Absolutely not. Our ego gets in the way.

Instead of 'Presence' we substitute 'need for accolades', 'requital', 'cruelty', 'fear', 'wrath', 'rage', 'insecurity', 'doubtfulness', 'shock', 'self-righteousness', and/or 'reflex'. (not a complete list, and go ahead, pick a combination!)

None of the aforementioned are rooted in co-existence.

Even with this perfect mindset, it is difficult to assert Presence.

Presence requires the ability to be physically dominant. The good news is that does not mean we have to be bigger or stronger, we have tools at our avail that make us physically dominant even over larger persons. (dare I say precepts! Or even tangible armaments some disdainfully/errantly call 'weapons'; within Presence, a 'weapon' is just a 'tool')

So… Should you work hard and become a most accomplished practitioner of any empty-handed martial art, or practice with martial armament, remember to find Presence.

That you may be effective at restoring peace even to the point of becoming deadly, but never dangerous.

The difference between these words: deadly and dangerous, is enormous and should be contemplated. Presence is the divide.

Tori's Hypotenuse

We should probably call this one "Advanced Daito Hands", but it is not about your hands, so let us instead call it what it is, Tori's 'Hypotenuse'.

You just learned about the "Hypotenuse" in the "Where Uke is Weak" section of this book. We viewed 'the coat hanger' in Uke's arm as a source of weakness. More specifically, we learned to use Uke's arm 'Hypotenuse' line to efficiently lock Uke's shoulder ball joint.

That very same line, the Hypotenuse line, exists in you when you act as Tori.

Used correctly as Tori, the Hypotenuse can increase the power and leverage of your arm. Here is how…

Remember "Daito Hands"?

In 'Six Precepts' we learned how Index and Pinky Pivots in Tori help Uke 'fall off the fence'. All we needed was a little bit of 'connection' to create a 'synthetic gravity' that forced Uke to support themselves on our forearm; with just a twist of our wrist, we removed that support, and Uke was toppling.

Besides being a very impressive way to take advantage of the fact that everyone is susceptible to gravity (no exceptions), Daito Hands also taught us that the thicker the arm of the lever, the quicker it is to have a supported object cease to be supported; (i.e., put the conjunction point further from pivot to maximize the effect of Uke falling off of your fence).

Would it not be incredible if we could somehow temporarily grow the width of our forearms to create that thicker lever and amplify the 'falling off the fence' effects on Uke?

We can! Here is how…

We cannot inflate and deflate our forearms at a whim (that is likely a good thing), but we can partner our arm with a new pivot line to create the desired amplified fence falling effect!

Let us start by taking a look at the act of pivoting on our index finger line. (technically, on the line of our radius bone.) (See Figure 58)

Tori's Hypotenuse

The 'Index Pivot' action in Figure 58 depicts starting at a 'right palm down' position and rotating to 'right palm up'; but you can (and should) just as easily reverse the motion.

The action should feel like 'sharpening the pencil' with our index finger as the pencil. Note that the action is accomplished by swinging/orbiting the tip of your ulnar bone (pinky side of the forearm) around your radius bone.

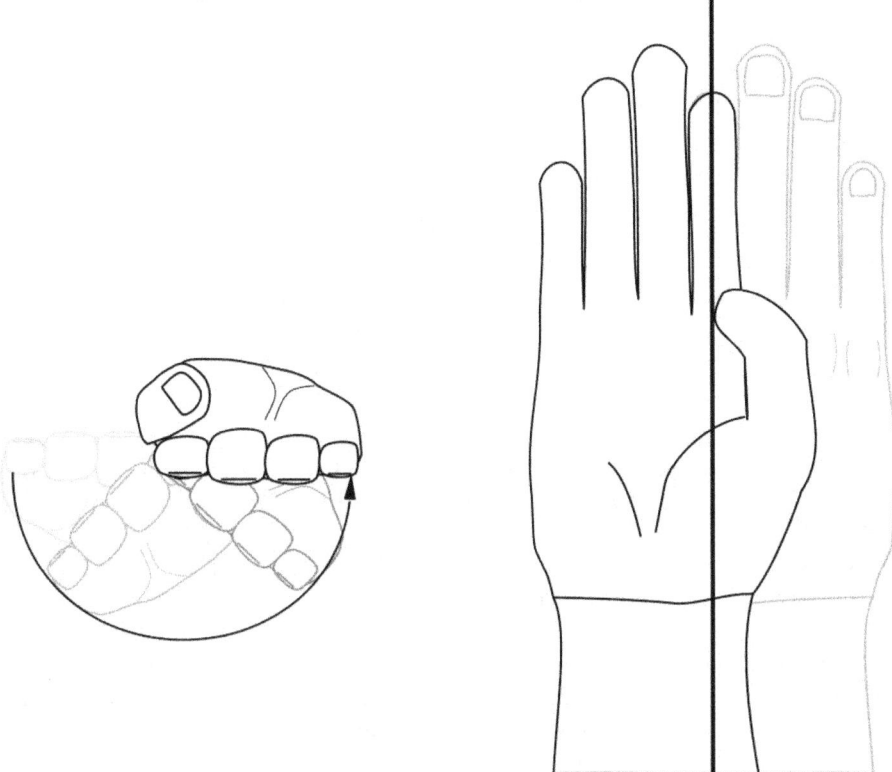

Figure 58: Index Pivoting from in front (on left) and looking downward at your right hand (on right); 'Sharpen the pencil' with your index finger

That gesture is more than enough to move an Uke that is 'connected' (i.e., is made to support themselves) on the pinky side of our forearm (typically by our resting our 'sword hand' on Uke, or when Uke grabs our wrist/forearm)

The point is that this action is happening solely within the space of the forearm; so, why not make it bigger?

Where We are Strong

Let us replace our forearm bones with the coat hanger!

See Figure 59 below, a comparison of your forearm to the 'coat hanger'.

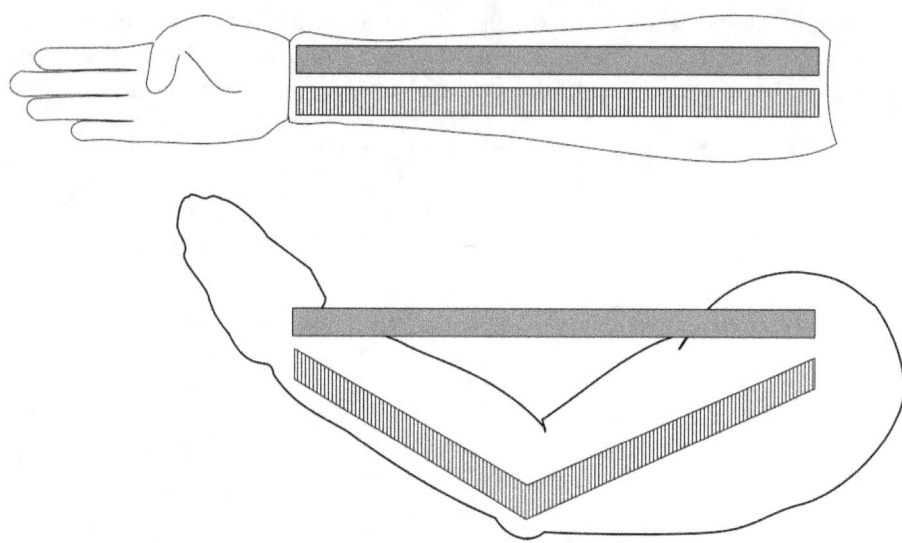

Figure 59: the forearm bones as a pair of rods (top), versus the 'coat hanger' (below). Top bone (radius) replaced by the "Hypotenuse", the lower bone (ulna) replaced by the entire arm.

When we bend our arm and use the entire length of our arm instead of just two forearm bones, two replacements have happened.

Your upper forearm bone (Radius), the one you just pivoted on, is replaced by the "Hypotenuse". Your lower forearm bone (Ulna) is now your entire arm. Note that the gap between the Hypotenuse and your arm is always at least as great as the width of your forearm.

In most places, significantly greater. Voila! A thicker lever!

Not impressed? You will be. Let us try and use it!

Set up the drill by extending your bent arm, finger pointed at Uke as in Figure 60. Have Uke grab your arm.

With no connection (i.e., without extension or compression on Uke) try rotate your hypotenuse; i.e., to swing your elbow to the sides, either in front of you, or to the side and upward.

Tori's Hypotenuse

An Uke of any decent size should be able to stop you or at least put up some major resistance.

Now, go find the biggest person in class to use as Uke, this drill is very rewarding when you see that size does not matter.

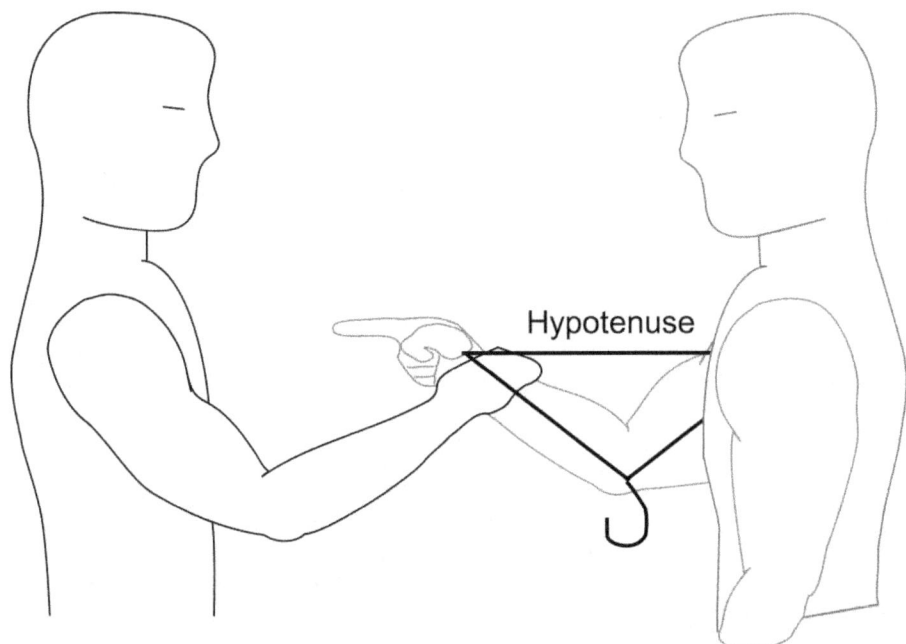

Figure 60: Tori's Hypotenuse Drill: set up with your bent right elbow, finger pointed toward Uke's Sagittal Plane. Uke grabs your right wrist.

This time, create compression connection by trying to physically touch Uke's chest at the area above Uke's COB; (i.e., chest height within Uke's 'Sagittal Plane'; our version of the Sagittal plane where it moves with the COB).

You will not get to touch Uke, Uke should stop you. (Poke Uke with force if they do not) Connection is created at the point that Uke stops your forward movement. (If you do not know what I mean, you have another book to read, and if you have read 'Six Precepts' and still do not know what I mean, you missed a major point: No connection, no Aiki.)

Effectively, you are asserting your Hypotenuse at Uke's COB.

Imagine that you are pushing your Hypotenuse at Uke. That is an important concept. It is not your arm applying the connection, it is your Hypotenuse!

Where We are Strong

Maintain that compression connection and swing your elbow in either direction; almost as if you were twisting your finger tip into Uke [with a bit of distance between Uke and your finger tip of course].

What happened? Uke has very little ability to impede your arm from swinging; i.e., cannot stop you from 'spinning your hypotenuse'.

Why?

Because Uke is 'falling off the fence'!

The connection has created a synthetic gravity. Uke presses against and is supported horizontally by your forearm. When your 'coat hanger' pivots on your Hypotenuse, your arm acts as the top of an upside down 'bent fence'; when the fence moves, so too does anything it supports.

As long as you keep the connection (i.e., keep 'touching' Uke's center [Sagittal Plane – area above, at, or below Uke's COB], wherever it may move to), you will have the ability to swing that coat hanger. That will in turn give you quite a good bit of control over an Uke that has grabbed you.

The great news is that it really does not matter if Uke has grabbed. This same drill can be performed when solely resting your forearm against Uke; (i.e., when Uke 'blocks'). The secret is that you first create connection by asserting and then spinning your hypotenuse.

Be sure to try this drill with Uke blocking; (e.g., Uke blocks your advance…

your advance could be a twisting punch or even a Wing Chun Lop Sau or Bong Sau. Instead of thinking of 'the Sau's' as 'deflection', think 'connection via asserting your Hypotenuse' and guiding Uke's resistance with your arm as you rotate your Hypotenuse. Keep your shoulders linked! [not so] Basic terms!).

Once you understand compression, try the drill extension connection.

Just keep your arm in that 'stiff elbow - unbendable' condition, draw your Hypotenuse away from Uke's Sagittal Plane (your Hypotenuse moves away from Uke like an arrow until Uke stops you), and again relax as you spin your Hypotenuse by moving your upper arm with your torso.

This drill also offers an excellent opportunity to practice using your body to rotate your Hypotenuse (instead of using your shoulder muscles).

Tori's Hypotenuse

Using the motion of your hips/torso to spin your hypotenuse feels a lot like 'using your chest to push your upper arm out of the way' or twisting to allow room for 'your elbow to drop toward your solar plexus'.

You might say we twist our body to 'move' or 'fall' into places that move our arm in the desired manner.

Relaxing into these spaces reduces the amount of effort we need to exert (we get comfortable!), it spares our shoulder muscles undue stress (less chance of injury to ourselves!), and leverages our weight instead of muscular strength (gravity is our friend!)

Such is the path to generating power from your COB.

Leveraging your Hypotenuse also pairs well with Daito Hands (Yes! You can perform both at the same time!) and even has some interesting advantages not found in Daito Hands.

Consider how much further you move Uke with your hypotenuse turn compared to a small Daito Hand wrist twist? (In defense of Daito Hand, it can be done in close quarters and is a more concealed movement.)

Another key advantage of your Hypotenuse turn: all of the rotation is centered within an imaginary line.

The benefit is that Uke has little to no chance to impede the rotating Hypotenuse because practically every point along your arm is part of the 'falling fence'.

You cannot stop a fence from falling if that fence is your support.

Use this drill to test that notion by connecting Uke to any part of your arm besides your wrist or shoulder (the only points on your arm that are actually on the hypotenuse).

Your results should always be the same: Uke falls off the fence.

One last (very important) consideration...

Another, somewhat tricky, and sometimes even troublesome advantage, is that, we can bring this 'falling off the fence' effect to our hand (instead of our arm)

Where We are Strong

This is important when we factor in that our hands are often our point of contact with Uke.

Let us take a look at the benefits of alternate hand positions in this drill.

Let us first take a look at our original 'pointing at Uke' pose and realize this is no different than how we punch.

In our initial 'using our Hypotenuse' drill set up (Figure 60), our index finger was in line with our Hypotenuse.

Assume the pose without an Uke, and turn this finger pointing into a punch by leaving everything else in your body in place as you close your hand into a fist.

With the slightest wrist adjustment, your index finger's 'palm bone' (metacarpal) should be in-line with your radius bone, and that base index knuckle is ready to partner with your middle finger's base knuckle to strike something.

All you need do is push that index knuckle outward along the line of the hypotenuse. I liken it to 'stretching' or 'lengthening' our Hypotenuse. Such is the essence of a straight punch. (and technically, also a push up and a bench press)

Capture that thought: a straight punch travels toward and presses against its target on the line of your hypotenuse.

Does that mean we can turn a punch into a "falling off the fence" situation by rotating the hypotenuse as we extend our knuckles along the hypotenuse line?

Yes, but you have to create and maintain the compression connection.

It is a little tricky, but the best I will offer for now is that it is accomplished by asserting the Hypotenuse toward Uke's Sagittal Plane (area in-line with Uke's COB) before and as the strike is made.

You can break through many a block if, as you meet resistance, create compression connection and then swing your elbow around the Hypotenuse (an eighth of a turn or a bit more if you desire).

This is one Aiki example of how a punch is a throw is a punch, is a throw, is a… it also represents the rudiments of Hsing-I 'Drilling' fist.

Tori's Hypotenuse

That punching/pointing hand position is just the first of three hand positions we can use when rotating our Hypotenuse.

It is the least effective at putting the 'falling off the fence' effect into your hand because the hand is in line with the Hypotenuse pivot; (i.e., it can be done, but not easily. It is really relying upon 'Daito Hands')

We improve our ability to create 'falling' by using the second hand position: the 'rising hand' (fingers to the sky), also known as the 'Sword Hand'.

The bottom of Figure 59 is a decent representation (although the fingers could be pointed higher.). It is helpful to also consider Figure 61.

When in this Sword Hand position, the hand is not on the Hypotenuse line, it is on the other side of the Hypotenuse's pivot compared to your forearm; (i.e., Your hand is above the hypotenuse and your arm is below. See figure 59, the hand is above the straight, grey Hypotenuse, the elbow sits below.)

Because our hand is not on the Hypotenuse, our entire hand shares the same 'top of the fence' aspect as the rest of your arm. (Short fence!)

Thus, it also shares the power generated when we use our torso to rotate the Hypotenuse. (Power in lieu of strength)

The secret is keeping the pivot in your wrist and not letting that pivot drift to where our hand is in contact with Uke; (i.e., the 'Conjunction Point' – anywhere we come in physical contact with Uke)

In this 'Rising hand' shape, it helps to use the focus/intent of trying to touch the pinky/ulnar side of our wrist to Uke's Sagittal Plane as we twist our Hypotenuse; (i.e., assert your hypotenuse toward Uke's chest above their COB, thus, create compression connection; spin your Hypotenuse).

Easier said than done sometimes.

We spend a great deal of our daily life working with our hands as the focus of the action. The sensation of contact between Uke and our hands can be irresistibly distracting and we lose focus on asserting our wrist/hypotenuse (and, instead, spin at the CP)

You can experiment with "Rising Hand" by setting up for the drill as before (Figure 60), but this time presenting Uke with a 'Rising Hand'. (Uke grabs your hand (as if about to start an arm-wrestling match).

Try rotating your Hypotenuse without first creating connection with Uke.

After experiencing the frustration of Uke having a great deal of influence on your ability to rotate your hand/arm along the Hypotenuse pivot…

 create compression connection by asserting your hypotenuse chest high above Uke's COB, (i.e., trying to touch Uke's Sagittal Plane with your wrist) and then rotate your hypotenuse. Freedom!

Try again, this time using extension connection (draw your Hypotenuse away from Uke) and, after you succeed, take another look at Figure 61.

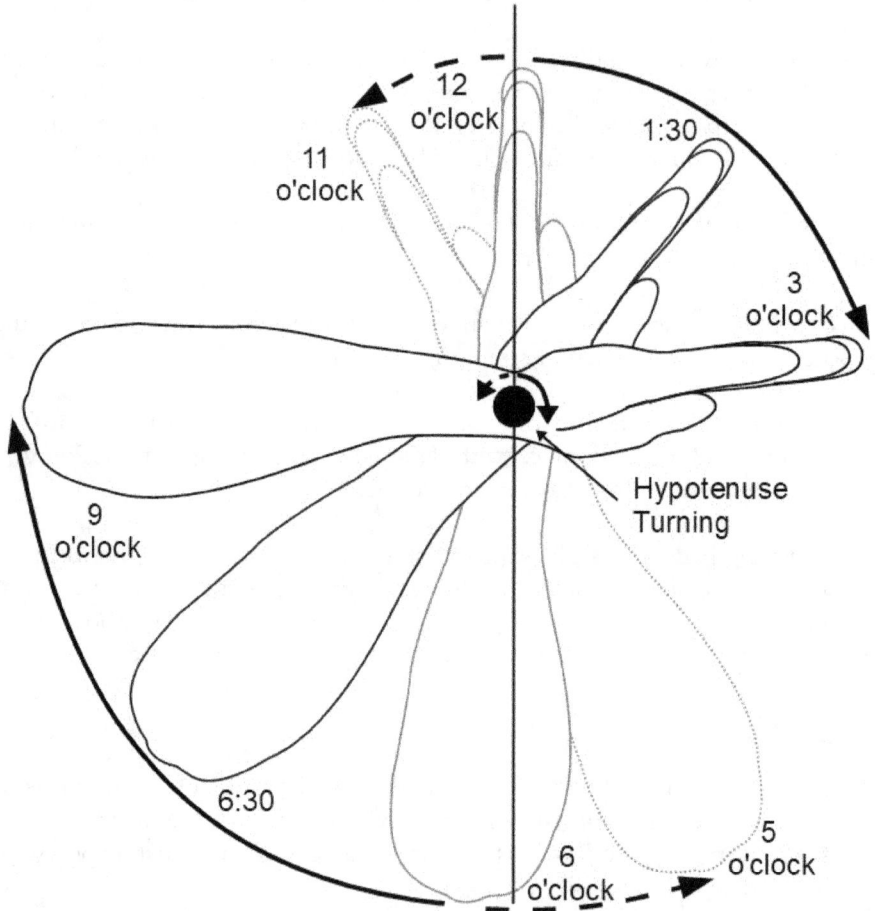

Figure 61: Front view (what Uke sees) of your right "Sword Hand" as you rotate your Hypotenuse. Note that the hand moves to the opposite side of the centerline in relation to the elbow.

Tori's Hypotenuse

Take notice of which direction your hand travels in relation to your elbow when in 'Rising Hand'/'Sword Hand'.

Elbows and hands are, of course, rotating in the same clock-wise or counter clock-wise direction, but they are on opposite sides of your wrist, thus, your hand (with Uke's hand in it) will move opposite of your elbow.

For example, as your and Uke's hand move leftward of your Hypotenuse, your elbow will be moving rightward; (i.e., Figure 61: hand at 12 moves to 3, elbow at 6 moves to 9 simultaneously)

It is important to note that, as your hand moves downward, your elbow will move upward.

This important observation helps us avoid the other common mistake: pivoting at our elbow (instead of the CP), which basically looks like, and is, an arm-wrestling maneuver. (Elbow stays in place as the hand orbits the elbow pivot)

Do not arm-wrestle, use your Hypotenuse (spin at your wrist!)

Remember I said that the advantage of changing our hand shape can be troublesome at times? Here is the last of the hand positions and the little catch that can be a hurt as much as a help if you are not aware.

Most persons are naturally (unconsciously) aware of the elbow and hand moving toward opposite sides of the Hypotenuse pivot (they easily avoid the arm-wrestling, or spinning at the CP/hand) and thus, they have no issue using a rising hand and this motion to deflect Uke's strikes at their face.

No problem so far, moving Uke's strike off the line and downward, especially when paired with us stepping out of the way, is almost always a good thing.

Trouble comes when Uke strikes low, and to intercept it, we drop our hand into the third position, called 'Falling hand'.

When defending ourselves below our shoulder, we typically drop our hand into the third position: a 'falling hand'. 'Falling hand' is otherwise known as 'Ulnar Deviation' (ala what we learned about in the joint locking section.)

Take a look at Figure 62 and again take note of which direction your hand travels in relation to your elbow.

Your hand follows the same direction as your elbow.

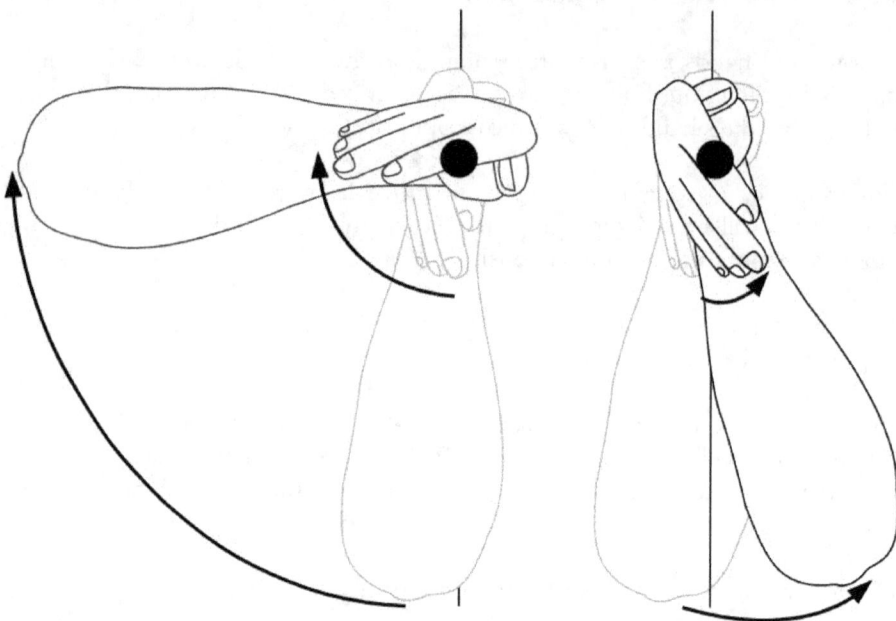

Figure 62: Front view (what Uke sees) of your right "Falling Hand" as you rotate your Hypotenuse. Note that the hand moves in to the same side of the centerline in relation to the elbow.

You get the same 'fence falling' effect on Uke, but you have to be wary of how you are moving Uke or you may be leading Uke's strike into you as you side step.

'Falling Hand' or 'Rising Hand' show up in practically every Aikido technique.

For example, 'Rising Hand' will present itself in Oshitaoshi (#5) (Ikkyo), and 'Falling Hand' in the beginning of Shihonage (#14).

Be sure you are asserting or retracting your Hypotenuse (compression or extension connection respectively) and make Uke 'fall off the fence' in those moves.

As revealed so far, this little secret in our arms called the Hypotenuse is already a great tool to move Uke with, but our situation gets even better...

The trouble is that we need some other background before we can go there.

Tori's Hypotenuse

So, for now, be sure that you truly can spin your hypotenuse even with the largest of Uke's; (i.e., get good at creating connection by asserting or retracting your Hypotenuse toward/away from Uke's Sagittal Plane and then spinning your Hypotenuse) Practice using all three hand positions. (See figure 63) and with differing points of contact; (e.g., Uke or you grab or rest a hand at different points along each other's arms.)

It will pay dividends to get comfortable with using your body, not your shoulder muscles, as the source of Hypotenuse rotation.

Doing so is a lesson in expressing the power of your COB through your hand (and avoiding having to have strong arms). Strength is in muscles; power is in the application of pivots and supports (aka fascial alignments).

When you are truly comfortable, experiment to discover Aiki hypotenuse rotation can be made to work with any hand position and, although not detailed here, is demonstrated in many other martial art forms.

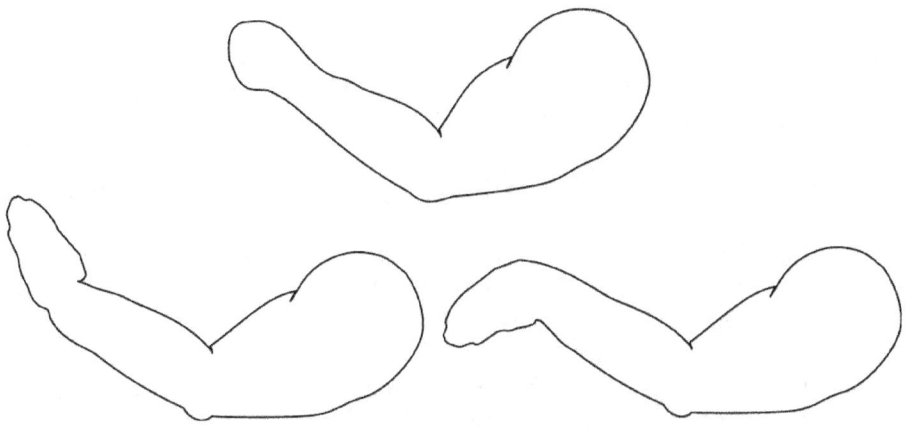

Figure 63: three hand positions we used in our 'Tori's Hypotenuse' drills. Top: punch; Bottom left: Rising Hand (Sword Hand), Bottom Right: Falling Hand (Ulnar Deviation)

Suggestion: Find as many Wing Chun or push hands videos as you can and look for the hypotenuse in action.

One of my favorite non-Aikido style expressions of the Hypotenuse turn is a move in Pentjak Silat called the Suliwa. (While you are at it, look up some Aikido videos and draw some conclusions!)

When watching, take note of the practitioner's ability to create connection and maintain it for the purposes of Aiki (control of stability and balance), but remember, not all arts bother to create Aiki, they do not need to. (No excuses for the Aikido practitioners, they must.)

Let us now explore another 'element' about 'where we are strong' before we start putting the major pieces together.

For all of the food lovers out there, I introduce...

Fork and Knife

It happened during my first few of months of Aikido training that I was Randori sparring with my, at that time, primary Aikido partner-in-crime Tim when I stopped to ask Sensei Carlisle, "When should I attack Uke?"

Sensei replied, "When he stops moving"

If you have read the preface of 'Six Precepts' you know where that discussion ended and the journey began.

Since then, I have been shown many thousands of times what Sensei Carlisle meant by that, (shown by many persons, Moe, Chris, countless others) but it took almost fifteen years until I was finally ready for the lesson.

Coincidentally, it was another Tim that finally got the lesson through to me.

Sifu Tim Wolfe is a masterful Aikidoka in his own right, but it was his demonstration of Black Panther Gung Fu that, very literally, hit me in the head hard enough to see what was going on.

I was practicing in the weeks between seminars when this concept became clear to me; it appeared as an instance of how a 'Fork and Knife' interact.

I could not wait to get back to Sifu Wolfe to share what I figured out while I was away and working on some of the material he had shared in the previous seminar.

When I shared, he smiled and said, "Yeah, you get it" and then proceeded to rattle off four more instances in modern western living where you can find this principle (four cases I had not even thought of. The experience was just one of many fine examples of why I go to him to study, not the other way around!) Those instances are his to share and if you ever get to Eastern Ohio, you should visit with Sifu Wolfe. Till then, here is my take:

Fork and Knife

The concept I share here is not specific to any style, it is as rooted in Aikido as it is on your dinner plate, and serves as another way to perceive the duality of nature, the relationship of Yin and Yang (they are complements, not opposites.) It is the hidden essence of the striking arts. Let us reveal it.

Stop for a moment and imagine your favorite, must cut meal; (e.g., For me, it is a steak, (~1 inch thick, cooked slightly less than medium rare! Yum! A 'well done' steak is 'rare'!) but if you must, I guess you could use a large stalk of lightly steamed broccoli that you would also find on the other side of my plate of steak. (Yummy too!))

The waiter messed up and gave you two forks with your meal. Not going to work is it? Before you give up and use your hands, ask for a knife.

The waiter takes all the forks and replaces them with, you guessed it, a single knife. (No tipping this waiter!) Not going to work well is it? Why?

In order to cut something, the object has to move (roll! [more on this later]) along the blade or the blade move along (roll) the object.

The movement is all relative. This means the knife stays in place as the object moves, or most commonly on our dinner plate, the object stays in place as the knife moves.

Thus, the fork serves to hold, control, and set up the conditions for the knife to come split the object of our intentions.

They work together, simultaneously, and as complements to each other (each doing their specific role, not trying to do the other's; thus, achieving a different, combinational, and greater goal. A 'marriage' of efforts.)

Sounds pretty simple, but this 'everyday miracle' is an important lesson in how we should be approaching our techniques.

Most commonly in Aikido, Fork and Knife are expressed in separate hands.

The first of our hands to come in contact with Uke facilitates 'the Fork' as it redirects Uke's attack and simultaneously applies weight onto Uke in a way that burdens Uke into a (sometimes momentary) rest where Uke's vulnerable target(s) are exposed at a perfect range for our other hand to become the knife and complete the desired response to Uke's aggression.

(Phew! Yeah, that last paragraph is a mouthful!)

All the while, our other hand, much like a perfect football 'corner of the end zone' pass, where the ball is in the air long before the receiver even turns to catch the ball, becomes 'the Knife' and has been on route to its target while the fork is still performing its job; and with the confidence that the fork will succeed, the knife arrives right on target and on time.

Contrast this with a game of 'whack-a-mole', where you never know when or where the target will appear.

A single-handed mode of attack missing as much as it is hitting.

What is the point..? The amount of skill required to hit a moving object is enormous compared to hitting an object you hold still.

I wasted many, many years sequentially applying a fork and then the knife, never truly realizing the roles, thus never combing them, I would more often than not, completely let go of the hold provided by the fork, switch to using solely the knife, and reverting to the luck of trying to hit a moving target.

Could you easily hit a fly with a hammer? (please do not try) Could you hold and hit a nail with that hammer?

When you answer honestly to yourself and understand the difference between those 'hammer' situations, you have the insight to why we should learn to simultaneously effect the two roles of Fork and Knife.

How many forks and knives do we have?

To answer that question, you have to consider that the fork and knife are not always in your arm/hand, your legs and hips can also act as either. (Find your local Judo master for a lesson in 'grasp Uke's gi and use your hip as the fork, leg sweep as a knife'.)

You also have the option to do as Sifu Wolfe so often does, perform both of the Fork and Knife roles within a single hand/arm. Combined with entering into Uke's attack with a Cat Stance (or San-Ti Shi), the technique is incredible.

One fork, more than one knife.

Your fork is additionally useful to 'listen' to Uke, to determine if you have Uke in a fully 'checked' position; to sense if additional control is necessary, and what Uke might be planning to do next. Sifu Wolfe dispatches me so quickly that I know he does not need to listen for a long time, but he can just

Fork and Knife

as easily hold me there as long as he desires. I guess if you are really good at creating a fork, it may remove the need for the knife. (Keep Uke 'checked')

What is the lesson? That you have two jobs to perform simultaneously!

Your contact with Uke must fulfill these two roles from the inception of your response to Uke's aggression to the very end of the 'technique'.

Where is the drill?

Well, I could tell you that an 'X-block' should no longer be seen as 'two hands blocking'. You could find that one hand receives Uke and applies the Fork, while the other hand is up front and already moving toward the strike. (I really should have paid more attention in Tae Kwon Do class!)

Alternatively, I could tell you that a good many Wing Chun practitioners seem to be displaying excellence in fork and knife. Drill by finding a Wing Chun practice dummy and see if you can derive a sense of Fork and Knife in the Sau's and strikes. (Note: Sifu Wolfe practices on a Wing Chun dummy!)

Instead, since the audience that would most likely read this book is Aikido and Aiki Jujutsu students, my suggestion is, 'go with what you know'.

You bring the drill; I will bring some tips.

Start by researching your absolute favorite Aikido/Aiki Jujutsu move. You can almost guarantee you are using a Fork and Knife effectively.

Take note of where you are leveraging the Fork and Knife, and continue progressing the investigation into every technique you train, especially the ones that are not your favorite.

What might be missing, and could make the change from frustration to success, is the awareness of Fork and Knife.

Here are the tips for when you are analyzing or 'bringing your utensils' to a new technique.

The fork always comes first, if not, you are just hitting a moving target.

Go slowly, work solely on the fork, focusing on willowing Uke with it, even if just for an instant. Complete the fork in the 'Tick, Tick, (pause)"

Once you are comfortable and consistent with the fork, start bringing in the other hand as the knife (simultaneously!).

A hint about creating the knife: I suggest that you might ditch the 'kata' aspect of your practice.

In many ways, the basics are taught as 'basics', and what presents as 'good kata display' is often not conducive to finding practical application; i.e., Tomiki Aikido practitioners: give up the 'demonstration' and truly practice your seventeen kata as Randori!

This is when study outside of your art makes a great impact! Sharper knives!

Find the techniques that resemble, or might even be the root of, your Aiki technique.

Insight to the throw exists in knowing how to convert it to a strike, break, joint-lock, etc. (alternate "Whoosh!")

Remember:
The duality in your technique is your strength, both roles created at the instant of the first 'Tick' in 'Tick, Tick, (pause), Whoosh!'…

and oh, what a 'Whoosh!' when they finish together!

Complements moving together, instead of one hand clapping!

Ligamentary /Fascial Grasping

Having trouble applying that 'kime' attack on a pressure point? Finding yourself tiring in your hands, wrists sore when working with a bokken or jo? Getting accused of using too much force when you grab, and/or cannot seem to hang on well enough to perform that wrist lock?

Ligamentary/Fascial 'Grasping' is the answer!

Grasping is also an excellent introduction to 'internal power'. Please be certain to 'grasp' (pun intended) this lesson as it will make the following lesson (on Zhan Zhuang) all the more fruitful.

So, what is Grasping?

Ligamentary/Fascial Grasping

It is learning to attach your hand to an object (e.g., a bokken) using the inherent structure of your hand and forearm instead of 'grabbing' an object by flexing the muscles (in your hand and forearm).

'Grasping' an object has many advantages over 'grabbing'. The primary benefit is an increase in friction between you and the object you want to attach to. I like to describe it as 'hand glue'.

Additionally, the lack of muscular exertion improves your stamina, allows you to absorb more impact (improves durability), provides greater range of motion and it allows a more direct transfer of the power of your hips through your hand.

So how do we 'Grasp'?

Sounds silly as I write this, but have you noticed that your body does not fall apart when you relax? Many never stop to put this awareness to use. Those that do are called 'internal martial artists'. Let us join them for a bit.

Take a moment to sit at a table and lay your right hand on the table, palm up, fingers pointed away from you (no bend in your wrist). Relax your right hand and forearm. (Truly relax them, not 'rested', make them flaccid.)

Did you notice that your fingers do not fall off? Even with a fully relaxed forearm, you should notice that your fingers still do not lay flat on the table.

There should be, at minimum, a subtle curve to your fingers.

Keep that right hand relaxed, wrist loose, and use your left hand to, one by one, grab each of your right-hand fingers by the tip and lift your entire right hand off the table (bend your right wrist, Figure 64); set it back down again.

Again, your hand does not fall apart does it? Turns out that you cannot easily rip your fingers off, but what is holding your hand together?

The 'things' that maintain the structure and connectedness of your fingers, hand, and even the wrist (your whole body for that matter) are the sacks of flesh (skin, muscles, connective tissue, etc.) that encase each other and ultimately wrap around the bones (or in case of muscles are attached to the bones via tendons), the bones kept linked by ligaments.

Lift your hand by each finger again.

Sense the tension it creates in the joints and the stretch in your flesh; through your fingers and even into your forearm.

You are stretching, and at the same time, loosening ligaments and fasciae.

This 'distortion' of the fascia, is best described as 'unrest'.

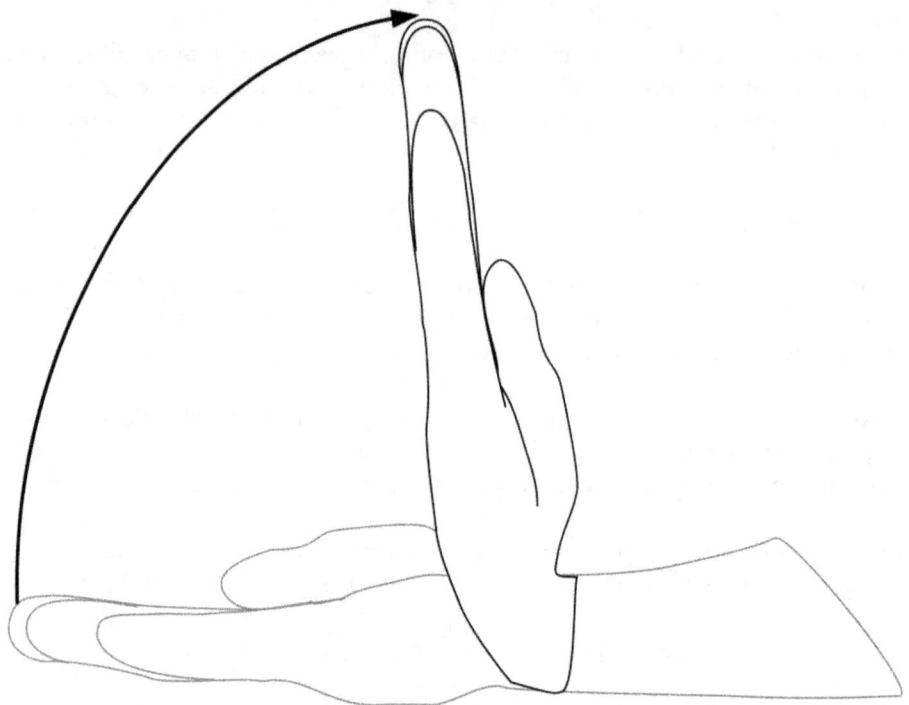

Figure 64: Resting your right hand on a table, use your left hand to, one by one, grab a right hand finger-tip and raise your right hand as above

There is a springiness to this stretch, maybe not as overt as a rubber band, but in its own subtle way, very much the same.

Of course, letting go of your finger will allow gravity to bring your hand back to the table, but what also brings your hand back to the table, is that, once you let go, just as a rubber band will, your hand will retract and come to rest again.

The flesh and joints of your hand and forearm have a natural shape they want to assume, and in this desire, they mimic a rubber band trying to return to rest.

Ligamentary/Fascial Grasping

An incredible amount of power lies within this aspect of your anatomy. All we need do is understand how to, and what happens, when we unrest, then rest (distort and restore) our structure instead of flexing our muscles.

Let us start with improving our understanding by adding even more ligamentary/fascial tension into our right hand.

Same right-hand palm-up, fingers out position as before, but this time not against the table, instead, simply out in front of you. (Like holding a platter/plate full of food.)

Keep your right hand supple and rested as you use your left hand to come under to the back of the right hand and use your left hand to flex your right-hand wrist to finger-up position (just like Figure 64 again)

In other words, the left hand pushes up on the back of your right hand to put the palm facing you, fingers to the sky. (Use your left hand to knock the imaginary plate of food out of your right hand and onto your forearm).

The point is what happened to your fingers: they straightened out without muscular exertion!

The subtle curve in your fingers that was there in the starting position of holding a platter has gone away.

Keep your hand pushed to its limit of motion and you will find that curling your fingers is difficult, almost impossible.

Interesting if we wanted to relax someone's grip on an object (disarm!), but not very conducive for grasping is it? Let us try the other direction.

Back to 'holding that platter' position with the right hand, supple, but rested. (note, and you will read this many times: Rested, Not Relaxed)

This time use your left-hand index and middle fingers to push downward on and extend your right wrist. Think 'drop the platter on the floor in front of you' (Fingers downward, the gray, shaded bottom start of the Figure 65 'moving image')

Take notice of your right-hand fingers, especially the pinky, ring, and middle finger. They curled rather naturally, and again, without muscular exertion.

Go ahead, try to straighten your fingers, it is not easy.

Your right thumb wants to come closer to the palm too. Now we have something to apply to grasping, but we are not done!

Complete the drill by keeping the extension in your right wrist (hold that right palm down with your left-hand fingers) and roll your right wrist over so that the forearm bones are crossed, your right-hand fingers are curled but knuckles upward, closer to the ceiling; (i.e., pronate your right wrist; put your hand in position to push open a door [but your fingers will be curled] Figure 65).

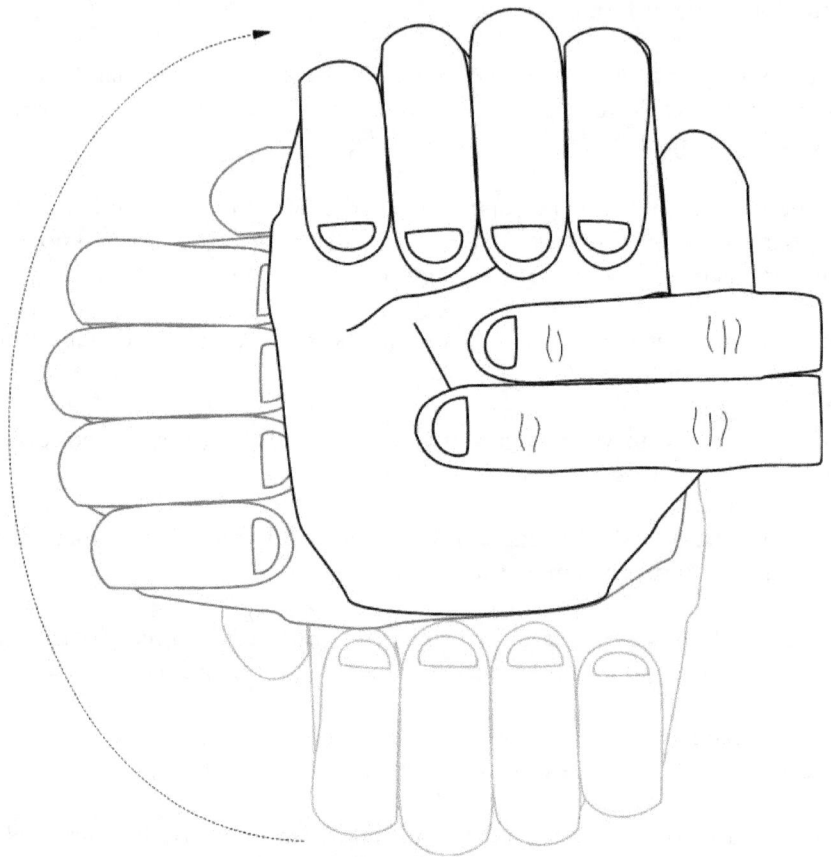

Figure 65: pronate your wrist from 'dropping a plate' to 'pushing on a door' (with your fingers curled); note the increased curl at the finish (fingers up) compared to the initial 'fingers pointed down' position

Did you sense your fingers curling even more? This is the awareness you want to develop, as distorting and returning your structure to rest is the basis of internal power.

Ligamentary / Fascial Grasping

What have we accomplished so far?

What we have done is confirm that our hand/forearm wants to return to rest and we can use that particular desire to shape our hand in a way that can aid us in attaching to an object without exerting our muscles!

Why did our hand shape itself accordingly without flexing, activating, or exerting any muscles in our right hand/forearm?

The response in our fingers was due to our mechanically (not muscularly) stressing the tendons and 'meat sacks' that curl and extend our fingers.

Those tendons connect the bones to muscles. So, when we shift our bones, we inherently tension the connected tendon(s), and effectively begin to stretch the associated muscle(s).

The muscles have their desire to stay at rest.

So too do the layers of 'fascia' that surround those muscles (e.g., the layers of connective tissue that wrap around the muscle and lie between the muscle and the next layer of flesh; repeat that 'wrap and in-between' pattern again and again until you get to the skin [epidermis layer]. Yes, even your skin wants to maintain a particular shape.)

Every layer adds their desire to keep their shape and thus, together, they all assist in curling the fingers without muscular exertion.

In this particular case, something had to give, and it was the joints of the fingers that had to curl.

So, let us put what we have learned to good use. How about we learn to 'grasp' a bokken?

Do not just sit there, go get a bokken!

Now pick up the bokken with your left hand with the 'blade up' (blade toward your left palm/the ceiling), grip the bokken just above the handle area (above the Tsuba), and get ready to place the bokken handle in your upturned right hand.

Prepare the right hand by putting it again outward as if holding a plate, but this time with a slight ulnar deviation. Just a little, not quite as much as we demonstrated in the Kotegaeshi lesson. (see Figure 66)

The objective is to line up the right hand's index palm bone (metacarpal) with your radius forearm bone; i.e., just like a punch. (note the thin line that cuts through the middle of the radius and that index palm bone [metacarpal] in Figure 66)

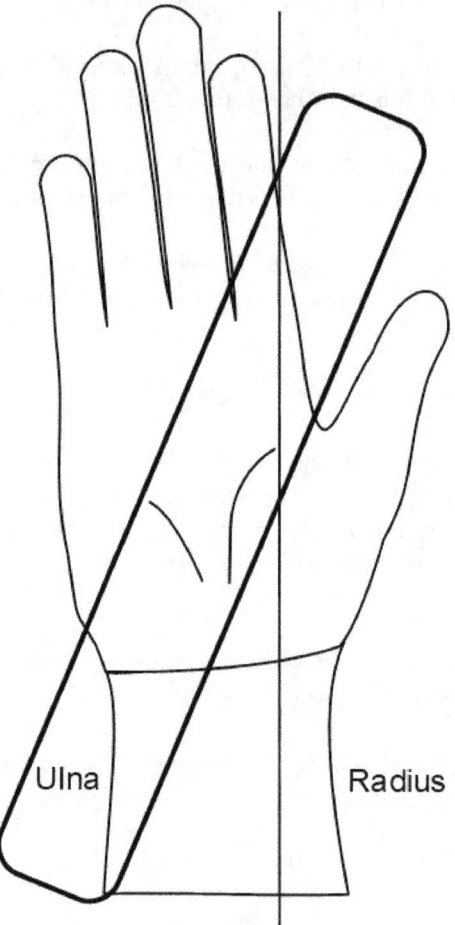

Figure 66: Right hand palm up, slight ulnar deviation, index palm bone [metacarpal] lined up with the radius bone (thin line); upper portion of the bokken handle across the hand from base index knuckle to middle of the palm heel

With right hand in position, your left hand crosses over the right hand to place the top part of the bokken 'handle' (i.e., at the normal position that you would have your right hand grasping) across your right hand, blade upward.

The idea here is to 'wrap' our right-hand fingers around the bokken.

Ligamentary/Fascial Grasping

Do this by pointing all of your right-hand fingers (including your thumb) upward and stretch them as high as you can, keeping that stretch as best as can as you lay your fingers (and thumb) over and around the bokken handle.

Once your right hand has enough grasp to at least keep the bokken from falling out of your right hand, you can cheat a little here by using your left hand to, one by one, help get the right hand fingers stretched across the bokken handle. (left hand pulling on the right hand finger tips as before; this time, stretching each finger before setting it down to rest.)

I like to think of it as 'wrapping' my right-hand fingers around the bokken, much like we are wrapping leather around a baseball bat handle.

Pretty good grasp so far huh?

Your right hand should be comfortable (rested!) at this point. It has to be, because we are going to improve the grasp by making it even tighter.

The objective now is to allow the friction between your skin and the bokken to hold the bokken in place as you truly relax and allow your right wrist to extend; i.e., allow the tip of the bokken to drop toward the floor as your forearm/wrist stay in place; your right hand folds over.

This should feel just as before, when you pushed your right hand down into the 'drop the plate' pose, but with left-hand replaced by the bokken and it drawing/extending your right hand downward (back of your right hand toward your right elbow).

You should have felt the pinky, ring, and middle fingers tightening around the bokken. Depending upon how well you stretched and wrapped your fingers around the bokken at the very beginning of this drill, you may have sensed additional 'stretch' in the fingers.

That stretch is desirable.

If you were too tightly wrapped in the beginning, you might have found you could not extend your wrist at all. If so, reset and give a little slack.

If you are lucky, you got it correct the first time. Most of us are not that lucky as we have not placed our hands in just the right spot on the bokken, so our grip might be a little left or right of center, or too close or too far from the tsuba.

Where We are Strong

Try a couple of times until it feels comfortable, then just as before, take the last step of rotating your right hand upward. Do this by using your left hand to grasp the bottom (butt end) of the bokken and rotate the bokken with the pivot in your right palm; stop once you reach the upward 'ready position' of your sword. Your right wrist should have rotated and your right forearm bones should be crossed.

This last action of rotating the wrist should have again brought a sensation of your right hand hugging the bokken handle without any sense of flexing your hand and forearm muscles. Your left hand should be 'grasping' too!

When you feel you have it right, keep that rested tension in your right hand, release your left hand, and shake the bokken a bit. The sword should not be lifting away from your palm. If it is, you need to try again.

For those that are getting it, ask yourself: how difficult would it be to have someone pull the sword out of your right hand right now? Do you sense the 'stickiness'? How supple is your wrist? Can you sense the extreme level of impact you can withstand when locking bokken with Uke? How much direct impact could your hand take and not release the bokken?

This is the condition you want your hand to be in when delivering force through a weapon (or any tool for that matter, a wrench, a hammer, a screwdriver, fork, knife, etc.) or even through the empty hand itself (fist).

To further that empty hand point, you have not only gained insight into 'rested tension' and how it is used in grasping objects. You have also learned the core secret to turning your empty hand into a striking device (but alas, striking is not the subject of this book, but there many other books out there!)

I find the books regarding Okinawan Karate take the most time to reveal the multitude of fascial/ligamentary tensioned shapes your hands can display, along with the contact points of your hand in these positions, as well as, the targets they are meant to impact.

Unfortunately, the fascial/ligamentary aspect is not often noted, so be sure to take it upon yourself to note of how the hand positions create tension in the fascia and ligaments of your hand/forearm to support the contact point.

There is a clinical, scientific reason that you do not want to be on the receiving end of strike delivered by a Karate master. They have perfected this lesson and demonstrate it effortlessly, devastatingly.

In Aiki, our elegant use of fascial/ligamentary tension in our hands is displayed when we grapple or use weapons.

Could this drill be applied to the Jo? Of course! (and any other weapon for that matter; e.g., knife, stick, whip, Tai Chi sword, etc.)

Could we apply this method to joint locking? Absolutely! and you should! Make it your goal to 'let go' of muscular exertion when 'hanging on'!

Now that you have taken the time to understand the difference between 'Rested' (very desirable!) and 'Relaxed' (the wrong advice in 99.44% of all contexts within martial arts) you are ready to contemplate applying 'Rest' to the rest of you.

Zhan Zhuang (According to Bill and Joel)

I hesitate to implicate Joel in what you are about to read, but our constant desire to one-up each other has led to a study of Zhan Zhuang that goes outside the book explanations we started with.

Yes, you read that correctly, so let me be the first to point out that, outside of a few experiences with a friend named Rob (who coincidentally, unintentionally set Joel and me on this journey to find the martial application of Zhan Zhuang), I am completely without an instructor directly transmitting to me the secrets of Zhan Zhuang.

So, if you would like, call me a hack, skip this section, but you might find that you have missed out on another very important introduction to internal power. (Even if it is coming from a book smart, experimental novice such as myself)

Still reading? Excellent!

So, what are we going to investigate?

At the simplest manner of explanation, we will learn more about creating structure, which is requisite for learning how to accept and project force with minimal muscular exertion. It will result in improving our posture, make us more durable, add to our stamina, and generally improve all of our martial abilities.

It all starts with standing, aka Zhan Zhuang, but it goes so much further.

Where We are Strong

For the uninitiated, Zhan Zhuang is the Chinese art of standing meditation (a form of Chi-Kung mixed with martial conditioning).

From what I have researched, Zhan Zhuang is foundational to Hsing-I and I-Chuan. That makes a lot of sense, because Zhan Zhuang seems foundational to all 'internal' martial arts; (e.g., Ba Gua, Tai Chi, and yes, although not formally an internal art, even Aikido [Especially Aikido!])

The martial secret of Zhan Zhuang is managing your fascial alignment at all times. Fascial alignment is first learned when standing and is then practiced when stepping. From stepping, we add it to everything we do. (standing, sitting, jumping, striking, kicking, meditating, etc. Everything physical is better with Zhan Zhuang!)

So, where did I read these things?

My 'instructors via print' in this craft (in no particular order) are Master Lam Kam Chuen, Waysun Liao, Jan Dipersloot, and Mark Cohen (all masters in my book!).

Find their books to know the formal written instruction I have received.

Which book?

All of them.

I own almost all written materials by these authors prior to 2015. I will suggest, based on my appreciation of their previous work, that any post 2015 work would prove just as valuable (if not more valuable! Internal arts get better with time/age.)

I was blessed to have some informal instruction too.

Informally, Sensei Steve Carlisle technically started me on this path long before I knew it was a path; when he first threw Master Lam's books at me while teaching me the basics of Chi Kung (my fault that I never went past the basics) and in how Sensei Carlisle constantly integrated his background in Chinese martial arts and traditional Chinese medicine with Aikido.

Two decades later, Rob came into class and cemented our desire to explore the possibilities by literally running over all of us like a juggernaut; claiming it was his Zhan Zhuang practice that enabled him. The rest of the clinical study was as test dummy for Joel Copeland and vice versa.

Zhan Zhuang (According to Bill and Joel)

Enough background, let us get to it.

Here is a graphic reality about most of us...

Although we are standing erect, we are not standing 'aligned', meaning we are not relying on the fascia of our torso and legs to keep us upright. As such, we are using muscular tensions to compensate for what equates to bad posture and all sorts of subtle damage is being done to ourselves. The damage gets less subtle when we opt to compensate and then engage in martial arts.

Let us learn to correct that situation and generate some power

Mountain Pose – the first alignment
Here is the proof that we are typically not aligned.

Ask anyone you can find to stand in front of you with hands in position as if they were going to deliver a pizza. (Upper arms at their sides, elbows bent, their forearms outward, and hands upward) You might liken it to putting their hands in position to carry a number of wooden logs to your fireplace.

Now, without actually touching them, feign dropping all of your weight on their hands.

What happened?

If they did not flinch, it was because they are one of three types of people.

The first type cannot pay attention to anything and did not realize there was a large amount of weight about to drop on their hands; (i.e., awake, but not really).

The second did not see you as a threat. Could have been your fault, maybe your feint was not believable, but lump these persons in with the 'too hard to work with at this moment' people.

The last type, and the most unlikely of all, was the one that stands ready to accept that weight at all times. This kind of preparedness can be made to be natural and non-stop, but it is not likely. (ask them if they practice Zhan Zhuang!)

The rest of us flinched. Those that did are actually the lucky ones, for there is a deep secret held within that flinching; a secret your body knows, and your consciousness needs to become aware of: fascial alignment.

You can teach yourself quickly if you can get your mind out of the way and watch nature take its course. Here is how.

Try that simple drill on yourself. You do not even need someone to perform the feint of dropping weight on your hands.

Hold your imaginary pizza box and then imagine someone unexpectedly dropping a large carton of eggs onto your hands (dropped straight down from the height of your chest, not thrown at you)

Emphasis on unexpectedly. (if it helps, put yourself in the same mindset as when playing the hand slap game. You know what is coming, you just do not know exactly when.)

You can let your ego tell you that there was no need to flinch or even adjust as you imagined catching the eggs, or... you can join the rest of us that felt a significant number of subtle shifts within our torso (and potentially our legs) that were trying to prepare us for bearing additional, external weight. (even as little weight as a dozen eggs!)

If you are not sensitive enough to feel the shifts, stop now, skip the rest of this chapter as you will be wasting time and risk getting hurt.

Try again later, but do not rule out that it could be that the solo exercise just needs a real threat to get you started.

So maybe, before you give up, go find a friend to feign dropping the weight, or actually drop a carton of eggs on you (We should all have friends that would throw eggs at us! <laugh>).

If you still cannot sense any shifting in your body, skip the rest of this chapter, take the eggs, and make a really nice omelet to share with your friend!

If you are still with me, you are sensing some adjustments in your torso.

Perfect, because that is all you need to know.

You could stop and search the internet for a bunch of diagrams and other explanations for what fascia is, or you can realize that your body already knows and it just revealed fascial alignment to you.

Yes, the answer is already inside of you. You just had to pay attention to your body.

Zhan Zhuang (According to Bill and Joel)

The subtle changes you felt are your bodies way of aligning the 'meat sacks' that lie next to or surround your bones and through the natural intent of these sacks trying to remain at rest (Not relaxed! At rest!)

(I am going to say it again: AT REST! NOT RELAXED!)

These sacks can keep your bones in position to support the (un)expected weight gain; (i.e., the imaginary carton of eggs dropped onto your hands).

Not everyone gets this drill correct at first try. The most common error is to lift the hands and tense up (flex), especially in the arms and lower back.

That has to stop. That is 'you getting in the way of your body'.

What are you tensing up for anyway?

This initial drill uses imaginary weight to avoid the muscular flexing situation (and to avoid injury!) There is literally nothing there to actually resist, and would you really use all that force to catch a dozen or so eggs? The intent is to keep the eggs intact, not break them.

Instead, 'get comfortable'; for 'Comfort' is your guide from here on out.

Catching imaginary or real eggs should be as comfortable as someone passing you a basketball.

Keep replaying the drill and catch those unexpected imaginary eggs until you find you are not rigid in the reception, and have adjusted to the point that you no longer need to, or can make no more adjustments.

Your elbows might bend a little, but you should be in condition to safely catch the egg carton should someone actually drop some on you.

When you feel you have it down, tidy up the alignment even more by changing the image of a carton of eggs to a larger and heavier box, one with dozen egg cartons (do not break any!).

Feel free to adjust your feet on the floor to make the catch comfortable.

Once comfortable again, stop for a moment and feel this body alignment.

When done correctly, you should find your upper body alignment is as in Figure 67. Looks familiar does it not? It should, it is a forward C!

Where We are Strong

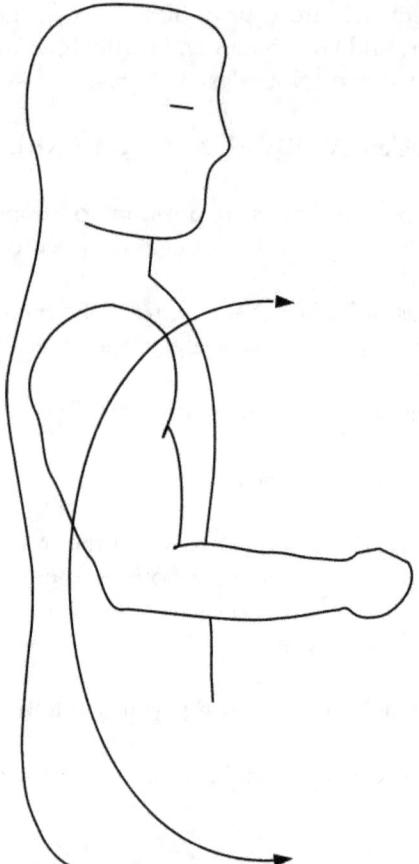

Figure 67: The Forward C is the basic Zhan Zhuang pose. Use the drill to 'assume' this position instead of 'posing'

Pay close attention to the lower back. In Zhan Zhang, that curve of your lower back diminishes greatly. (Please go and look at the five Zhan Zhuang poses on the Internet)

The lower back is an area of intense focus in Zhan Zhuang. I can almost guarantee that with minimal research you will find someone pointing out the lower back position in the photos.

If you found your comfort via the drill, you can drop your hands fully to your sides (drop the eggs/pizza) and stand in the first Zhan Zhuang pose (the 'mountain pose') for five minutes with relative ease.

Trust me, it is not as easy as it sounds!

Zhan Zhuang (According to Bill and Joel)

You just skipped past MANY hours of practice at Zhan Zhuang by doing this drill and 'aligning' your fascia with a conscious task of catching an increasing, albeit fictional, weight instead of simply 'posing'; i.e., mimicking someone else; (You are welcome, because no one helped me like that!)

During that five minutes of exercise (yes, standing still in mountain pose is exercise!), we will further improve our alignment by imagining that someone is adding even more weight to our shoulders.

Stay comfortable, but as you imagine more and more weight pushing down on the shoulders (as much as you desire, a whole planet if you want to go that far), what you should be sensing is your body making ever more fine adjustments to support the imaginary weight. No tensing up, stay comfortable, let the body align. You are doing great, but we need to address something first.

Before we take this to the next step, let us stop for a second and talk about the mountain pose and the lower back.

It should be easy to 'hang' your shoulders with a slight spread in your shoulder blades. ('Linked' shoulder: shoulder blades flat against your rib cage, collar bone [clavicle] parallel to the floor and also against the rib cage)

I have not heard anyone complain much about their shoulders. What does come up from time to time is a discussion about the lower back.

Some might argue that we need that curve in our lower back to generate power and that a flat lower back puts a person at risk of injury.

I am not going to tell you one way or the other. You have to make up your own mind on what is safe and what you should do. You should not have had any weight applied yet, so if you are even remotely worried about injury, stop this drill, skip this section, there are plenty of other topics in this book to learn about.

For those that are continuing, again, do so at your own risk and be sure to warn others long before you share this practice. As we said in the disclaimer beginning of this book, in every word of this book, we are solely preserving a historical aspect of the martial arts... So, if you are with me...

In preserving that martial heritage, let us pick things up again by describing the relationship between your abs and that significantly straightened lower back.

Where We are Strong

The lower torso area lacks a true bone structure (aka no rib cage). The lumbar bones are only a small part of the circumference of your midsection, so…

If you were doing this according to plan, your abs are going to be somewhat flexed (I did not say tensed!). The abdominal muscles help join with the oblique muscles (your sides, 'love handles') and your lower back muscles (your QL's too!) to form a bubble that supports your lower torso.

There is a balanced tension between these muscles that keeps you structured but not rigid. It is easy to find comfortable balance from side to side (obliques against obliques). The trick is finding the balance between your frontal abs and the lower back. Your key indicator of success is your breathing.

When performing this pose correctly, you should be able to breathe abdominally. In fact, abdominal breathing is easiest when you assume this aligned pose; i.e., you should sense all sides of your lower torso trying to puff outward as you breathe in, and draw inward a bit as you breathe out.

You can expect to sense pressure changing in your 'pelvic floor' as well, but the front abdominal wall does the most moving.

The secret is that your diaphragm is the impetus for the puff and collapse your 'abdominal bubble'. Your abdominal muscles remain comfortable, at rest, and supple, like the outside shell of a balloon. A balloon that responds and reflects the motion of your diaphragm.

The point is, if you are too tense in your abs, you are cutting off your ability to breath, and that should not feel comfortable. When you get it correct, this is a fantastic method to intentionally perform abdominal breathing (and create diaphragmatic support for singing, speaking, etc.)

Be sure to take great heed to what you just read, as the awareness of how to fascially align your body and how that structure is affected by breath is the foundation for higher level skill in internal arts. (You might even say 'critical', as in, either you understand and practice it, or you might not excel)

Back to the pose itself.

This mountain pose position, with a comfortably aligned torso, creates a larger 'bubble'; one from your shoulders to your hips. The aligned torso does not twist much, and in that fact lies the intense support of the spine.

Zhan Zhuang (According to Bill and Joel)

Thus, your torso has become an excellent conductor of force from the hips to the shoulders and vice versa. Your 'box' frame, shoulders over hips, can be described as a balloon, or even more descriptively, much like a well inflated ball, e.g., a football (American or European, either works), or even a basketball. (Others describe the spine as a 'bow' when in this condition. Either way is fine, I prefer the idea of entire torso in lieu of focusing on the spine, but the condition/goal is the same. Un-authoritatively, I believe this condition to be called 'Song' [in the Tai Chi classics]. Rested, not relaxed.)

How much energy is there in a sports ball when you put pressure on it?

Welcome to 'at rest structure' in your torso; energized, structured, but not rigid. This is how the internal masters apply one of the 'six harmonies': hips harmonizing with the shoulders.

'Your torso as a ball' can receive and bounce back energy, which makes you much more durable. Force against your torso will be displaced and shared across more of your frame so you can better deal with impact and reduce risk of injury. Including the impact that you are about to create as you drive force into whichever object you choose to impart pressure upon (aka kick, push, Hsing-I: Pi, Tsuan, Beng, Pao, or Heng Chuan, Ba Gua palm changes, etc.)

You should realize a significant improvement in your posture. Forget that 'support your head like it is tied to a string' explanation. That explanation does not describe even a tenth of what you are trying to do which is: align yourself, specifically your fascia, and add breath support, to bear weight. (some 'traditional' instruction is very misleading/devoid of detail)

Just like our fascial grasping, the lack of intense muscle flexing improves our ability to keep this functional pose, so we do not tire as quickly. Not to mention, all of our internal organ functions have chance to be improved as we do a better job exchanging air with our lungs and basically massage all of our organs against the walls of our torso a as we breathe abdominally. (Look up the proposed advantages of diaphragmatic breathing on the web if you do not already know what I mean.)

Abdominal breathing is a key element of how Tai Chi and all the internal arts focus on their 'center'. (Just one element of many, but a key one)

You are in your best condition to perform most martial practices.

That is a lot to be said for a simple standing pose. I suspect this is why the Hsing-I masters reportedly valued Zhan Zhuang so highly.

Let us go (much) further
The Zhan Zhuang we have investigated thus far is an excellent study in posture, potentially gaining health benefits, and prepping for Chi Kung. Most stop here, but there is more...

So, what else can we learn from this first Zhan Zhuang pose beyond 'putting on a shirt'? (which is how Joel and I often describe the fascial alignment in the torso).

To be blunt, you do not get the martial benefit of this training until you actually feel some force against your alignment. Let me show you some solo and partner methods to train your alignment against force. Plus...

If you tried any of the other Zhan Zhuang besides 'mountain pose' (aka 'Wu Chi'; the first Zhan Zhuang pose), I will bet you would also like to know how to position the arms so that they do not tire so easily. This lesson reveals a method to unite our arms with our body (via a little rant about 'unbendable arms'), so let us cover that too, but wait, there is more...

We can then also do a quick investigation on mind/body awareness and how Zhan Zhuang assists in improving our mind/body unification skill.

This will have us take a moment to investigate what 'No-mind' is so that we leave the discussion with a basis for understanding Zhan Zhuang's meditative aspects (something that you will certainly be coming in contact with as you read more about Zhan Zhuang and any/all of the internal martial arts)

At minimum, expect to at least come out the other end of this discussion having a deep understanding of how to 'move with alignment'.

Improving the alignment for Martial benefit: push and pull
We gain martial benefit from Zhan Zhuang by applying fascial alignment and breath support to the rest of our martial practice; i.e., we learn to receive and project force while 'wearing our Zhan Zhuang shirt'.

Keep in mind, a pivotal aspect of every martial training is learning how to receive and project pressure, but it typically is learned indirectly while doing a myriad of other things over many, many years; (e.g., kata, sparring)

Because this ability to receive and project pressure is learned indirectly, progress can be slow. Due to lack of focus or recognition as a 'thing', proper fascial alignment is typically described as having 'good posture' and passed

Zhan Zhuang (According to Bill and Joel)

over as something you either have or do not have. This is how many aspects of the arts (and sports!) fail to be passed along.

Instead, master that simple 'how to align your torso fascia and assume a mountain pose (forward C)' lesson, as that is itself both an achievement and a pre-requisite for practicing what you are about to read. Not everyone is aware of their body to the degree required to 'put on their shirt'. You have quite a blessing if you can. Congratulations! Here is where we go to next.

You could immediately run to the dojo, 'put on your shirt', ask others to start pushing on you in every dimension and you will gain from it, but that ability to handle external interaction is actually pretty advanced.

Gaining skill in Zhan Zhuang takes time, and if you have not guessed, it involves some extreme subtlety. Let us grow the root before we try to make ourselves strong.

The best way I know to improve this mountain pose is with solo exercise.

Where? In my favorite place to practice: near a door frame.

Why is a door frame such a great place to work? It has a great deal more ability to withstand pressure than your typical wall (should you press in the areas without the stud behind it). I also suggest that everyone has access to a door or at least some form of entryway that is large enough for us to move around within. Could you use a pole or a tree? Sure, but not everyone has a stationary pole rooted well enough to press against, and trees are outside! (I like to work out in air conditioning! Ha!)

Get started by finding your favorite door frame and performing the double-weighted mountain pose inside the door. Do so with one foot in each room that the door separates, the door behind you, we want to be looking at the side with the latch plate, not the hinges.

Choose a hand, and 'with your shirt on' (that comfortable alignment in your torso), push against the door frame.

If this is your first time, all that 'comfort' likely became a whole lot of tension and/or you pushed yourself away from the door frame. A good many will find their alignment crumble and completely lose that sensation of being like an inflated ball.

It is normal, but let us fix it.

Where We are Strong

The point is that we have to build our ability to stay 'at rest' in our fascia as we direct the force into the door frame.

We nurture this ability by exerting an extremely light amount of pressure against the door frame at the start and slowly building the pressure in gradual steps, readjusting (re-finding comfort) at every point along the way.

So, get in your mountain pose inside the door again, and this time, simply touch the door frame without even pressing on it. For some, even this will cause them to lose their alignment, but even if you maintained your 'shirt', stop for a moment, and while maintaining contact with the door frame, re-align yourself; (i.e., re-Rest yourself!).

It will help in this situation to imagine holding back a train instead of catching dozens of eggs.

Feel the forward C coming back as you replay the 'stopping a train' imagery as many times as it takes to re-align (get comfortable!).

Once you are aligned, press a little harder against door frame. "How hard?"

It is best to stop just before you feel your alignment break. Give yourself some time to sense the near loss of alignment and then allow yourself to find comfort once again. It is no issue if you go too far, 'lose your shirt', or push yourself away from the door jamb. Just reset and try again.

What you should begin to feel is that it takes a good bit of practice to truly 'commit to gravity' instead of committing to the pressure that you are exerting on the door.

What that means, in so many words, is that you are channeling that force against the door frame down through your legs/feet and into the floor.

In my experience, it required a great many gradual increases in pressure to get the knack of it. Push and align, stay there for a while (15 – 30 seconds) and sense the comfort; then add slightly more pressure, realign, repeat, again, and again.

I sometimes imagined myself as falling through the floor. That falling through the floor imagery helped me find 'at rest comfort' in my torso and even down through my legs.

Ah yes! The legs!

Zhan Zhuang (According to Bill and Joel)

You should expect to sense the fascial alignment in your legs as you discover the equilibrium that balances the pressure on your hand with the pressure you exert on the floor.

In my feet, I feel most stable when the weight is focused in my arch, in the centerline, nestled close to the heel (but still in the arch).

The legs feel as though all the muscles along the inseam are activated (from inner thigh to the feet), not tense, rested and comfortable. This 'yin' portion of the legs balances the 'at rest' tension in the outer portion of the legs.

The force toward the floor seems to pass straight from my hip sockets, through the air, directly to that spot in my arch I just mentioned.

In other words, through my leg hypotenuses! (roughly, the line between your hips and your ankles)

It will help to scan your body as you practice. When everything is in synch, take the time to feel the different parts doing their part.

Sense the construction of your 'Forward C' (i.e., feel the 'shirt' of your torso). Feel the intent of, but not the actual motion of, spreading your 'linked' shoulders as you maintain your hips in a forward CAM position.

Do you sense that the forward C projects at both the shoulders and the hips?

You almost cannot help but push forward. Sense the near effortless springiness in your legs. You are not trying to bend or straighten your legs; they just seem to keep their shape; both leg hypotenuses staying their comfortable length.

Press as hard or as soft as you can against the door frame, nothing changes in the legs, they comfortably maintain their curvature.

Your arms might be a different story right now. There is likely a lot of tension there, you might even be tiring. We will fix this in a grand way later, so pay no heed to any discomfort in your arms for now. Instead…

Keep this sensation in your torso and legs as you completely let off the pressure on the wall.

Where We are Strong

Now, as best as you can, maintain your perfectly aligned pose as... with one hand preferably, two if you must... grab the edge of the door frame and pull yourself toward the wall.

Oops! The structure crumbled again did it not! (laughably, I almost banged my nose on the door jamb when I first became inclined to try pulling)

Hopefully your 'shirt' stayed on, but the point here is that, your body should not have to shift much to accommodate the change from pushing to pulling.

Just as your legs and torso were not changing much to adjust between soft and heavy pushes against the door frame, they should not have to adjust much when making the complete shift between pushing and pulling.

At first, this change in direction can be befuddling. It can help to stop the physical pressure and get back into the imagery.

Leave the door way, go anywhere safe, and assume the mountain pose, carry the weight of the world on your shoulders, then remove it.

Next, imagine standing chest high in a rushing river, water flowing toward you as you face upstream.

Speed the water up and slow it down. This is just another version of 'stopping a train', the difference is the lack of involving the arms.

When you think you are stable, reverse the flow of the imaginary water, have it come at you from behind.

If this is your first time, and you listen to your body closely, you will likely sense quite a shift in your fascia.

The most drastic change is typically felt in the placement of the weight on your feet. The weight will want to drift toward the balls of your feet.

Therein lies the objective. We have to find our 'sweet spot' within our feet to wear the weight that works for both frontward and backward approaching force. (I suggest it is in the aforementioned place in the centerline of the arch, closest to the heel. [not the 'bubbling well' [Yong Quan Kidney 1], but I should point out that this is a static pose. Things change when moving.)

We find that spot in our foot by continually toggling the direction of the imaginary pressure and re-discovering the 'comfort'.

Zhan Zhuang (According to Bill and Joel)

Each time we toggle the direction of the imaginary force against us, we should have to move less and less to keep aligned, and eventually settle into comfort.

Comfort is the goal, so take as long as you need between each shift from frontward to backward to regain the alignments.

Within a short period of time, you should begin finding the 'true alignment' of your torso fascia and very soon after that, a combination of alignment of torso and leg fascia.

Try imagining the water coming at you in short bursts from random directions, find that comfortable pose that contends with all those imaginary forces, and you are ready to start with the wall again.

Assume mountain pose in the door frame, this time grasp the door frame with one or both hands and shift from lightly pushing against the door frame to lightly pulling on the door frame.

Between each shift, re-align from shoulders all the way to the floor (torso, inseam, and placement of weight within your feet). Each time you toggle from push to pull, you should sense you require fewer and fewer adjustments necessary to keep yourself in place and that you can use ever increasing amounts of pressure.

Improving the alignment for Martial benefit: shifting weight

These static double-weighted standing exercises are excellent, but for it to become martially (occupationally) functional, we have to take the alignment into stepping (and eventually into all of our motion).

So as not to bore you, I will forego a lengthy explanation of modifying this against the door frame drill by shifting your weight in increasing amounts onto either foot until you are finally performing the drill on one foot. (Captain Obvious says, "Practice standing solely on one foot then change and practice on the other!" Thank you, Captain!)

When you are able maintain 'your shirt' on each foot… maintain a push or a pull on the door frame as you practice complete shifts of weight; (i.e., stepping) from one stationary foot to the other; i.e., Change your nickname to 'Dances with Wall' and move your unweighted foot to a new position each time you place it down; (i.e., step in, out, and all around the door frame).

Just be sure that you maintain the push and pull actions as you focus on 'down, over, and up' shifting weight between your feet.

Attaching to the wall with one hand will widen your options for stepping and facing different directions.

Challenge yourself by toggling the push and pull in the middle of your step; as best as you can, face away from the door jamb (as if getting pushed and pulled from behind)

Stepping in Zhan Zhuang is important practice. (The traditionalists out there are likely spinning on their heads right now. 'Stepping in a standing still art!' Kung Fu Heretic!)

There you have it, now you can move away from the wall, and work with a partner.

Before you dive into the deep end and practice in every move you know (and ever hope to know!) try starting like this…

 integrate your Zhan Zhuang with forward drop-stepping by standing in front of Uke. Have Uke place their hands on your shoulders, and start pushing them backward by drop-stepping forward.

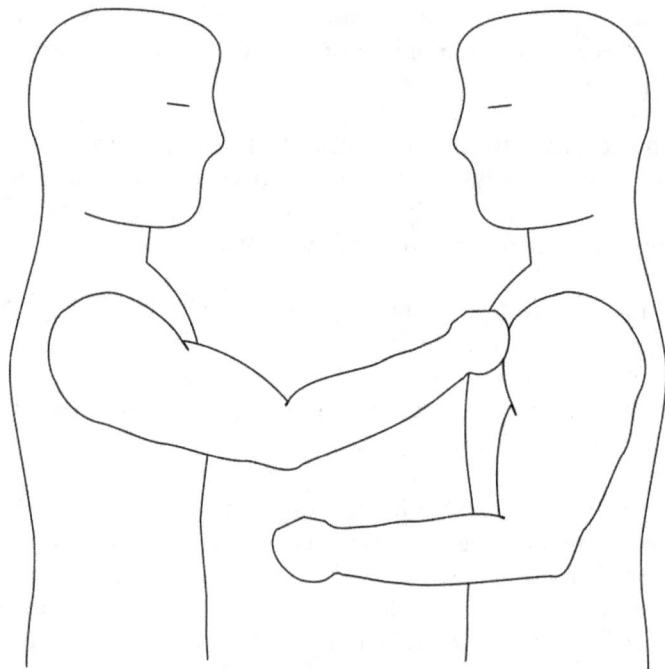

Figure 68: add picture of Uke on left, Tori on right pushing forward with their shoulders; arms in 'cradling a stack of wood' pose

Zhan Zhuang (According to Bill and Joel)

Pushing Uke along in this fashion will keep your arms out of the mix and let you focus on your 'shirt'.

I have seen many succeed when they assume the 'delivering pizza/holding wooden logs' pose as they step through Uke (Right side Figure 68). Their hands and arms are not in contact with Uke in any way, but the intention of holding the pizza helps put the 'shirt' alignment in place.

Have Uke add resistance as you improve. You should be having zero issues maintaining your forward 'C' and Uke will be feeling the full force of your 'hip swing' coming through your shoulders. This exercise is essential to bringing the force of your COB (center) into your hands. Do not skip it.

In no time, you will be using your Zhan Zhuang when practicing Aiki. 'Wear your shirt' at all times, and the rest takes care of itself.

Everything you do physically, in and out of the dojo, from this point onward, should be 'with your shirt on'.

Walking, running, leaping, lifting, throwing, etc. and especially sitting! I often use the desk I am sitting at to replace the resistance provided by the door frame. It helps me re-establish good sitting posture.

When the risk is minimal, e.g., waiting for a red light to change, I have used light pressure on my car's steering wheel to help me realign.

How about some tips on traditional Zhan Zhuang?
What I appreciate most about Zhan Zhuang is that it teaches how to align your body (namely, fascia and ligaments) to engage in accepting and projecting force in a way that requires minimal muscular exertion.

What is the benefit of minimizing muscular exertion? Stamina!

Using your muscles requires effort. Lactic acid builds, and muscles tire; your ligaments will not. Neither will the basic desire of your fascia to hold their shape.

My basic frustration with the traditional practice of Zhan Zhuang is that it takes the path of teaching these alignments via exhaustion.

Stand in place for 20 minutes (with your arms outward!) and you will soon agree, the traditional focus is exhaustion bringing about the necessity of finding a better way. (Do our front shoulder muscles ever recover?)

I much prefer the methods I just described. (I wish someone had shown them to me instead of having to discover them for myself.)

Taking the path of exhaustion often turns people away from the practice of Zhan Zhuang. It is hard enough to have people look at you strange for 'exercising by standing still'. Why should it hurt too? There is no need.

So for those who desire the full spectrum of the Zhan Zhuang 'kata' (form, sequence of moves, poses, whatever), please allow me to describe how to leverage the 'Aiki Joint Locking' we learned a few sections back as the path to significantly less painful Zhan Zhuang (and tie this into creating 'Unbendable' arms)

The basic Zhan Zhuang 'kata' has five poses (See Figure 69, but this is a good time to go to the web to find other examples of these poses. Although each pose is basic, and the names pretty much explain what they are, it will help to see them on video as well)
- 'Wu Chi' a.k.a. Mountain pose (traditionally, palms toward hips)
- 'Holding the Ball'/'Hugging a tree'
- 'Holding your Belly'
- 'Standing in the Stream'
- 'Holding up a Balloon' (I call it 'holding up the world' or 'volleyball set [for a spike]')

There is a progression here.

'Wu Chi' or 'Mountain' pose does not engage your arms. Why?

My guess, because it is hard enough to learn how to stand without adding the arms.

If you can 'put on your shirt', standing still in mountain pose for 5 minutes should be a cake walk. I find it necessary every now and again to 'hold a pizza box' for just a moment as I re-discover comfort, but alignment is easily regained and I can return to practicing Chi Kung (poorly! <sigh>)

When you move to the second pose 'Holding the Ball', you begin engaging your arms and the fun begins.

Now, more than ever, will you discover the essence of 'teaching via exhaustion'. You will also quickly discover the meaning of 'linked shoulders', because any other position in your shoulders will tire you out quickly.

Zhan Zhuang (According to Bill and Joel)

Figure 69: the five Zhan Zhuang poses

The problem is that, even with your clavicles and shoulder blades dropped and resting against your rib cage, you remember that you have a ball joint and will soon find your front and potentially middle deltoid muscles tiring.

My advice? Start by using your Hypotenuse! Here is how…

Get in Mountain pose and then bring your hands to your front, elbows dropped; be sure your wrists are at the correct height and position in front of you.

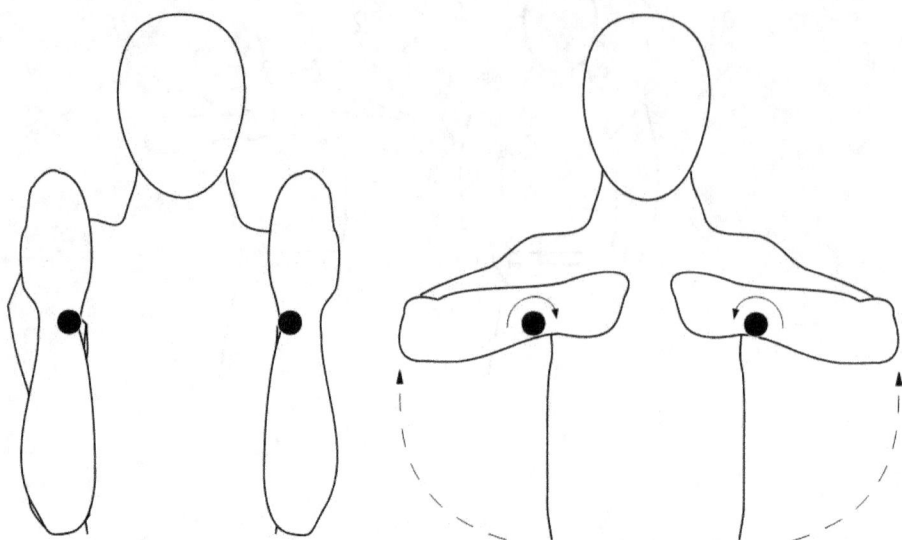

Figure 70: Rotating your arms on your Hypotenuse to achieve the 'Holding the ball'/'Hugging the Tree' pose using ligamentary tensing

Next, rotate your arm hypotenuses and intentionally bind your ball joints forward (i.e., 'close your shoulder kuas') as your hands rotate into place.

The weight of your hands should be maintaining the twist in your ball joints; i.e., 'hang' your arms so that your shoulders lock.

You should recognize this feeling in your shoulders from the Aiki Joint Locking Kotehineri practice; i.e., you were just now your own Uke and closed/locked your own shoulder kuas!

This action puts your attention on using the near limit of your range of motion in the shoulder ball joint and removes the reliance on muscles.

Stay put in your shoulders and let us now put your hands in position.

To get those hands where we want them, keep your elbows exactly where they are (i.e., do not move your shoulders) and simply apply a pinky pivot wrist turn to yourself; (i.e., pinkies in place, flop your thumbs over and forward [away from you])

Nurture a constant intention of your thumbs falling forward and outward over your pinkies; (i.e., apply a Kotegaeshi outside wrist turn on your forearm.)

Zhan Zhuang (According to Bill and Joel)

It will take some practice to do this mix of twists for longer periods of time, but you will be progressing a lot faster than the trial and error method of exhaustion.

The guidance of twisting some tension into your joints in order to stiffen them is the basic theme for the rest of the poses as well.

In the third pose, 'Holding the Belly', use the same Kotegaeshi, pinky pivot, outside wrist turn on your forearm, but with your hands down around your belt line, open your shoulder kuas to keep your hands away from your body (Try this pose with closed shoulder kua's as well)

If feels subtly like you were going to toss a ball to someone using a hip-pop!

The fourth pose, 'Standing in the Stream' is best explored as opening the shoulder kua as much as possible before allowing your thumbs rotate forward to hang and cause a pinky pivot in your forearm.

In other words, put your arms out waist high to your sides, rotate your thumbs backward as far as can, leave your shoulders in place, and then 'pinky pivot' rotate your thumbs forward again; let your thumbs 'hang' forward.

That fascial tension should keep you up for a long while, just be sure to keep the shoulders linked (for some reason I have the tendency to try and bring my shoulders blades together. This tends to lift my shoulder blades from my rib cage. That causes a lot of unwanted tension and I tire quickly)

The fifth pose, 'Holding up a Balloon' is actually a lot of fun. I like to think of it as shooting a basketball the incorrect way, with two hands. You could also think of it as 'setting' a volleyball (for your teammate to spike).

I enjoy this pose as you get to settle into your legs a bit by shortening your leg hypotenuses and opening up the leg kuas a bit. (all the other poses seem to be more accepting of a subtle twist in the hips that closes the hip kuas; i.e., not quite, but edging/intending toward pigeon toes.)

The intention of opening the hip kua will rotate your knees out a bit and create comfortable tension in the knee and ankle joints.

The arms are arranged so that the shoulder kuas close as you apply an index pivot (where your intention is to have your pinky's falling over your index fingers on toward your face); i.e., simply apply a Kotehineri to yourself with palms to the sky!

All four of the Zhan Zhuang poses that include your arms are excellent for teaching you torso alignment. They all mimic the 'holding a pizza box' aspect of helping you find the fascial and ligamentary alignment in the torso and legs, but each arm pose gives a unique pressure against your frame. Thus, they each help you find comfort in alignment in a different way.

You will receive advice from other texts such as 'like having a ball under your arm pit', or like 'sitting on a ball'. To be transparent, those descriptions did not help me in the least, and if not for Rob running over me, I might never have returned to improve my Zhan Zhuang practice.

So, sure, do as the books say and take a couple of months in each pose trying to figure out fascial/ligamentary alignment via muscular exhaustion or take my advice and speed yourself along so you can begin your work on the martial aspects/benefits. (and complete the foundation for focusing on the mental and health benefits of Chi Kung!)

Either way works. The only importance is that you practice Zhan Zhuang!

And on that martial aspect of Zhan Zhuang topic…

Unbendable is a 'twisted' concept
Now is the best moment to bring a quick side channel discussion to the forefront. I will keep it brief, but indulge myself a bit of "Flame On!".

I am absolutely certain that you have seen the 'unbendable arm' demonstrations where Tori lays their arm on Uke's shoulder as Uke hooks their hands on inside of Tori's elbow and attempts with all their might, and in vain, to pull downward to make Tori's elbow bend.

Remember that remark I made in the beginning of the explanation of 'Presence' about 'non-discernable forces flowing from our fingers'?

Well, if you had not guessed, I do not accept the thought that some intangible energy is flowing so strong through Tori's arm and out their fingers that it somehow prevents their joints from bending.

I am calling shenanigans!

How about an explanation supported by something tangible, measurable, repeatable, teachable, and rooted in science that says the lesson you just learned, about how to stiffen your own joints by (mis)aligning the structure of your fascia and also in the lesson of aligning/pivoting your hand to prepare

Zhan Zhuang (According to Bill and Joel)

for Karate strikes can together make a big difference in your ability to hold your elbow joint in place in this drill (and other situations!)

In defense of those that purport to 'use the force' to create an unbendable arm, they are not always purposefully trying to sell snake oil.

If you do not know exactly what you are doing because you simply have never had an issue performing the trick, you might not ever notice the subtle manner with which your joints are being supported by a combination of small bits of muscular exertion and a subtle twist of the fascia. It might even feel like "Get a bucket and a sponge! I am spraying Ki out of my fingers again!"

Since the dawn of time, what man cannot explain with science gets explained 'spiritually' or 'mythically'. (Are 'Six Precepts' and 'Codex' attempts to dispel Aiki myth with science? You bet ya!)

Most spectators have no idea of what to look for anyway. Nuff said!

What we are going to do is take a quick moment to re-analyze our Zhan Zhuang pose(s), and revisit the impact that closing or opening our shoulder kua has on our ability to rotate, in this case, drop, our horizontally outstretched upper arm bone (Humerus) downward to our side.

Why?

In this traditional 'unbendable arm' setup, Uke cannot bend our elbow if they cannot rotate our shoulder to make our upper arm bone point downward; (i.e., put our humerus vertical; aka 'at our side')

In other words, we simply cannot bend Tori's arm in this setup unless Tori's ball joint rotates.

Go ahead, put your arm horizontally forward in front of you. Now, drop your elbow!

What happened in your shoulder? Interesting huh? 'Dropped elbows' are very much an aspect of our shoulder joint. (not so obvious until you try it.)

So, acting as Tori; i.e., being the person exhibiting the unbendable arm…

Simply open or close your shoulder kua (in my opinion, it is a bit easier with open kua) when demonstrating and it should be enough to resist most Uke. (Yeah, it really is that easy to explain, and perform (with almost no exertion!)

Where We are Strong

You are also being assisted by Uke's body being in the way of you dropping a straight arm to your side; i.e., if your forearm stays aligned (i.e., straight-ish elbow), Uke is actually helping you keep your upper arm in place!

If you still need some help, include some additional support in your elbow.

Those 'fingers flexed outward'? Yeah, they can lend assistance too!

Lay your right elbow into your upturned left hand. Wrap your left-hand fingers around your right-arm elbow as well as you can.

Now stick those right-hand fingers straight out, bend them, straight again...

Did you notice the tendons and muscles tensing (tugging) on all sides of your right forearm nearest your elbow?

This 'stretching of your fingers' helps find that equilibrium between the muscles that flex and extend your wrist and it will collaterally have effect on supporting your elbow.

Relax your fingers for a second and shift your right hand from neutral to 'sword hand' (rising hand, radial deviation). Can you feel extra tension being created in the crook of your elbow?

Straight fingers, rising hand, even Pinky and Index pivots, they all add support and tension to your elbow and it only requires a small amount of muscular exertion. It is very efficient and extremely subtle.

Couple the rested tension in your elbow with the rested tension in your shoulder, and with Uke literally in the way, Uke has almost no chance.

Nice parlor trick! And look! No messy Ki cleanup!

Although I am being sarcastic (caustic? Bad Bill!), there is an important lesson in usefulness here that goes beyond the mythical.

The Zhan Zhuang poses teach us this 'unbendable' skill set, and with it, we can better unite our arms with our 'shirt'. (In actuality, we are unifying elbows with knees, which is the second of the physical 'harmonies' [in the six harmonies of the internal arts] This particular harmony is actually a focus on your kuas, both hip and shoulders)

The result? Instead of a 'sleeveless T-shirt', you put on a 'long sleeved shirt'.

Zhan Zhuang (According to Bill and Joel)

Your arms become an integrated object with your torso, and every inch of your upper body is driven by and reflects the power in your hips; e.g., your 'down, over, and up' is efficiently delivered to your elbows. (we have one more lesson to learn before it reaches our hands)

Now, do I mean that you should constantly keep your arms out, shoulder locked like some kind of Frankenstein's monster?

Come on, you know me better than that.

There is a time and season for everything. Aiki, like all things, requires a mix of soft and hard, yin and yang.

Why does the Karate practitioner synchronize the twist in their arm to the impact and follow through in their straight punch?

They do it to assist in creating an unforgivingly solid transfer of their hips and legs into their hands; but only for an instant. An instant is all that is necessary. Timing is everything.

We do similarly in Aikido. Let us prove this with a drill we can call the 'sideways unbendable arm' drill.

Test your ability to unify your arm and body by placing an arm directly in front of you (or even slightly to the side) as you might to perform Gyakugamaeate (#3). (See Figure 71)

With your palm up, no Zhan Zhuang in your torso, and a neutral shoulder kua, ask an Uke to come at you from the outside position and push your arm across your body.

Do your best to resist your body being rotated, and keep that arm right where it is at in relation to not just you, but the entire room. (Figure 71)

Chances are you could not hold your position at all, but if you could, you could not hold your position for very long.

You most likely caved at the shoulder ball joint as the arm moved across your body.

Try again, but this time, put on your Zhan Zhuang shirt, and in your arm, apply to yourself a bit of Kotegaeshi.

This means that you should open your shoulder kua to keep your shoulder in place in relation to your body and add a bit of pinky pivot to help support your elbow. (Let that thumb hang down as if someone was applying Kotegaeshi to you; i.e., use a subtle pinky pivot)

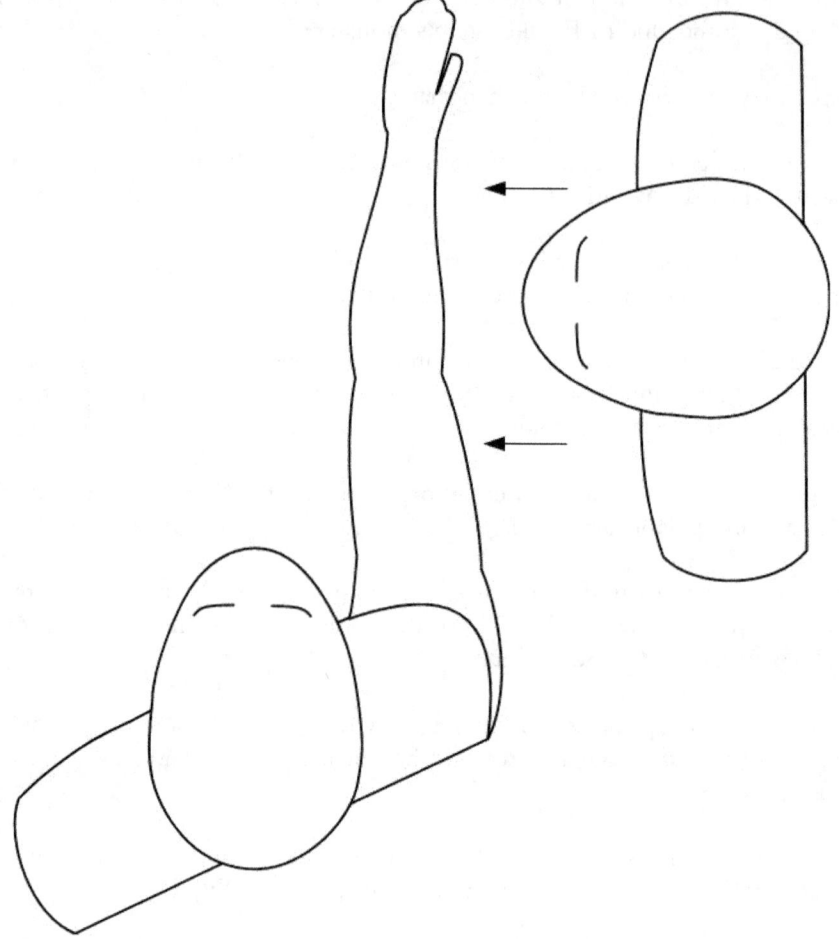

Figure 71: Uke pushing sideways on Tori's arm attempting to rotating the arm in front of Tori

Am I asking you to completely lock out your joints and completely relax your muscles? No, there is danger in that.

You should be competent and confident in finding that subtle middle ground where your fascia supports the muscles and vice versa.

Zhan Zhuang (According to Bill and Joel)

Ask Uke to try moving across you again.

You should find you are significantly better at staying put (and acting as a clothes line for anyone that dared not get low enough to limbo through)

Take the time to repeat with a Kotehineri inside wrist turn (close your shoulder kua and index pivot in your hand).

The good news is this works, but the bad news is that both methods are actually super close to becoming Uke applying a joint lock upon you.

Thus, our weakness can become our strength and our strength become our weakness. What a wonderful world we live in. Paradoxical at times.

Flame off! After the Ki embers cool a bit, let us talk a bit about mind/body connection…

Mind/Body Awareness
These Zhan Zhuang drills strengthen our alignment in two ways.

First, as we just learned, by physically aligning the fascia and ligaments, but most importantly, by strengthening our Mind/Body awareness.

What exactly is Mind/Body awareness anyway?

It is the integration of your cognitive and reflexive guidance systems.

"Huh?"

Put simply, allowing mind and body to guide each other simultaneously.

Obviously, there are two sides to this coin: your body, which is a 'feeling' and 'reflex' organism, and your mind, a 'sensing' and 'directing' agent.

On the body side is your muscle memory. Actions happen as a reflex, where your movements follow a set pattern regardless of additional input.

On the mind side is 'you' receiving input, processing it, and adapting to it.

This feedback loop between what you perceive and what is routine is happening all the time at various levels of efficiency and detail.

Ever fail to throw Uke?

Where We are Strong

If you have missed a throw (or two), could you answer to yourself what went wrong? or were you moving too quickly along an ingrained pattern of motion to even notice? Is there any 'mental residue' from the entire experience that could give you a hint to what just happened?

If not, that is a whole lot of reflex and very little mind interaction.

How great would it be, if instead, your mind was conscious of not only the routine, but additional factors that contribute to the success or failure of the technique? So much so, that you could replay events in your mind and pinpoint where adjustments needed to be made!

And even then...

What happens on the next try when we attempt to make those planned adjustments and things still go wrong?

Are we listening to our body with enough detail to assess if we actually performed the planned change or is yet another, altogether different correction necessary? Did we notice Uke do something different? Do we have the efficiency to make an adjustment on the fly?

As you develop Mind/Body awareness, you become more aware of the aforementioned and the more spontaneous you can become. Dealing with random situations may still be awkward, but they will be manageable.

How much mind/body awareness is necessary to become masterful at Aiki?

Tough to say, but I know this much, not only does the limit of your ability to execute an appropriate response depend greatly upon this mind/body awareness, so too does the speed of progress in learning more complicated techniques.

"Great pep talk! How does Zhan Zhuang fit in?" Great question!

Zhan Zhuang builds Mind/Body awareness through the practice of 'listening to our reflex'. It is this focus upon allowing the body to communicate with your conscious that opens the door to improvement.

Everyone has a body that is already aware of the dangers of misalignment, but the body itself has bad habits. The task is bringing that awareness to forefront of your mind by 'listening to' and 'watching' how your body reacts to misalignment and then correcting/improving the situation.

Zhan Zhuang (According to Bill and Joel)

Continually focusing on your alignment will forge a stronger reflexive habit of staying aligned. You will not even have to 'think' about it. Posture becomes more than just a pose, it becomes a feeling; an alarm that goes off in our head when our body strays from that comfort, and at the same time, the habit of alignment will eventually result in your body attempting to reflexively correct bad posture in all situations.

This is especially true when Uke tries to misalign us. Our 'spider senses' kick in and the body sets off on its own reflexive auto-correction. At the same time, it sends to our brain a 'record' or 'log of sensations' that we can replay in our minds and analyze afterward; if we listen.

In short, we are developing our kinesthetic awareness and memory.

The conversation between body and mind is slow at first. It requires a good deal of time learning 'how to listen'. Your body is a distraction of its own; there are a multitude of things going on inside of you.

For your martial sake, the first 'body voice' you want to become aware of is the fascial alignment. Become so aware of your alignment/posture that you can hear its voice above the rest of the input vying for your attention; i.e., listen to your alignment as often and in every situation you can; (e.g., sitting, standing, running, cooking, fighting, etc.)

The next goal is to build coordination. This is where you re-build the larger set of martial reactions we call reflexes. This is no longer Zhan Zhuang, this is called 'practicing' your art.

The trick is, we have to consciously 'notice' the body's feedback, not necessarily engage with it.

In most cases, we do not want to actively, consciously engage with the motions, we simply observe. We practice slowly so we can 'experience' (listen to) the motions, for if we try to get our mind actively directing every little nuance, we will certainly fail.

As you slow down and remove distractions; i.e., 'focus', on particular aspects of your desired skill, keep your mind 'listening' to your body. You will find a wealth of information at your avail as your body and mind discover together what did or did not work.

That conversation of success/failure will initially come after the situation, (experience comes after you need it!) but eventually becomes instantaneous.

Speed will increase as the familiarity grows between mind and reflex.

This is trained by constantly practicing at the 'top speed' you can while still performing your action correctly and with 'mindfulness' (active observance).

You should find that your 'listening' might keep you very slow for a short while, but as your mind learns what to expect, you will 'get out of your own way' and find a deeper mastery of your action.

The goal is to eventually direct our bodies with our mind; i.e., initiate reflex from the basis of awareness of all the things around us, but we cannot do this until this conversation between mind and body is a real-time discussion that happens in the background (automatically as a habit).

Let us end this introductory discussion with an admission that I am a big stickler for awareness. My experience is that, the more I engage my mind in my actions, the better I become. Awareness begets appropriateness.

On the surface, Zhan Zhuang can seem like a superfluous, perhaps even ridiculous practice of 'standing still', but it actually leads to a great deal of insight about mind/body awareness.

Practice often! Here are some more words on how…

How can I be "Mindful" when I have "No-Mind"?
Zhan Zhuang is a fantastic way to practice Chi Kung.

I do not claim to be extremely good at the practice of Chi Kung, but I do practice (via Zhan Zhuang), and since the discovery of alignment, the Chi Kung practice is improving.

The significantly reduced physical effort created opportunity to focus more upon the mental aspects of Zhan Zhuang, one of those aspects is 'No-Mind' otherwise known as Mu-Shin. (or 'Mushin no Shin': the 'No-mindedness of Shin (Mind); i.e., 'Mind without Mind', or best: 'Quiet Shin')

How do we train our Mind/Body connection, and at the same time, strive for 'No Mind'? Is it a Zen paradox or is it proof that 'Mind' in each of these phrases is being utilized differently?

The correct answer is the latter. English lacks precision when it comes to our mental construction. This means that we have to define two different translations of Mind if we hope to understand both concepts.

Zhan Zhuang (According to Bill and Joel)

Let us start with 'No Mind' and unlock the true meaning of Mu-Shin

Mu means 'without' or 'lacking'. Easy enough. The Shin, on the other hand, translates as 'heart/monkey-mind' and that is where the trouble begins: the words Heart and Mind.

Shin Mind ('Shen' in Chinese), your 'monkey mind', is your 'thinker'. It triggers your emotions, is poorly translated as 'heart', better translated as 'your feelings' or 'your spirit' (in the sense of, for example, 'being in good spirits' i.e., happy or 'being of low spirit' i.e., sad.), and for all intents and purposes, is your ego.

It is the filter that taints everything you come in contact with.

Why is the Shin Mind there? It is a self-preservation mechanism, born of the consequence of lacking omniscience; (i.e., not knowing everything)

"Huh?"

To avoid getting deep on the subject, let us say the Shin Mind tries to help us stay alive (exist) in a world we have little understanding and control of.

The Shin-mind (ego) is centered on the belief that if we can comprehend the order of this world (or at least fabricate a sense of order in our Shin mind), then surely, we can interact with it in our best interest (which at its base, is to survive).

We have simply to know ourselves (our identity), in order that we can identify and avoid those things that are not in our best interest; gravitate toward those things that are. (We [errantly mostly] assign 'good' and 'bad')

All sorts of advantage and disadvantage comes from this 'thinking'.

As an advantage, you begin to identify what will physically end your existence and wonderful reactions such as fear happen. Your 'fight or flight' helps you survive.

Another example of advantage is falling in love and continuing not just the species, but, in some small way, continuing you (through your lineage, again, a kind of survival!)

All sorts of good comes from having an identity. The problem comes when we 'think too much'.

This happens when we stop adapting to the situation at hand, and instead work toward trying to predict and control outcomes based on past experiences.

In this 'state of mind', as your Shin Mind seeks control, it triggers powerful, but ultimately distracting emotions that give a sense of validation to these day dreams.

How effective are you at anything if you are day dreaming?

Not very, let me assure you. (trust me, your Sensei agrees!)

Pacifying your Shin Mind (stopping the 'over-thinking'; stopping the 'day dream') offers your faculties (sight, smell, taste, hearing, touch, your reason, reflex, intelligence, and yes, even your sense of identity) the opportunity to receive as much, and as many, types of input as is possible for you to become aware of and act upon in the present situation.

This implies that, since we are not contending with past and future, Mu-Shin, or 'Quiet Shin Mind', lack of distraction, leaves us directly connected to the only moment there ever is: now.

Some call it 'being in the moment' or 'in the zone', still others call it 'mindfulness' (and the inclusion of 'mind' in 'mindfulness' is super confusing! They are referring to the second type of Mind! The one we are about to discuss.)

You might consider it 'performing at the best of our abilities' because we are 'bringing all we have' to the task.

So next time you find yourself feeling anxiety about the future or concern about something in the past, especially when you are in the middle of practicing Aiki (e.g., that little voice inside your head asking, "Will I ever get this technique correct? I have not been able to all day"); know that you are letting Shin Mind get in the way of you doing your best.

"Just quiet the Shin mind. Emotions come under control, and we do our best"

Sounds easy does it not?

I hope it is easy for you. Inner peace is a phenomenal gift to give yourself. I strive to experience more often.

Zhan Zhuang (According to Bill and Joel)

The question to ask at this point is "What is left if I 'Lose my Mind', 'Lose my ego'?!?!?!"

You are left with the Mind of 'Mind/Body awareness'.

The internal martial artists call this Mind the 'Horse Mind'.

In Chinese it is 'Yi' or 'I' {capital letter I, pronounced 'Yee'}, the very same 'I' of Hsing-'I', which is exactly the same 'I' of 'I'-Chuan. (and is the same kind of 'I' as in I-Ching, but that particular 'I' belongs to/is the Tao!)

In English 'I' is called 'intent' or better yet, it is recognized as 'your Will'.

It is your 'true heart', the one that lies behind this 'passion heart' [Shin mind]. (Oh yeah, English mixes up the word 'Heart' too!)

It is the very same 'Will' we describe when we say that you 'will' not walk across a room until you 'intend' to, no matter the circumstances.

If your will is strong enough, it sets your direction in life, ultimately, your destiny.

Your Will is the manifestation of your individuality (your 'divinity') in the sense that it makes you separate from everyone else (and if you believe in one, distinct from, but in the image of [a] God.)

You can align your Will/Intent with the Will/Intent of others, but you never 'share' the same Will. (There is no 'Hive Mind') (compare this to the noun form of the word 'identity'. You can identify [verb] with others, and 'share' feelings, but you cannot 'be' the same person. English is a mess.)

Experiment for yourself and introduce yourself to, well, yourself. Your real self. The part of you that can, if you let it, perform on an infinitely higher level than you can being driven by your Shin Mind.

Here is how.

Find a nice quiet place, meaning somewhere with little distraction, and, of course, where you are physically safe. Close your eyes. Quiet the Shin mind; i.e., no thinking. Not even the music that plays in your head even when there is no sound.

Not so easy at first is it? All sorts of thoughts start popping in.

Where We are Strong

The advice you are most likely to receive at this point is to 'pay attention to your breathing'. They give that advice a lot of different ways. It is great advice, for, if you can keep yourself from adjusting your breathing, paying attention to an autonomic behavior in your body helps you see the natural function of your body. (the part of you that works best when you are not getting in the way)

The problem for me is that focusing on breath never really worked because it is/was boring and, in full transparency, I have a hard time leaving my breath alone when I am watching it.

Instead, I like using the following drill…

It is called 'Rapidly replacing your ego/identity'

With your eyes closed, imagine yourself as someone else. Be sure to include the environment around you; i.e., change, as best as you can, your surroundings and the role you play in it.

It is good to be as detailed as you can in this imagination. Bring all five senses to the task. Mentally, become someone else. For example,

Be a surgeon about to save someone's life. Can you feel the mask on your face, the cold scalpel in your hand, see the patient on the table, the other doctors there to help, sense confidence as you know you have done this procedure literally over one thousand times before, the smell of the antiseptic, that you feel a little thirsty, hear the beeping of the monitors…

After a minute of that, become a caveman running from a prehistoric predator; scared out of your wits, sweating, breathing heavy, but still more to give, adrenaline pumping, grit under your feet as you lead this monster to the group of other hunters ready to take it down, so hungry, you will be glad to eat again, even though this particular lizard tastes gamey, …

Then just as quickly, become an astrologer discovering patterns in the sky; be a monarch, a mailman, Rockstar, or whatever situation you feel you can quickly give incredible detail to.

It has to be detailed, you have to assume the identity/role whole-heartedly, and for not much more than a minute or two in any identity.

After about four or five mental scenery changes, stop! Sit there and wait for the next 'scene' to begin, but do not let another scene start.

Zhan Zhuang (According to Bill and Joel)

Sit now for a moment between 'scenes'. Who are you now?

The same person you were throughout all of that imagery, that is who. The indelible part of you that comes along in each of those speed journeys of 'self-induced amnesia, waking up in a different environment'.

You can think of it as 'the blank slate of 'You''.

This 'entity' is your will. The 'principles' center of your being. The part that reasons, 'I want to be this kind of person', but you fall short because your desires (emotion, Shin-mind) get in the way; (e.g., fail at dieting)

Say 'hello' to yourself! This might be your first time alone together.

I have known others to call this part of you 'The Watcher'. It is another great way to describe your 'Horse Mind'

'The Watcher' was able to experience the new ego situations and not only perceive the new story, but adapt to it (important!). By buying-in to the 'role' in each one of those 'scenes', you have effectively replaced (displaced) your 'ego', and that is why it tends to loosen the grip of your current one.

There is a good chance that you are wondering (with expletives!) "How does this have anything to do with Mind/Body, Aiki, and/or Zhan Zhuang?!?!".

It does in a very important way: you effectively cannot perform anything, let alone Aiki, when distracted by the 'Shin Mind'. The problem is, you have likely been distracted and you did not even know it.

You see, in the exact same way that you just used a drill to place yourself into an 'alternate situation from what is truly happening', your Shin Mind has been replacing your current moment with a much more subtle delusion, a near irresistible one that you can really buy into. You call it your past and your (potential) future.

The reason your Shin mind is so hard to quiet down is that it knows you really, really well. It knows your fears, your joys, your likes, your dislikes, your sadness, everything. In many cases, the Shin mind nurtured those preferences/feelings to help keep your attention (keep you distracted)

Are you completely focused on your situation with Uke, or are you sharing time dealing with your Shin?

Where We are Strong

Ever see a boxer dangle one hand in the air to catch their opponent's attention and then hit that opponent with the other hand?

What does it mean to 'get in your opponent's head'? - stoke the flames of your opponent's Shin to create strong emotions to distract them, so that you can blindside them (e.g., taunt your opponent to get them distracted by rage, intimidate them to generate fear, mock them to incite embarrassment, etc. Fear, doubt, worry, regret, sorrow, anger, embarrassment, guilt, rejection are just a few distractions we can influence in our opponent [and ourselves!])

When the Shin mind is in full effect, there is no opportunity for you to be 'engaged' in your actions. All you have left is blind reflex.

Blind reflex is 'mindless' in a bad way. (mindless is NOT mu-shin, instead it is 'without any mind' at all.) Your ability to perceive what your situational receptors are telling you is near eliminated. In a few words, you are dulled and unable to adjust quickly, a.k.a. you cannot adapt what you know to meet the need because you cannot assess the need. You lack awareness.

In contrast, with your 'Thinker' quieted, the 'filters' are gone and your 'Watcher' is able to receive directly the reality of the current moment.

It is very liberating, for we are no longer a victim of emotion. We become an entity of focused (pure) intent.

The question now should be, "what do I do with pure intent?"

The answer: act!

This is the intent, the Mind, the 'I' that should direct your actions. (The 'I' of Hsing-I and I-Chuan when we allow our awareness to guide and initiate our responses/reflex based on the 'reality' of a situation.)

The wise ones will tell you that you are in the perfect condition to 'do nothing', also known as 'Wu Wei'. Better said as, 'let your action arise as natural course of what is happening around you'. What do I mean?

Did you notice how well your 'Will' adjusted to your game of 'rapidly replacing your ego?' It was easy. Every time you changed scenes, you did not have time to build a detailed 'back story' (i.e., past) or set complicated expectations of outcome, you imagined and without hesitation, performed what was natural to do for the situation in that moment of 'now'.

Zhan Zhuang (According to Bill and Joel)

This ability is there for you 'here and now' if you turn off all the illusion of the past and potential future.

Short warning though. Coming into awareness of your 'true self' is sometimes a scary thing.

It is often difficult to practice 'losing the ego', for you have relied upon your ego for survival for as long as you have been alive. For me, there still exists a 'letting go of the side of the pool' type hesitation to relying solely on my Will, but the benefits are spectacular.

Rest assured that your ego will always be there. You cannot eradicate it. Ego has a purpose and you want it around. Be sure to tell yourself that your ego is only taking a short break, or if it must be active, it must be strictly aligned to your Will (the first of the three internal harmonies! A big part of what it means to be 'Centered')

One last caution before we close

Now even though you have quieted your Shin-Mind and are exemplifying Mu-Shin, there is still room for refinement when it comes to your Will.

Adapting quickly to situations is still rooted in your principles, what you believe as a sense of morality and righteousness.

In other words, Principles govern your will, and allow/disallow you from performing actions that typically end up being discussed in ethics and legal debate.

Your principles can be affronted by the actions of others. If you are not strong of will (full of intent) you can find yourself shocked, and stopped cold in your tracks.

It is not an emotional reaction. It is based on a jarring discord between your principles and those of others. Let us call it, "I cannot believe they just did that! It is unholy. I could never…that is an absolute atrocity"

The actions of others are not the only aberrations you will come to wrestle with.

Your principles are also vulnerable to discord between your principles and how they align with the 'nature of things'.

I will offer just a hint of explanation: this is a personal subject and you 'will' have to discover what you feel is 'Your Truth', and thus suffer/enjoy the alignment or discord with 'THE Truth' accordingly.

In my experience, there is only one 'Truth' and I am not its progenitor, merely its beneficiary. As far as I can tell, this realization is as close as anyone could ever hope to come to as to define of the word: humble. Staying humble (and kind) is a grand goal.

So, before I go off the deep-end and get too preachy...

It is far beyond me to suggest that I know which principles are correct. The ethics and legality debates are as old as man, but I will suggest this much as I believe it will aid you greatly in martial settings...

should you succeed in removing your distractions; discover yourself in the 'here' and 'now', as you become engaged with 'this moment'; you will find strength, solace, and serenity, both during and afterward this event if you 'Will' let your intent move in the direction/manner of Presence.

With that, it is time to close the discussion on Zhan Zhuang.

We are ready now to take all of this internal strength we just refined and deliver it unto our hands!

The Transmission

How do we get the internal strength into our hands?

We use 'down, over, and up' to generate force. We might even add to that force by twisting our hip as we step!

We carry that force through our torso by aligning the fascia (Thank you, Zhan Zhuang!).

We channel that force into our shoulders and elbows with a little bit of fascial torque and a slight bit of muscle (Thank you again Zhan Zhuang, this time for uniting our shoulder and elbows with our torso and motion!)

Something is still missing. We get some of the benefit of the force created by our body, but what is keeping that force from amplifying and passing into and through our hands?

The Transmission

Do we bend our elbow as we do in the gym?

What good will bicep curls and triceps extensions do for us in a fight?

The answer is that bicep curls and triceps extensions might actually be helpful.

We could bend solely at the elbow and drive a good amount of force with a triceps extension as most persons do when using a hammer. We might even give Uke a good upper cut as we curl our arm, but we can do much, much better.

How about, instead of relying on a rather weak set of upper arm muscles, we put power into our arms by letting our legs/body do the work?

It is easy, once you know how; that know-how is called 'The Transmission'

The Transmission is a study in simple machines; in particular, the lever.

(Aiki is largely about placing pivots in the correct spot!)

We are going to point out how, by the most common (obvious) usage, your arm functions as a third-class lever (when you curl it/pick something up)

With a bit of strategic pivot placement, we are going to convert our arm instead into a first-class lever.

All manner of great things can happen once we have this skill.

Some of which we will discuss in this section (e.g., getting the force generated by our hips into our hands/arms) and the rest will be explained in 'connection builders' section (e.g., how to use The Transmission to create Aiki, not only in your hands, but in martial armament/tools [a.k.a. weapons]!)

Before we get into the application and since our focus is going to be creating levers in ourselves, it helps to actually know a bit about them.

Sure, levers are simple machines, but they are not by themselves simple.

Levers may only have two parts: a rigid length of material (lever arm) and a pivot called a 'fulcrum', but there is a very important magic within them called "Mechanical Advantage"

What is mechanical advantage?

In base terms, it is a force multiplier! Something that can take, for example, one pound of pressure coming from you and turn it into many times more pressure on the object you hope to put pressure on.

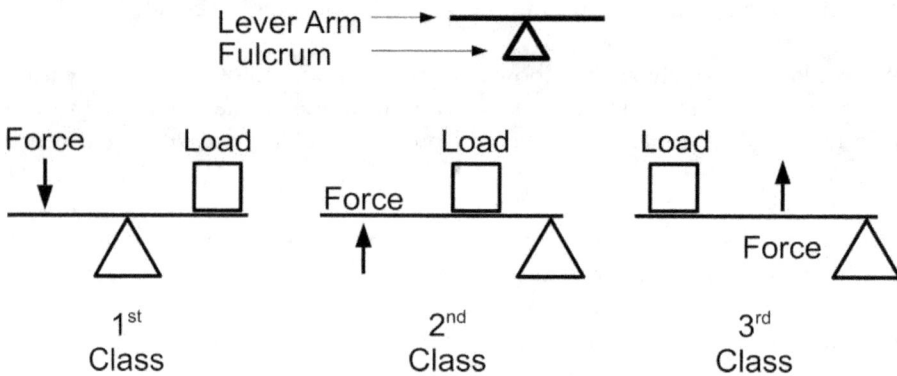

Figure 72: Lever components and three classes of lever

So, although I suggest that you can skip the short technical side conversation that we are about to have, I also suggest that you might miss an opportunity to become an Aiki Engineer!

Fear not, we will make assumptions and imply 'perfect world' conditions so we can explain this in a very straightforward manner. Give it a try…

Let us get started by looking at the first-class lever in Figure 73.

Looks kind of familiar huh? It should, it is just a teeter-totter.

With the fulcrum at the exact middle of the teeter totter plank, when we rotate (e.g., counter- clockwise from '9 and 3 o'clock' to '8 and 2 o'clock'), both ends of the lever move at exactly the same rate; i.e., they travel exactly the same distance in exactly the same amount of time.

The necessary acceleration and speed to get to those positions in that time is also identical. (The acceleration is going to be important in just a bit)

Not bad, but not much mechanical advantage. (None actually!)

The sad fact of this situation is that, in order for the teeter-totter to be of use, the weight on either side must be the same or you just get lop sided. For all intents and purposes, a teeter-totter acts like, and is, a scale.

The Transmission

You learned this around the age of five when you noticed that you had to find someone the same size as you if you wanted to effortlessly enjoy a teeter-totter.

It was effortless, because you and your likewise weighted partner 'balanced out'.

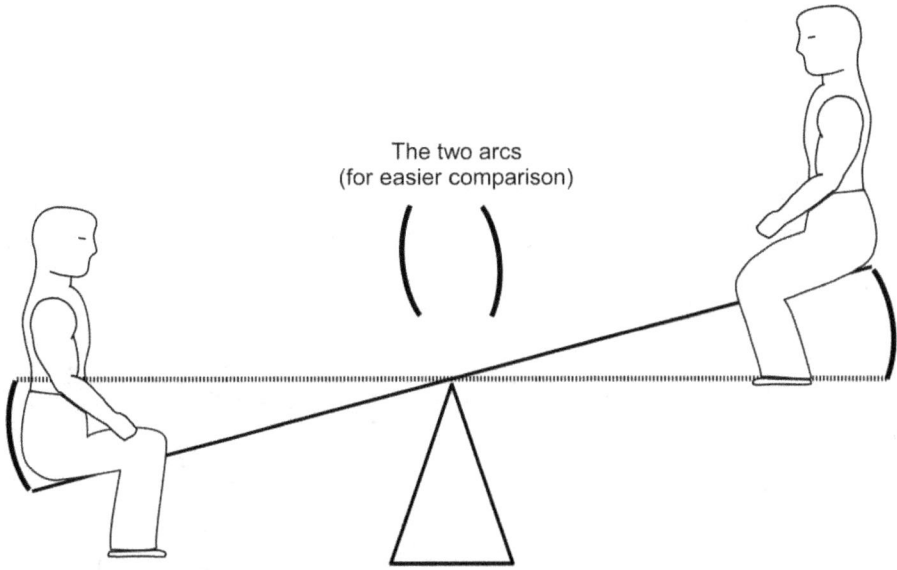

Figure 73: A teeter-totter (First Class Lever)

The cancelling out of each other's weight meant both of you could ignore gravity and, with minimal effort, push against the ground to rise upward.

The 'weightless' effect is incredible, both having equal effect on the teeter totter means neither of you were more advantaged than the other.

In battle, we do not want to be equal to our opponent, we want every advantage we can get so we can handle even those much larger than ourselves. Mechanical advantage to the rescue!

The mechanical advantage we seek appears when we shift the lever arm in relation to the pivot.

Leaving out a whole lot of detail about Figure 74, it will be enough for us to know that the mechanical advantage is calculated by comparing the length of each side of the lever arm.

To be specific, longer side length divided by the shorter side length. If the longer side is 10 feet long, and the shorter side 1 foot (10 divided by 1; which equals 10), you will get ten times the amount of force applied to the object on the short side as compared to the force you are applying on the long side.

Thus, same weighted persons will produce a lop-sided lever. (Same two persons from Figure 73 are in Figure 74, and the Teeter-Totter is slanted.)

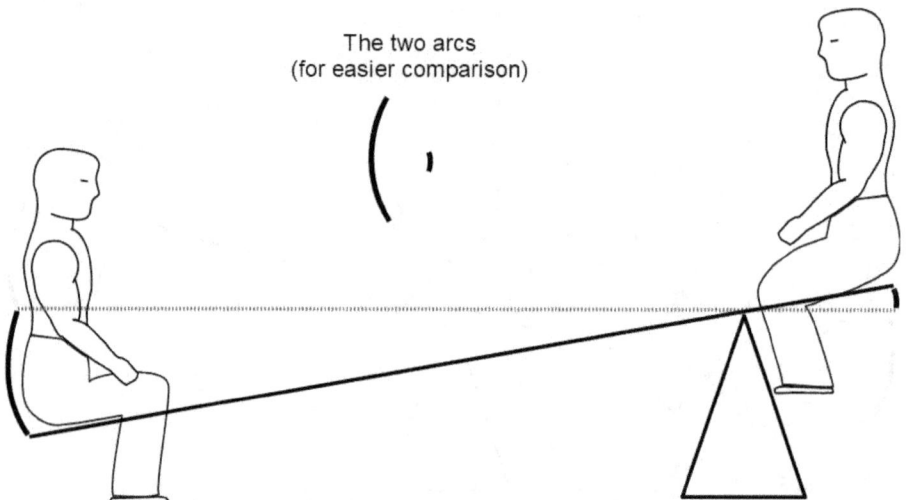

Figure 74: shifted lever pic, moves one hour, but different distances

{Quasi-math-y part that you can just skip!

Why (and with a smidge of detail please)?

Again, keeping things extremely simple, if the short side moves 'one hour', 'one number' on the face of the clock [i.e., one twelfth of a circle [1/12], or 30 degrees], so too has the longer side.

The difference is that the tip of the longer side moved ten times the distance in the same amount of time. Each side moved = 2*Pi*Radius/12, so if short side radius = 1, long side radius = 10; round Pi to 3 and you can estimate the short side moved about a half foot, the long side moved 5 feet; i.e., ten times as far. (see the 'arc comparison' in the middle of Figure 74)

The only way to do that in the same amount of time is to by having ten times the acceleration.

The Transmission

That is interesting because Force = Mass x Acceleration.

Archimedes figured the mass part did not matter so much when considering the mechanical advantage, it was the ratio of the movement that mattered.

We could put any amount of mass on the long side and the force would be amplified on the other, regardless of how much resistance the short side was offering.

I hope that made sense, if not, forget it, ten times the length on one side equals ten times the force on the other, skip to the next paragraph.

Quasi-Math-y part over.}

In figure 74, for the weight to balance out, we would need the person on the short end (high-side) to weigh ten times the other (e.g., a 50 lb. playmate on the bottom of the teeter-totter could partner with a 500 lb. gorilla! A 100 lb. person could balance a 1000 lb. cow!)

This wonderful advantage is a result of the mechanics of a first-class lever.

At worst, when both sides of the first-class lever are equal (ala a teeter-totter), there is an even exchange of force (One pound of pressure on opposite side for each pound applied to the other; e.g., a scale)

At best, we gain advantage when we make one side longer, and push on the longer side. The pressure is amplified to the shorter (according to the ratio of the length of the sides.)

This means that it is typically in our best interest to create an unevenness in the lever and stay on the longer end of the stick!

Take a look at Figure 75 and what do you see?

Your forearm is only 'half a teeter totter'.

Your elbow represents the fulcrum (for now!) and your forearm bones are one side of the lever arm.

What would it be like to have only half a teeter-totter and try to pick up our friend by pushing or pulling upward on the plank just a short bit away from the fulcrum?

Pretty darn impossible right?

That is because the mechanical advantage is working against you. This situation is very similar to pushing downward on the short end of our adjusted first-class lever in Figure 74; (i.e., trying to be the 500 lb. gorilla)

Study Figure 75 to understand that your biceps (and a muscle called the Brachialis that sits below the biceps) are positioned to do the work of curling your arm.

These muscles connect to your forearm just a short distance away from the elbow fulcrum, and worse, they do this between the load and fulcrum (just as a third-class lever does)

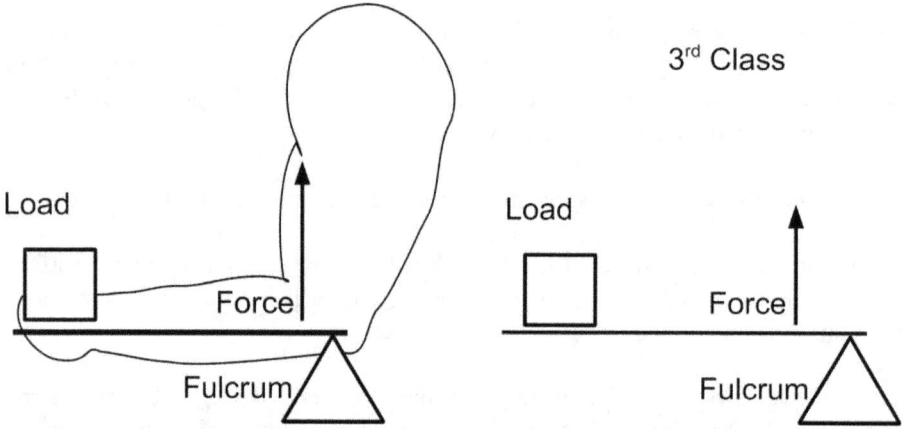

Figure 75: pic of the arm to reveal the 3rd class lever

Everyone is a bit different in shape, but for the sake of explanation, let us suggest that those muscles connect one tenth of the way from your elbow to whatever you are supporting in your hand. (this is an approximation, but it is not that far off from reality for the average person)

Why, oh why, if our arm is at a serious disadvantage in moving objects around, do we so often try to 'bicep curl' Uke when it is ten to one against us? (i.e., why do we exhibit OABD? – Over Active Bicep Disorder!)

I am sure there are a lot of reasons, but one good one is due to the lack of awareness that we can do something else...

Let us convert our arm into a first-class lever! It is quite simple and very useful. Again! Aiki(do) is all about pivot placement.

The Transmission

To prove it, we are going to move the pivot of our arm from the elbow to an alternate position: the middle of our forearm.

Doing so will reveal that your body can generate more force than your bicep and brachial muscles.

Let us start with the third-class lever experience, with pivot in our elbow, and perform a bicep curl, here is how.

Find a chair with a back about as high as your solar plexus.

Stand behind the chair and hang your right arm downward such that your elbow is lower than the top of the chair, back of your upper arm braced against the head-rest area. (Right side of Figure 76)

Now curl your arm.

In most gymnasiums they would call that action an 'isolated bicep curl' because this exercise solely engages the arm curling muscles.

On the highest end, I have seen persons curl as much as seventy pounds this way. Most never achieve sixty and those are the 'stronger' persons. The rest of us are never getting to fifty pounds without training.

That point is why we use a 'hip-pop' when trying to unbury our hand when Uke has our hands pinned to our hips.

You could try this curl exercise with some weight if you so desire, it should be obvious that you cannot move a great deal of weight, but what I want you to notice is which muscles fired. (hint: solely your upper arm muscles)

Let us reposition and try again.

This time, place a folded towel or a pillow on top of the chair (for comfort!) as you again stand behind the chair and place your forearm across the top of the chair with the top at about the middle of your forearm

Let your hand drop a bit toward the chair seat.

With the middle of your forearm staying in contact with the chair, bend your knees, which drops your elbow, and let your hand rise.

Voila!

Your forearm just became a teeter totter; i.e., a first-class lever! (Left side of Figure 76)

Your hand did not rise quite as high as when you performed the isolated bicep curl, but it does rise!

Given that we 'halved' our forearm when we decided to make the chair our fulcrum, we are not gaining any mechanical advantage, but each pound of pressure is 'transmitted' to our hand. (hence the term 'transmission')

Take notice of which muscles were engaged...

That is correct, none! (at least none in your arm)

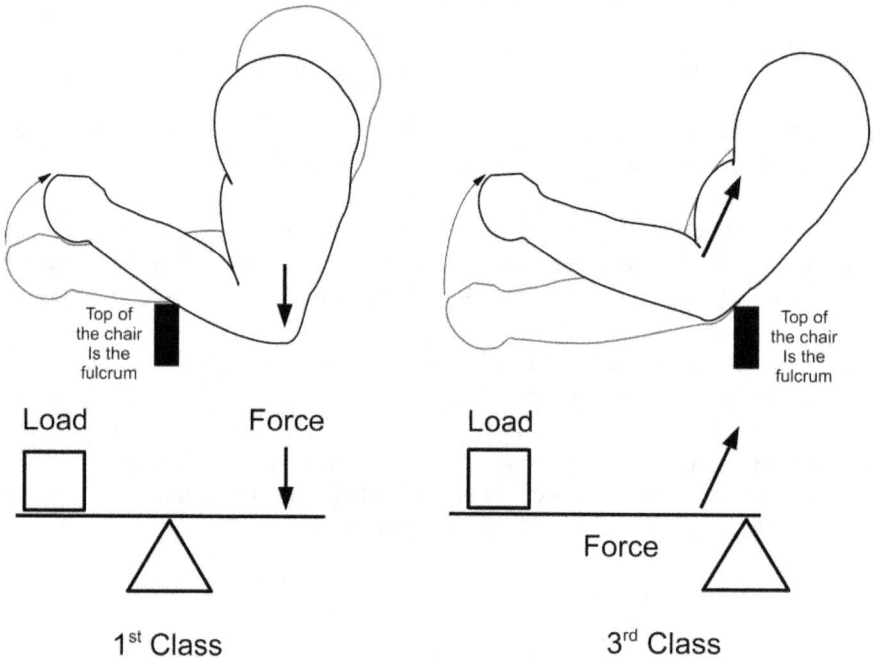

Figure 76: Your forearm as a third- and first-class lever

All you did was bend your knees a bit and lowered yourself.

Nearly all of your weight was applied to the elbow side of your 'forearm teeter totter'.

The Transmission

Applying weight versus exerting muscles opens the door to using 'power' instead of 'strength'.

You should trust me that you can move a great deal more weight this way.

I also suggest you forget about trying to test your limit, for it is likely that your forearm bones will break long before you reach your new max weight. Do not test it, just know you have become drastically more powerful than you were using a third-class lever arm.

I can hear you saying, "Thanks! Awesome! Will you help me carry around my Aiki Chair everywhere I go?!?!" (and under your breath, "dummy!")

Answer:
We could travel around with something to generate a fulcrum with, but then again, do we really need that?

Remember, levers only have two parts: a rigid length of material (lever arm) and a pivot called a 'fulcrum'

The two-part lever recipe is magical when you realize that the fulcrum does not really have to be a 'thing'.

"What?!?!"

It is true, think about it. What does a fulcrum do?

The fulcrum has one role: to enforce a position in space (a stationary spot), upon which, the lever arm can turn upon.

Can the lever arm not rotate against itself without external support directly upon the pivot point? Sure, it can!

That spoon you use to bring soup to your mouth; that is a lever. So too are fishing rods and pens (and tennis rackets, baseball bats, etc.)

Some will argue, "The fulcrum in those objects is in my hand, therefore the fulcrum is tangible!"

My retort?

True for most common usage, but does the pivot have to be at the point of contact in our hand?

Where We are Strong

It does not, and, just maybe, we can get 'outside the box' and use these items a bit differently than is typical. (Let me explain later in this book. We are not ready to talk about martial armament yet [i.e., the tools of martial arts that some call 'weapons'])

As you read, remember this secret: all a lever really needs is a point in space to stay in place as the lever arm turns around it, the key is to build coordination so we can perform just this task.

Let us drill! (incorrectly at first!)

Place a hand on the wall and, while maintaining that contact, loose wrist, walk around some.

That is not Zhan Zhuang! That is called 'truly dancing with the wall' and you look silly. Stop it! (I hope no one was around to see that! <wink>)

Unfortunately, when we put our hand on Uke instead of the wall, this seems to be a very common thing to do. There is no pressure against Uke using this action. Let us fix it.

This time, find a door frame and while standing inside the door frame (a foot in each room), let us put our arm in front of ourselves parallel to the floor as if holding a shield. (bottom of Figure 77)

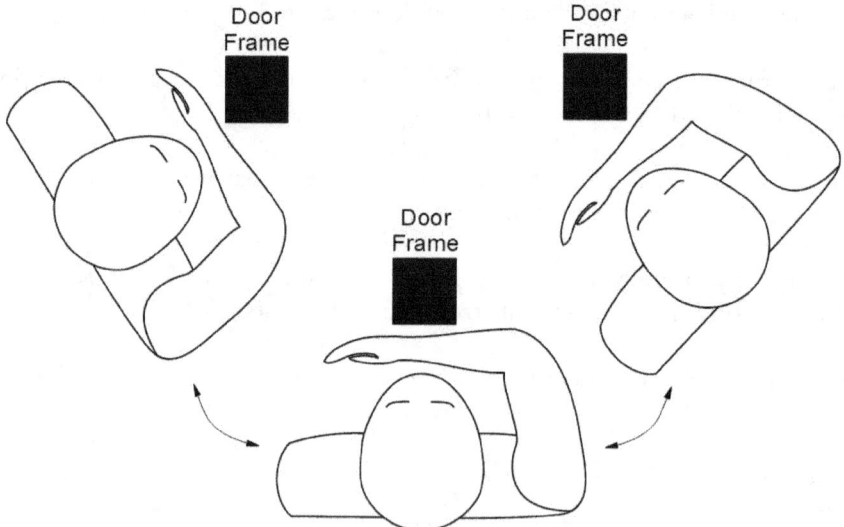

Figure 77: Basic transmission drill; moving entire body around the door frame

The Transmission

Place the middle of your forearm against the latch side of the door frame, and while maintaining contact between the frame and the middle of your forearm, move your body such that it 'teeter-totters' your hand and elbow into and out of the adjoining rooms. (suggestion: step around! In and out of both rooms as you stay facing the door frame)

Imagine as you perform this, that Uke has grabbed your wrist. (important, Uke grabs your wrist, NOT the part of your arm in contact with the wall. The effect on Uke should be obvious. See Figure 77 again)

Get creative with your stepping, face in any direction you wish, but always maintain the contact with the wall and the middle of your forearm as you teeter-totter your elbow and hand in and out of the rooms.

Next, (not depicted) stand perpendicular to the door frame (i.e., in one room), use the outside middle of your right forearm to touch the wall (like 'holding a pizza box', but just one hand reaching into the other room) and bend/extend your legs and see if you can create the vertical teeter-totter effect you created with the top of the chair example in the left side of Figure 76. (Laugh if you will, but this does take some coordination!)

If you have mastered these basics, move on to creating an hourglass type path for your hand and elbow (with the middle of your forearm as the 'skinny' part of the hourglass).

The hourglass effect takes a bit of coordination. The Tai Chi Classics call it 'harmonizing the knee and elbow', I call it a hip turn. See for yourself.

Use a pole as a guide (or stand in the door way, someone hold a Jo staff, etc.) with right arm in 'shield holding' position again, place the middle of your forearm against the pole, and get into a basic 'cat stance' with your left leg back and heavily weighted, right leg forward, lightly weighted, with only ball of the right foot in contact with the floor. (Bottom of Figure 78)

Maintain the weight in your back leg (left leg) at all times as you rotate your left hip counter-clock-wise and start to turn your entire body leftward; (i.e., leave your left foot in place and close your left hip kua; which should result in your body facing leftward.) (i.e., Figure 78 - Moving from Start Position [bottom] to Closed Kua Position [upper-right side])

Allow your right foot to pivot in place on the ball of your right foot. I like to think of this as allowing my right heel to follow the right elbow, but technically, it is the right knee following the right elbow.

You should sense your right elbow moving forward past the pole, your right hand coming toward you (contact between the middle of your forearm and pole maintained).

At the limit of your 'Closing your left kua', reverse the motion and return to the start position; (i.e., Turn your body clock-wise)

Continue moving clockwise past the starting pose, until your left hip kua opens and your hips are facing rightward as much as can; (i.e., keep rotating clock-wise through the bottom starting position of Figure 78 and move into Open Kua position in upper left.)

As you move, notice that your elbow retracts as the hand begins to move past the Jo/door frame. Your right heel swings around the ball of your right foot, and at the extreme of your motion, will be pointing at or maybe a bit beyond your left ankle.

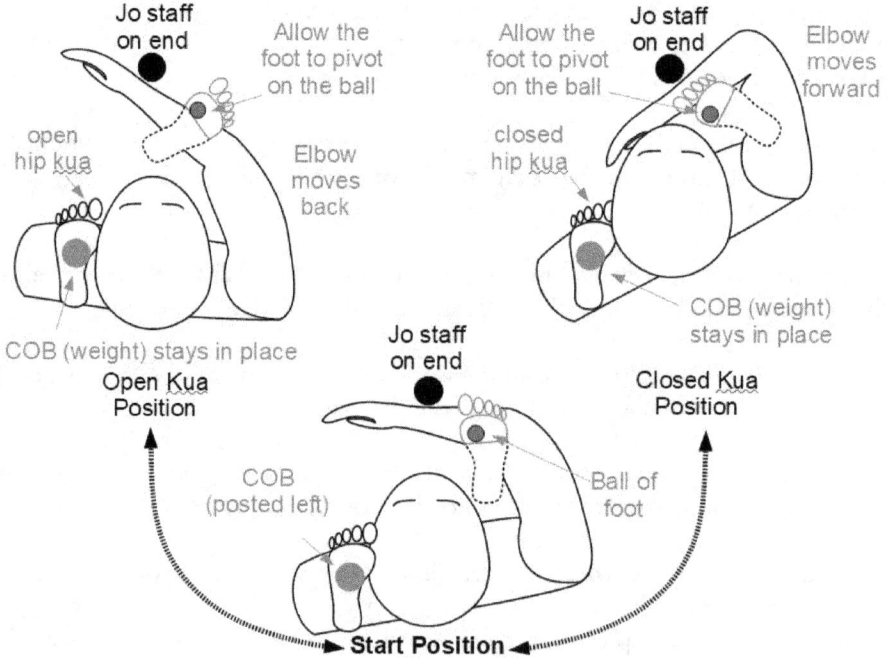

Figure 78: Hip turning to create the first-class lever; dropping and raising elbow to create an 'Hourglass'

I have been doing this a long time, and it has become easy, but not so for many when they first try it. Expect to need to go slow at first.

The Transmission

Once you have the knack, you can begin to drop and raise your elbow as you twist and change this horizontal back and forth into an hourglass.

Again, this is likely to be a bit tricky at first.

Eventually, this drill is nothing more than dancing 'the twist' with your arm against an imaginary door frame.

I have trained others by holding a jo staff upward with the bottom in contact with the floor as replacement for the door frame, but it would be just as valid to use tip of your finger as a guide for them to stay in contact with (this is important later in the connection builders section).

You likely find that you cannot just use your finger until your student becomes a bit coordinated. (that is why I use the Jo; you find out quickly that the coordination is not easily created and the student pushes the staff all over the place. Door jambs do not have that issue, they stay put!)

You might say this action will be impossible to perform with an Uke holding onto your wrist, especially a resistant Uke, but truly it is not.

The point is that we do not need to be super strong to help guide Uke when they are moving and committed. We also have avenues to guide Uke when they are static (No-line and 20!)

Having a first-class lever coordination is going to make all the difference in the world when we are leveraging those places where Uke is weak.

In fact, I think you will find (at the end of the connection builders explanations) that there is no other way to deal with a resistant Uke.

This skill is useful even when Uke is not connected to us; when they are trying to strike us.

To quote Sensei Merritt Stevens, "Uke can only be strong in one direction"

Obviously, there are some conditions implied in that elegant wisdom, one being Uke's ability to maintain their path when punching. Redirecting a punch is easy (so too is being on the end of the punch but that hurts!)

You may not know it yet, but you have learned one of the key secrets to creating leverage in all physical activities.

Where We are Strong

This knowledge is a game changer for many, once they learn how to apply it.

With that thought, let us end our discussion of 'where we are strong' and move to 'Where We Are Powerful'; i.e., 'Advanced Connection Builders'…

[

"Wait! What about second class levers!?!"

Well, we do not use them as much in the dojo. There are examples, but not enough to warrant discussion right now. Time to move to 'Connection Builders' with a promise that we will cover them in just a bit.

"Wait! Can I mix up that 'hourglass' drill by shifting weight and pivoting on my lead foot?!?"

Do we really need more visits from Captain Obvious?

Move on to 'Connection Builders' and learn to use your 'Transmission'!

Go!

]

Advanced Connection Builders

Let us continue bringing power to our hands by emulating animals, common tools, shields, and learning a thing or two about circular Aiki!

The Albatross

The albatross is a continuation of the 'elephant arms' training introduced in 'Six Precepts'.

The base 'elephant arms' drill offers us two very important skills.

First, it teaches us how to follow our COB as it moves. This is an extremely important lesson, because the maximum amount of force that we can generate is almost entirely contingent upon drawing objects (e.g., Uke) toward or projecting them away from our COB.

We have to know where the COB is if we hope to best use it.

The second lesson is how we can 'cheat' Uke by lowering one side of the elephant arm. This not only reduces our effort/exertion, but also to allows Uke to 'roll into a void' which invites Uke to move in the desired direction.

This 'slanting of our hands' is in contrast to maintaining a 'parallel to the floor' position and directly supporting Uke's weight/pressure throughout the move. The parallel hands method works well, but why not take every advantage, and why not take the 'slanted hands cheat' to yet another level?

To get to that higher level, we will add a new angle of attack to the second lesson, 'slanted hands creating a void that entices Uke to follow our lead'.

With any luck, we will gain the ability to launch Uke without Uke-Do (*oo kay dough* - Uke making the fall happen to make Tori look good)

Better yet, we will learn from nature to do it!

So enters: 'The Albatross'

You might be asking, "How does an albatross have anything to do with Aiki?" It is all in the landing!

The albatross is beautiful in flight. With minimal visible effort, the albatross glides along air currents, rising and falling in altitude at a whim.

Advanced Connection Builders

Their rather large wing span is often credited as the reason for their grace in the air, but what is undeniable is the detriment those elongated wings bring to the poor bird when they land.

You might describe the common event as a rather abrupt stop. Think 'feet then face', for as the bird touches down with its feet, the momentum of the top-heavy torso often results in a 'face-plant'.

Obviously, these birds are not as seriously clumsy as I have described (or we would have a lot of flat faced birds!), but the visualization of this action, especially if you think of it in a cartoon sense, will become very obvious and extremely useful.

Let us do a quick review the standard elephant arms drill, then expand upon it to show you where the albatross fits in.

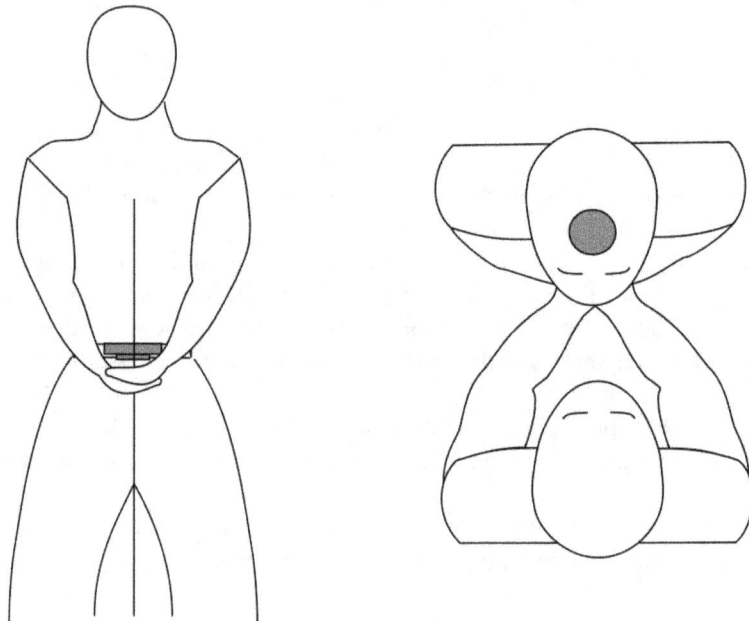

Figure 79: Set up for elephant arms drill; Center your weight and crossing hands; Tori and Uke lined up for elephant arms drill

Face Uke in a double-weighted stance (horse stance). Bring your hands to the middle and overlap them (I suggest you avoid interlocking your fingers, but whatever makes you comfortable). Have Uke put their hand or hands into yours and apply some of their weight by leaning onto your hands. (Not all of their weight, but most is fine. You may remember Figure 79)

Move your COB from one belt loop to the other (i.e., shift your weight), keeping the burden of Uke's weight as near your COB as possible. (Again, another déjà vu picture below: Figure 80)

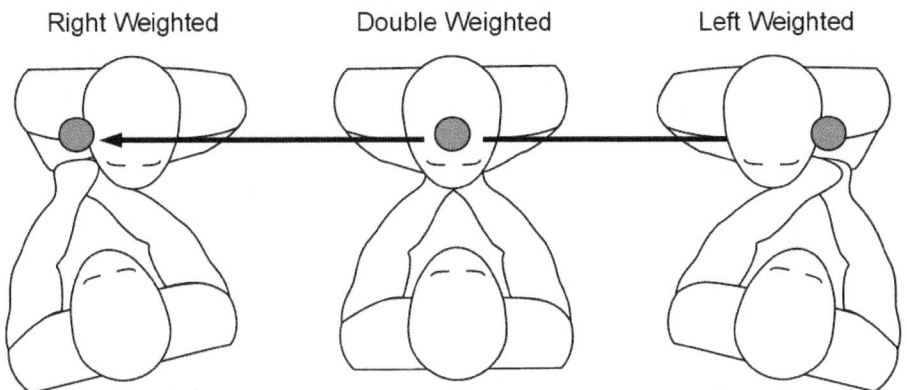

Figure 80: Elephant Arms Drill; Shifting the COB left and right and taking Uke's weight with you

I expect you kept both of your hands/wrists at equal heights at all times.

Great! For that is the initial drill that teaches you where your COB is.

Try again, but this time, use the 'cheat'; i.e., lower the side of your 'elephant arm' to create a downward ramp (declined plane) that 'rolls' Uke in the same direction you wish to move.

For example, if you are right weighted and have nowhere to go but to shift your weight to your left…, use 'the cheat' by dipping your left hand and allow Uke to roll into that void as you take the both of you in that direction. (Figure 81)

We learned in 'Six Precepts' that we could try lifting one side or the other, but that requires effort, so be sure you are relaxing and dropping the side that you wish to move Uke toward, not exerting to raise the opposite side.

Pay close attention to the action and relationship between yours and Uke's hands. Take notice of what is required to create the void that we want Uke 'roll into'.

This skill is the most overlooked aspect of creating Aiki. It is unfortunate, as it is arguably the very essence of Aiki (physically). So, take notice and know that we will dive deeper before this book is over.

Advanced Connection Builders

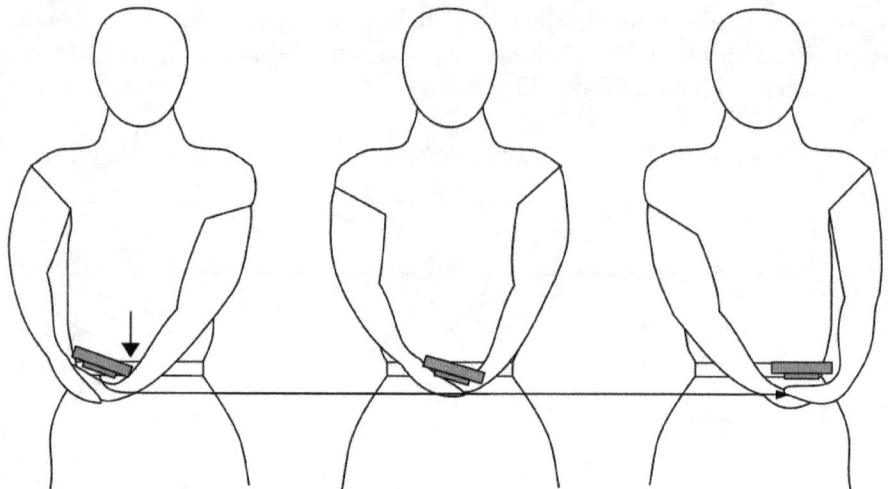

Figure 81: Elephant Arms Drill; Shifting the COB to your left; taking Uke's weight with you while using 'the cheat' (dropping the left side of the elephant arm to match the direction you wish to move Uke toward

If you have the 'swing' of things, let us expand our play by replacing the 'cheat' with an 'albatross'.

In the 'cheat' we dipped the elephant arm on the side in which we are about to move Uke. (same side/direction)

In the 'albatross', we drop the side of the elephant arm opposite of the direction we wish to move Uke.

This position is akin to the albatross coming in like a plane for a landing with Uke on their back. Think, 'feet down, nose up', no front landing gear!

The important part is that we again give Uke the void that invites them to 'roll' down our hands, but we do not let them fall off; i.e., keep Uke 'hanging onto' the albatross' back. Give Uke the 'thrown-back' feeling you get from a vehicle moving out from under you, but still staying seated.

It feels a little like Uke is pushing on our hands in a manner that would have our lowered wrist sweeping the floor if it could reach. (In the case of Figure 82, as if our right hand/wrist could make contact with the floor to sweep the floor, but of course, it cannot).

Let your albatross give Uke a ride on its back as it glides across in synch with your COB, and realize that this is a one-way flight!

The Albatross

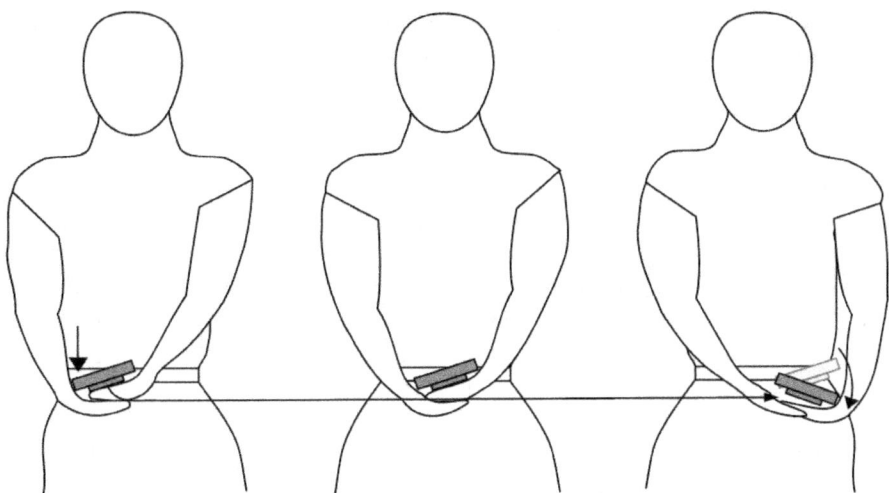

Figure 82: Elephant Arms Drill; Shifting the COB leftward and taking Uke's weight with you using 'the albatross' (dropping the side of the elephant arm opposite of the direction we wish to move Uke. Note the 'cheat' at the end

It ends at the point at which you have completed your sideways motion and have no further to go without stepping.

So enters the landing! (Crash actually! and if we are skilled, Uke's vault!)

Notice at the end of the motion, on the far right of Figure 82, is a shift from 'albatross' to 'cheat'.

This equates to that rather awkward landing (feet then face)

Where the left and center images of Figure 82 are likened to the albatross gliding across with head up, feet back and below ready for touch down. At the far side, the poor bird touches down and its top-heavy stature drives its nose downward while the feet stay in place. Ouch!

Two drops! No lifting! Drop one side of your elephant arm, then at the end, drop the other even lower.

Explained in words and in the context of this Figure 82 example...

at the start, drop your right hand and maintain the 'lowered' height of your right hand as you progress across your body. When you reach the end, again, keep that right hand at that 'lowered' height as you drop your left hand even lower to create the 'face plant' shift from albatross to cheat.

The fun part of this drill is feeling Uke launch off of your hands. I expect that you felt it. The success rate is quite high, even for first timers.

Two areas of focus seem to round out the lesson.

First, remember that it is just imperative to again create the void for Uke to roll into when you deliver the 'nose-dive cheat' at the end.

I cannot stress enough how important 'creating the void for Uke to roll into' is to every Aiki technique. Get used to this action/goal. It has been said before, but it bears repeating: Creating that void truly is the base ingredient of all physical Aiki technique and should be applied at all times.

Second, dropping to re-pivot your hands is much less exertion than lifting.

For some reason, many have a habit of trying to lift instead of drop.

You might not be consistent at this motion at first.

The albatross is a very counter intuitive move. Carrying Uke on the back of the albatross is weird enough, but it is that 'not letting Uke roll off, but still having a void for Uke to roll into' part that tends to be the challenge.

For your potential benefit, as an alternate metaphoric explanation, creating an 'albatross slant' in our hands can be imagined as 'popping a wheelie' on a motorcycle. Uke rides 'almost rolling off of' the back of the motorcycle (until the bike reaches a rock at the end and the front tire dives downward!)

The trick is to imagine getting into position with the 'back tire dropping' instead of the 'front tire rising'. (if that is too hard to grasp, go ahead and raise the front side, but you will have to exert some force to raise upward [instead of relaxing the backside to move downward])

Others have described the Albatross as 'walking backwards with a wheelbarrow', but again, I shy from that because it implies lifting the high side instead of relaxing and dropping the lower.

This analogy does work if you consider that you would want to keep the dirt in the wheel barrow as you walked backward (much like we want Uke to stay on our hands, but desiring to roll off)

The Albatross

It is arguable which simple machine we are using here (I ascribe to declining/inclining planes, but of course, what is a lever once it is slanted but just another example of a declining/inclining plane!)

I like the albatross analogy best (due to the focus on 'dropping the landing gear'), but use the motorcycle or wheel-barrow intention if that works better for you.

Whichever analogy you use, 'The Albatross' method of moving Uke must be mastered and, although optional, I strongly suggest that you master shifting from albatross to cheat by dropping lower in lieu of raising up; (i.e., learn to make the shift without additional exertion).

You will know you are getting it when you feel Uke launch.

Think through this 'Albatross to Cheat' action a bit further, apply it to your techniques, and you will find this secret showing its 'flattened face plant' in many places, especially in weapons.

Case and point...

Within Tomiki Aikido's Koryu Dai San kata (third of six 'core' katas) are a good number of opportunities to use this 'albatross to cheat' action to throw an Uke that has grasped the jo-staff. (the Internet awaits you to find and view a demo of that portion of the kata! [Although I cannot guarantee the practitioner is going to demonstrate albatross to cheat])

More often than not, that insight seems to beg another question, "Is a jo-staff the best way to learn to perform the albatross outside of the elephant arms drill?"

The answer? Maybe, but your forearm will do just as nicely. (Hint: the forearm areas on either side of your 'transmission' can 'albatross' or 'cheat'!)

That answer should lead us to another great question: "Are there scenarios in which I have been following these same 'albatross and cheat' principles in everyday life?"

That improved focus question leads us to a deeper understanding, one that reveals which everyday tools the 'Albatross and Cheat' are really emulating...

 tools you have been using all of your life and you might not have thought you were training the whole time...

Advanced Connection Builders

The Broom and Shovel

Chores have long been the Martial Arts Master's preferred method to instill martial skill in their apprentices. I can name many martial arts films that prove my point! (likely you can to!)

Why?

Well, chores get the job done in more ways than one.

Your student gets better, and your house or dojo gets clean! Win-win!

Now, I cannot claim to have had a maintenance engineer teach me the finer points of applying and removing wax from a car, smoothing large swaths of walking area, or even applying color to vertical outdoor surfaces <wink>, but my parents made sure I swept all the dirt I dragged into the house, and because I grew up in northern United States, I had to pitch in and shovel a whole lot of snow.

Now I am not saying that every parent is secretly a martial arts expert (although most seem to know how to perform corrections when kids misbehave. Mine did!), but all parents want help keeping the house clean.

What a blessing, for it prepares us for learning how to move objects with a tool! In these cases: a broom and a shovel.

It is an interesting twist that all 'weapons' (martial armament), including your hands, are nothing more than tools. In Aikido, a sword/bokken, a jo, and your hands are all tools we use to move our favorite object: Uke.

What is even better, is that brooms, shovels, swords/bokkens, jo's, arms, hands, etc. all share the same principles when we use them efficiently to create Aiki.

Let us focus on the universally relatable situation of moving objects using a broom and a shovel to better understand the art of moving Uke with the tools of Aikido!

The Broom
Have you ever stopped to realize that a broom is a hybrid mix of a lever and an inclined plane? (You have! We both need a life! Ha!)

The Broom and Shovel

When it comes to brooms, we have two kinds: push-brooms and straight brooms. Both types fit this lever/inclined plane combo description, but, for now, we will focus on the straight brooms. (push-brooms are a bit more akin to rakes [operate on a level plane; more like a 'push-rake'])

How is a broom like a lever? It has the requisite parts: the broom stick (rigid lever arm) and can be made to spin at a pivot point (fulcrum), but to complete the classic lever analogy, we still need an object we want to move (a load) and an external force (a leading force).

In this case, let us use the broom stick as the lever arm, the dirt on the floor as our load, and our hands will provide the leading force.

"What about the pivot?! Where do we place the pivot?!"

It is by examining the pivot that we gain martial understanding!

Let us begin by considering how we sweep a floor. We start with a quick bristle check! (for the detail-oriented persons out there)

We need the broom bristles to 'be in the way' of the load.

If bristles are too flexible, they simply pass over the objects (dirt) we wish to move and leave residue behind. We also need the bristles to be the correct width and density.

If the broom is not dense enough or the bristles are too thick, they will not 'be in the way' of the dirt. The smaller objects will simply pass through and only the larger objects get moved; (i.e., like a rake.)

Assuming we have the appropriate broom bristles for the job…

we need some pressure to keep the broom gliding along the floor and imparting force upon the dirt.

Keeping the broom in contact with the floor is achieved by working the broom as a lever and it is the pressure from our hands that determines which type of lever it becomes.

Pressure from our hands onto a broom is translated one of two ways, either as a first-class or third-class lever. Either way can be made to work, but they do affect the efficiency, and thus, the results.

Advanced Connection Builders

Using a broom as a third-class lever means our top most hand becomes the fulcrum (stays in place), as our lower hand swings downward to push on the middle of the broom stick. (See Figure 83)

This type of use typically 'splats' the bristles flat to the point that the part of the broom that affixes the bristles begins to come in contact with the floor.

As we begin to sweep, the 'splatted' broom bristles will actually roll along on top of the dirt. It should put you in mind of the broom gliding along on top of a bunch of marbles. (really small marbles). The bristles become elevated, and in a sense, lose contact with the floor. The bigger pieces of dirt hold the bristles up, and the smaller pieces of dirt remain untouched.

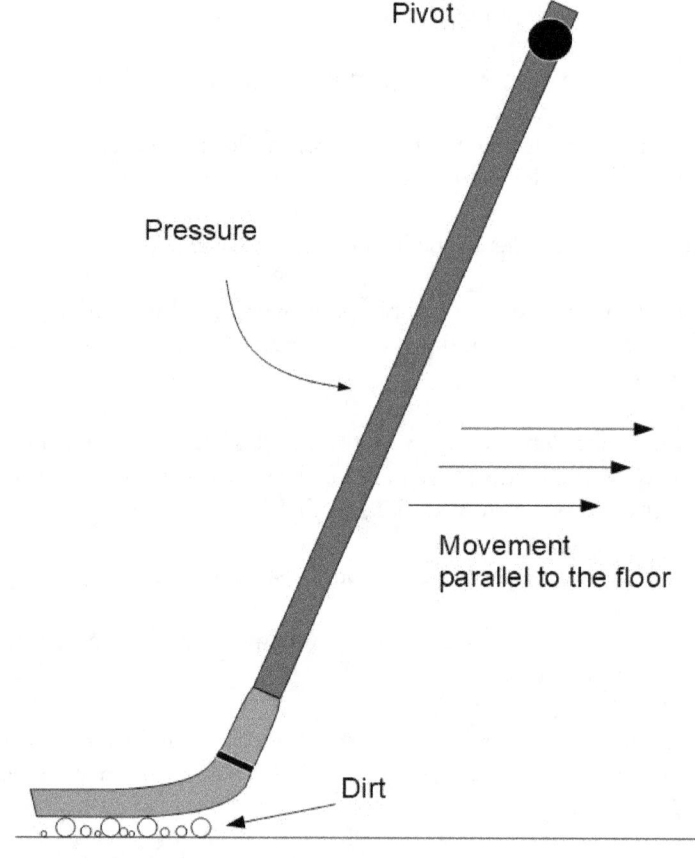

Figure 83: Third class lever broom. Bristles are 'splatted' to the floor

The stroke used in a third-class lever is awkward as well.

The Broom and Shovel

Keeping the pressure on the bristles in this manner requires that we move both of our hands horizontally, parallel to the floor.

It more closely resembles the stroke used with a rake where we hope to leave some parts of the load behind (the only difference in that we grip a rake with 'thumbs toward the rake head', broom grip is 'thumbs up' away from the floor.)

In this method, the tool handle (broom or rake) maintains a constant angle with the floor for the entire stroke. (See Figure 83 again)

Do not laugh, I have seen grown adults try to use a broom in this manner.

It is not an efficient use of the broom as a tool, but it is a third-class lever. (Because the lower hand is applying pressure to the broom stick between the load [the dirt] and the fulcrum provided by the upper hand.)

We can switch to using a first-class lever, but our dirt moving efficiency does not necessarily get any better.

We make the initial switch to first class lever by using our lower hand as the fulcrum and lifting the top of the broom with our other hand. (Figure 84)

This 'lower-hand pivot, upper-hand lift' method uses the broom as a first-class lever, but may suffer the same 'bristle misuse' if we lift upward too hard.

Lifting with too large a force will, yet again, splat the bristles; (i.e., bring the bristle binding close to being in contact with the floor.) Thus, it also suffers the same penchant for leaving material behind.

The broom stroke is again awkward, for the broom stick has to be kept at a consistent angle throughout the stroke; (i.e., the hands move parallel to the floor); fine for a rake [with an adjusted grip of course]), but this is not how we use a broom.

Both of these versions of using a broom are inefficient because we are using the broom solely as a lever.

In this case, we are overdoing it with the lever's ability to crush objects.

In these examples, the 'crushing' aspect was displayed as splatting the bristles against the floor and was an attempt to pinch/crush the dirt between the broom bristles and the floor.

Advanced Connection Builders

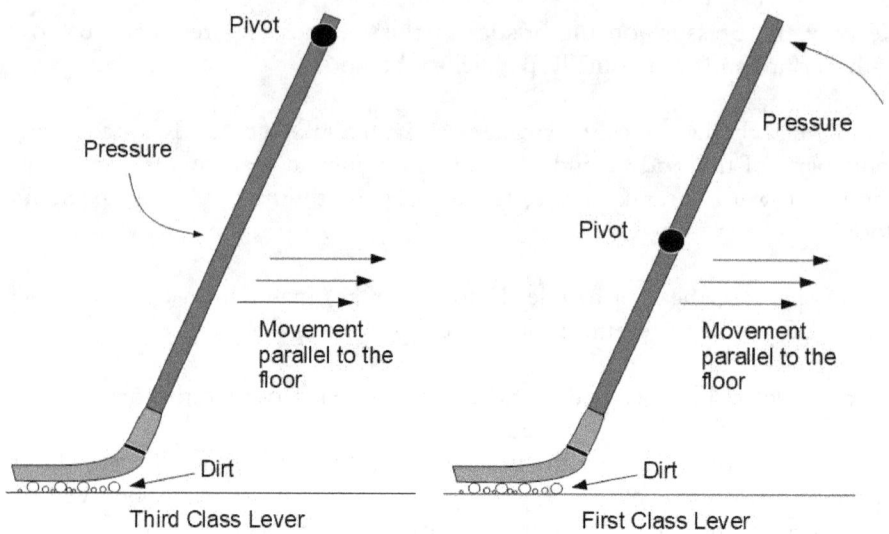

Figure 84: using a broom as both a third- and first-class lever; no big difference in the effectiveness

We do not want to crush the dirt; it would be far more efficient to 'roll' the dirt. All the dirt, even the smaller pieces, all moving/rolling at the same time.

What rolls an object? An inclined plane!

You know this if you ever watched a rock roll down a hill, or a ball roll down a slide. You feel this effect when you walk on a hill yourself. The only thing stopping you from rolling is that you are not shaped like a ball.

Having the broom stroke impart force to the dirt (load) as an inclined plane requires that the length of the bristles make contact with the dirt at an angle instead of flat (parallel to the floor); i.e., We want the bristles to contact these dirt 'marbles' at their side, instead of at the top.

The effect would look like the 'marbles' were rolling down to the end of the bristle as they simultaneously roll along the floor for a short length of distance. (Figure 85)

This action cannot be achieved using the broom solely as a lever.

To grasp how to create the inclined plane aspect of a broom, it may help to contemplate hoop rolling (hoop trundling), where you use a stick to elicit a hoop to roll down the street.

The Broom and Shovel

The stick in Hoop Rolling is nothing more than a lever/inclined plane; and by gliding the stick along the hoop, we indeed roll the hoop.

Any one that has worked in an auto tire shop or rolled their spare tire to the flat tire they were about to replace should understand perfectly. They were rolling that tire with their hand as the inclined plane.

The broom is just a big lever/inclined plane trying its best to roll a bunch of very small objects a.k.a. dirt.

Each broom stroke rolls the dirt from near the middle of the bristles, to their ends. (See Figure 85 again)

Once we reach the end, we reset the stroke by moving ourselves sideways a bit before positioning the broom at the edge of the dirt, re-contacting both the floor and the dirt with 'lightly bent bristles' (in lieu of an extreme 'splat'), the goal being to contact the dirt with the middle of the bristles; thus gaining a length of bristle to roll the dirt against again (the bottom length of the bristle to be exact).

This is the same with rolling a hoop, especially when we are directly behind or in front of the hoop. The stick is made to run up or down the length of the hoop, but we only have so much stick, so we have to reset.

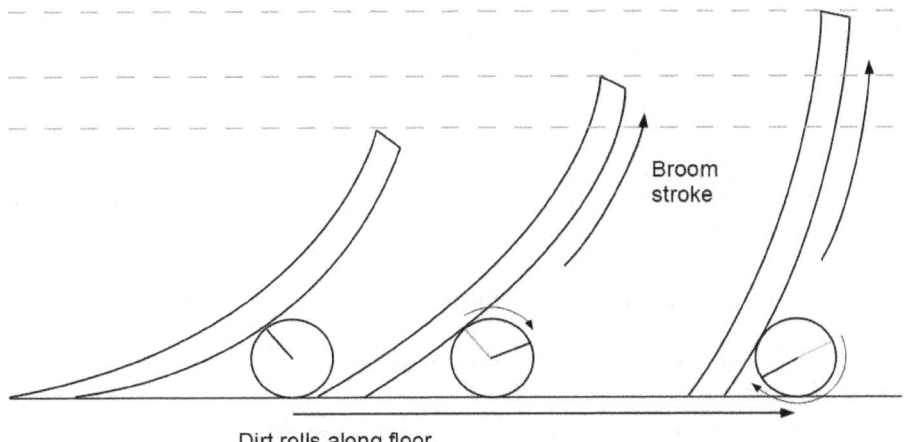

Figure 85: the broom meets the dirt with bristles bent (left). The broom stroke 'rolls' the dirt along the bristle as it rolls the dirt along the floor. The bend in the bristles moves/'relaxes' as the broom rises

Advanced Connection Builders

One last analogy to help illuminate the 'roll along the length of the bristle' concept...

Ever have one of the toy cars or motorcycles that you had to pass a length of plastic stick with gear teeth on one side through a slot that housed a true round gear (with the same kind of teeth), and then pull the plastic stick out quickly to turn that round gear and drive the motor within? (The car or motorcycle would then speed off as you put the wheels on the floor.)

In the Mojo, we call that stick the 'zipper ripper' as the gear teeth resemble half a zipper, and you had to 'rip' that stick out of the vehicle to make it go.

Again, you should see the effects of a length of straight material 'rolling' the rounded object it is in contact with.

The broom analogy is important, for this is exactly the same principle within a typical sword stroke and when you use your knife at the dinner table.

With a broom, you want to successfully 'roll' the object you intend to move.

In the case of a sword, the object would like to roll, but when the object cannot keep up or cannot roll at all, the blade begins to separate/reave the surface presented to it. On your dinner plate, the fork holds your food in place as your knife does its best to roll, but instead slices/severs the object.

We will dwell on the blade application in just a bit, but first...

all of this analysis brings into focus an important aspect of using a broom efficiently...

we cannot operate the broom as an inclined plane (with efficiency) if the fulcrum stays within our hands!

Where then should the fulcrum be placed?

The answer: the pivot must shift/move within the area between our hands.

That answer implies a lot so let us investigate the details.

The pivot (fulcrum) drifts; it moves along the broom handle. It starts at the lower hand (ala the first-class lever action) and drifts to the upper hand in a 'ripple' effect. This means that it moves to the top and ends in the Third-class lever position when the broom is completely vertical.

The Broom and Shovel

Not a simple answer, not one many think about in the first place, but it subtly explains one reason why many cannot cut well.

There is also an interesting side relationship to take notice of, for as the pivot drifts within the handle upward between the hands, the bristle bend ripples downward from the middle of the bristles to their ends i.e., bristles mirror the pivot (the pivot in the handle is the motivation for the action).

The point is that the bend of the bristles and the placement of the pivot both move outward toward the ends of the broom; away from each other. (See Figure 86: notice the pivot moving up as the bristles straighten)

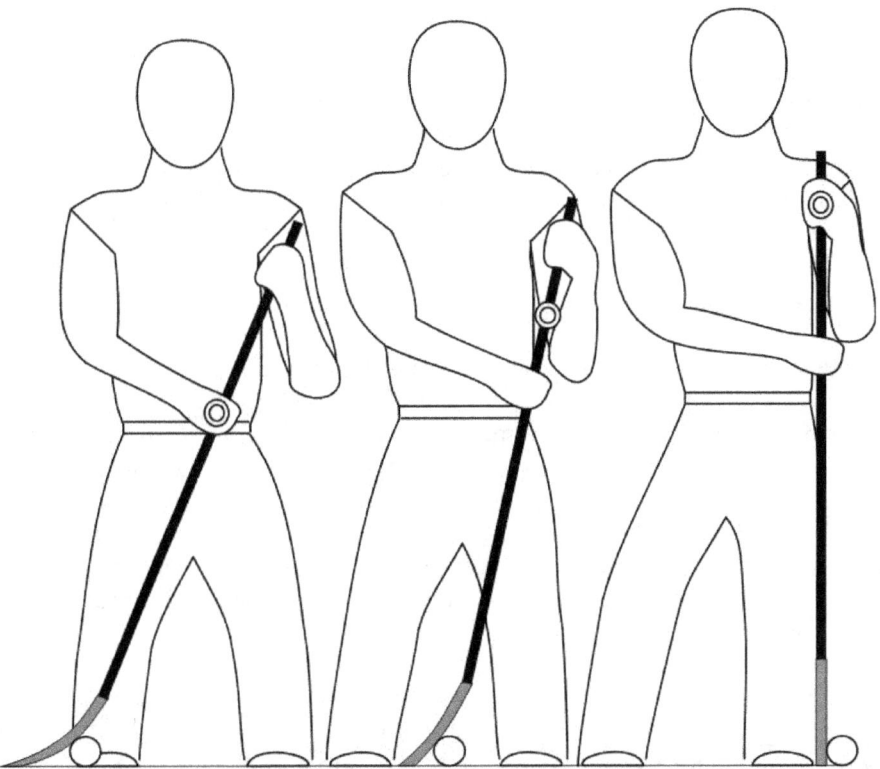

Figure 86: Sweeping a ball using our hips. Pivot in the broom moves from bottom hand to top; the bristle bend moves downward to the bottom

The 'rolling bend' in the bristle is easy to picture if you 'splat' your broom bristles to the floor (ala our earlier inefficient examples of broom usage) and then lift the broom handle to allow the bend in the bristles to undulate/shift down toward the end of the bristles.

Advanced Connection Builders

The bristles should look like you are lifting them from the floor as you would a piece of tape.

Slowly repeat splatting a broom and then lifting up to get a feel for how the bend shifts from middle to the end of the bristles, but as you do, pay attention to the 'lifting'.

The lift is one of two simultaneous actions required to keep the bristle bend in contact with the side of the dirt/balls (e.g., ten o-clock position), instead of the top. The other action is, of course, moving the broom across the floor.

The shifting pivot is how these actions come together to meet our goals.

Some will relate the movement of the pivot in the broom stick to how we construct (quadratic) Bezier curves in computer graphics.

The 'sliding dot' that makes the curve is the fulcrum of that line/lever. The 'bar' is the part of the broom stick between your hands. In most cases, you will see the 'broom' moving along between Point0 and Point1 (and shoveling happening at the same time between Point1 and Point2! Oops, not investigating shoveling yet, but we are getting there!)

I chuckle as I know some of your heads just popped!

That Bezier curve example is completely accurate. Regardless of your 'techie-ness', go ahead and look on the Internet for the moving graphics of how Bezier curves are drawn on computers. (Easiest to understand when watching a Quadratic Curve being drawn. It is an excellent example that might just make visual sense to anyone, even the non-geeks, and in a way that could never be captured in a book. What I hope is that it makes the shifting pivot visually apparent to you.)

What is more important than seeing the pivot move is that you are capable of replicating the shifting pivot and can find it in your own actions, so let us try performing Figure 86 ourselves in front of a mirror…

To sense this phenomenon directly, we switch out the dirt and roll something a bit bigger. In this case, we use a baseball (or similar sized ball)

When sweeping the ball from our right to left, our intent is that the bend in the bristles always meets the ball at approximately the ball's ten o'clock position. (go back a few pages to see Figure 85 again)

The Broom and Shovel

It requires that we keep that bend in the bristles undulating downward within the bristle as the bristle 'zipper rips' upward and makes the ball roll.

The coordination for the broom is actually fairly easy. It is not much more than a little lift and a little drag. (Do not tell your spouse/roommates or you will be the one doing all the sweeping in the house!)

The way I suggest to perform these actions is to have your top (left) hand solely lift the broom (vertically) as the lower (right) hand both drives the broom along the floor (horizontally) and is responsible for maintaining the bristle bend in just the right spot.

Broom head below on your right side, place your left hand on top, right hand roughly in the middle of the handle (thumbs upward on both), and bend your knees slightly as you drop into a double weighted horse stance.

Keep your left hand stationary in relation to the rest of your body as you shift your weight into your left leg (over and up! [we started at 'down'])

That is right, we want our body rise to do the lifting, not our left hand.

As you swing onto and straighten your left leg, your left hand, stationary in relation to your chest, will follow your body as it moves upward, thus lifting the broom handle; your right hand drifts across your body as it maintains a bit of pressure on the middle of the handle. You should not need much more than the weight of your right-arm to draw the broom across and maintain the bristle bend in just the right spot throughout the motion.

When you get the knack of things, watch yourself in a mirror and see if you can spot the fulcrum drifting from somewhere just above your right hand upward toward the top of the handle where your left hand is.

The primary point, besides teaching how a broom operates (important for understanding the sword!), is that we recognize that the pivot/fulcrum of our tool is not always in the same space as where we contact the tool; (i.e., not in our hands) [To be more specific, we might say: in 'efficient' usage, the pivot is hardly ever where our hand meets the tool.]

Try it a few times, and when you are ready to progress to shoveling, continue the broom stroke to the point at which the broom arrives completely vertical in front of you, likely in front of your left foot pinky toe or maybe just outside of your stance, and stay there a moment. (right side of Figure 86)

It may require that you raise your left hand higher than where you started, that is ok, just be sure that your right hand and arm come across your body.

This position will be the starting position for the next half of our lesson: the shovel. So, get a feel for it, but relax a moment, because we need to expand on a few concepts before we go to the shovel…

Here are some 'food for thought' statements:

First, to get to this stopping pose, our legs basically mimic the action of the broom!

This sweeping motion should feel a lot like the 'front wheel drive' drill; where we moved 'over and up' and learned to 'drag'/'sweep' our non-weighted leg along. Our non-weighted leg kind of looks like a broom if we allow our feet to come together at the end of the weight shift, correct?

It does! Coincidence?!? Not at all!

The words 'leg sweep' should be taking on new meaning for you. The leg sweep is actually another representation of a cut.

More thoughts…

Remember the comment about the pivot within a spoon, fishing pole and a pen not necessarily needing to be at the points at which our hand touches these objects? (last section of 'Where we are strong')

Consider what you just read and decide for yourself where the best place for the pivot might be…

For example, where would you put the pivot in the spoon if you want to lightly drizzle some liquid onto an object (e.g., when wanting to make even width lines of chocolate on a desert like a pro [no bags, no bottles])?

Hint: turn the spoon into a first-class lever: pivot between you and the load.

Answer: fulcrum goes right up against where the spoon blade meets the handle; i.e., closest to the load. The spoon head becomes the 'short side' of the lever and moves much slower than the 'long side' where your hand is.

You can get a much more even pour this way, because the spoon head tilts slower than the action in your hand, thus more easily controlled.

The Broom and Shovel

How about for achieving your new record for longest one-handed cast with a fishing pole? (with a spinning reel where your hand cannot be at the very bottom of the rod.)

Hint: third-class lever, leading pressure asserts between the pivot and the load (the lure/hook/bait), aka upside-down third-class lever broom.

Answer: start with pivot at bottom of the rod and let it drift to your hand as you cast) [we did not cover this, but a third class lever is a speed multiplier, the end of the rod moves MUCH faster than our hand; exact opposite of our desire with the spoon]

That last couple of paragraphs are extra credit work, but you will know if you have truly caught on to this lesson if those two situations make sense. (If not, blame the terrible explanation and move on.)

The example that is more than just an 'extra credit' exercise is the Pen. Calligraphy is a microcosm of swordplay, and begins to reward you for taking the time to analyze brooms/brushes in detail.

Is the pen mightier than the sword or are they complements?
Is the pen mightier than the sword? I have met some persons along the way that could have hurt me just the same with either! (No joke!)

Quasi-humor aside, let us take some liberties with the word 'pen' and bring this to calligraphy which employs a brush. (sorry, no ball-point pens today)

The relevance here is that a brush is just a smaller broom and, in this context, we will find that a flaw in our broom is the essence of excellence for a calligraphy brush.

What is this flaw in a broom that we so admire in calligraphy?

If the object we hope to move is too fine, too small, the broom will leave traces of the material we hope to move.

If that material is dirt, well then, we have to get out a mop to lift and suspend the material in a liquid, then move the liquid, or we get a vacuum to perform a 'dry lift', but...

If we have a 'hand sized' broom, the surface we are applying the broom to is paper, and, instead of dirt, that material is ink/paint... the rules are the same, and we have what we want!

Advanced Connection Builders

The drag of the calligraphy brush rolls, but leaves material (ink) as it moves mostly due to the bristle construction. The ink is not 'pinched' onto the paper. The calligrapher rolls the ink along the paper and uses the friction/absorption of the paper and bristle 'flaw' to leave ink behind.

The focus is not the ink, the trick is to learn how to manipulate the brush and leave traces of paint/ink as you go.

The artistic value of this action is subjective, but the objective aspect is that the artist's/calligrapher's results reveal their ability to guide the brush.

That ability is the eye/hand coordination and understanding of how the machinations of the lever/inclined plane effect the placement of the ink.

Consider the initial contact between brush and paper. Did the artist intend to leave a pool of ink?

As they drew, did they control the bristle bend well or was it shifting from splatting to balanced? Did the artist let the brush run dry? Did they lift the brush directly upward from the paper or leave the paper with the 'lift off' of a plane? Was the brush ever spun?

Was any of this appropriate, aesthetic, and/or intentional?

Calligraphy is not just text, it is a log, a record of a point in time, a tangible and interpretable proof of the artist's ability with a lever/inclined plane fused with their awareness and intent.

Should one have in-depth understanding of the machinations of a sword (levers!), would they not be able to translate those skills to fine motor skills?

Conversely, might calligraphy help them discover nuance not readily observable in the larger world of sword play; thereby enhancing their sword craft? Would the art work reveal progress over time (provide assuredness)? Could they find a new mind/body awareness (eye-hand coordination) and practice the same no-mindedness?

It is no surprise to me that Shodo would be a common practice amongst swordsmen.

Let us take a small step along in their footsteps by 'crossing sword and brush' and tangibly experiencing the physical similarities between a sword and brush…

The Broom and Shovel

Ever take notice of the Tai Chi swordsman's empty hand displaying Tai Chi Sword Fingers? (a two-finger 'Scout's Honor' salute)

It might feel a bit like finger painting (calligraphy with your fingers), but assume the 'two finger salute' and glide the pads, not the tips, the <u>pads</u> of your fingers (where your whorls are), downward along the wall.

Pay close attention to where you had to apply the pressure (the pivot!) within your hand to keep those pads gliding along. For me, it is somewhere between the middle finger knuckles and the base knuckles that connect fingers to the palm.

We can put the pivot even lower into the hand, wrist, or even forearm, the feel is the same, but just a different focal point.

It feels a lot like 'sweeping' does it not? Your fingers are representative of the Shodo brush's bristles. Think on it.

Hold onto this experience. It will come in 'handy' again in just a bit.

For now, remember to avoid fencing matches with great painters/calligraphers.

The Shovel
Ok again, back to where we were. Where were we?

Oh yeah! Standing on our left foot, at the end of the broom stroke.

Brooms are best suited to operate with the load being rolled behind the leading force (behind the force being applied on the broom stick).

In other words, the broom handle exits your outside door before the broom head reaches the door sill and rolls the dirt outside.

What is on the flip side? What is the 'Yang' to the 'Yin of a broom'?

Answer: The shovel! The broom's perfect complement.

When we reach the limit of motion for a broom; (i.e., when the broom bristles have fully 'caught up' to the top of the handle and the broom is completely vertical; a.k.a. where we ended our broom lesson) instead of resetting the broom for yet another pass, let us instead decide to keep ourselves in place, but find a way to keep the dirt moving; but how?

Advanced Connection Builders

By shifting our action from sweeping to shoveling! That is how!

Why shoveling?

It gives us a renewed mechanical advantage that allows us to continue moving our object, this time, without having to move ourselves.

If the shovel gives us mechanical advantage, it should be obvious that it is somehow a lever. It truly is, and what is more is that, it is also, just like a broom, a hybrid combination of a lever and an inclined plane.

If we understood our exploration of a broom as a lever, using a shovel as a lever should be fairly straightforward.

The shovel's handle is a rigid lever arm that requires a pivot, and we supply one with our hands; and again, we have the ability to create a first- or third-class lever.

Let us skip a detailed explanation of how keeping the pivot in either hand creates one type of lever or the other, but let us remember that a pivot in the hand at the top end of the shovel creates a third class lever; pivot on the hand in the middle of the shovel handle will generate a first class lever.

Who cares anyway, for keeping the pivot static in our hands impedes efficiency.

We do not want that.

Let us instead focus on the 'drifting' pivot that, this time, slides downward on the lever arm; i.e., starts at the top hand and drifts toward the one at the middle of the shovel. (exact complement/opposite of the broom!)

For the geeks, see Quadratic Bezier Curve generation and, again, watch for one side of the bar shoveling as, simultaneously, the other end sweeps.

For the rest of us (and the geeks like me too!), let us remove the broom head, and instead of tacking on a shovel head, let us just use the handle as a jo staff.

Do not break the heads off of any broom sticks, simply flip your broom over or swap out your broom for a jo staff and get back into that 'vertical broom' pose from the end of our 'broom' drill; oh, and go ahead and lose the ball.

We do not need the object to roll.

The Broom and Shovel

We ended the broom drill with left hand on top, right hand across our body, both hands gripping with 'thumbs upward toward the sky'.

Let us switch the hands, not the grip (for now). Put your right hand on top, left hand in the middle, both with 'thumbs up'; (i.e., keep the jo where it is, just switch which hand is on top, not the grip.)

We also want to have our weight ninety percent or better on our left foot; (i.e., 'posted left' position, ala right side of Figure 86; [just remember to switch your hands])

Our new goal is to 'shovel' that jo staff tip to our left and along the floor.

This is accomplished by our right hand providing the 'leading force' that drives straight downward the entire time (i.e., our right hand never moves leftward, it moves solely and directly downward) as our left hand subtly guides the jo staff outward along the left; the jo point glides along the floor as if being 'squeezed' out of the starting position by our right hand dropping and driving.

Do not move your body, but feel free to bend your left knee (as additional motivation for your right hand to drop).

There is purpose in staying in the 'post left', 'left side weighted' position.

We want to isolate our hands as the motivation for the jo jutting leftward.

Staying left weighted and shoveling outward to the left has the effect of giving your right hand no option to move to the side, thus forcing the downward motion we desire.

The shoveling action should reveal the pivot of the 'jo shovel' drifting from, at start, your right hand above, and dropping closer to your left hand below as the tip of the jo is moved further from your toes. This pivot shift reveals itself in the need to add more pressure in your left hand to keep the jo tip against the floor as your right hand drives downward.

Extend/shovel the jo tip as far as you can without stepping and when you reach the end, reverse the motion by picking your right hand straight upward as you straighten your left knee and allow your left hand to apply enough weight to keep the jo tip on the ground.

Voila! Your reset/'return to vertical' was a broom!

Alternate the rise and fall of your right hand; i.e., remain standing in 'posted left' position and 'shovel to your left, then broom back to the start' a couple of times as you try to sense the 'drifting pivot'.

Sensing the pivot drift is not essential, but it will enhance your awareness.

Now, we have done well to learn two new drills, but there are two very important aspects we must focus attention upon.

The first is the 'inflection point'. The point at which we change from a 'broom' to 'shovel'. The 'inflection point' in this case is the 'vertical broom' pose we left ourselves in at the end of learning about the broom, and the 'vertical jo' at start of the shovel.

With the vertical jo staff at that 'starting the shovel' position, you could just as easily begin 'shoveling' to either side of your body. All it would take is a quick shift in our grip.

Your stance can be double or single weighted, just be sure to raise and lower the top of the broom, like a piston, directly over the same spot in the floor.

The tip of the jo on the floor, as well as the rest of the jo, are the portions that move leftward or rightward.

We want to become incredibly aware of this piston action, so let us emphasize it by changing things up and performing the 'broom to shovel and back' drill parallel against the wall instead of vertical against the floor.

Find a two jo-length stretch of baseboard (where your wall meets the floor) and, in the middle, draw (or put tape or simply imagine) a jo-length line perpendicular to the wall; (i.e., the dashed line(s) in Figure 87)

Let us call the perpendicular line the 'piston line'; the 'baseboard line' is the one along the baseboard and it is divided into left and right halves by the piston line coming out from the wall.

Place your jo completely against the wall and covering the left half of the baseboard line. The inner tip of the jo should be touching the 'piston line'. (Position 1 in Figure 87)

From this position, grab the 'top' of the jo; the part touching the piston line, and move the top of the jo outward from the wall along the perpendicular piston line.

The Broom and Shovel

As you allow the tip of the jo in your hand to move along the piston line, keep the 'bottom' tip of the jo gliding along the baseboard line. (move through position 2 toward position 3 in Figure 87)

You should be sensing 'a broom' along the baseboard. (and if you want to, you could simultaneously sense a 'shovel' scraping along the 'piston line')

Stop once your jo is completely on the piston line. This is the 'inflection point' where we shift to shoveling. (Position 3 in Figure 87)

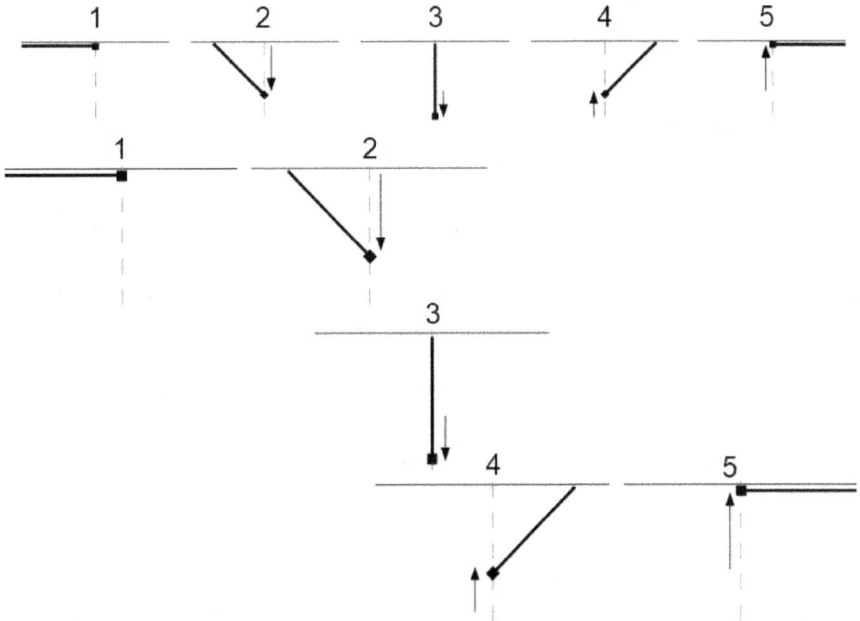

Figure 87: Jo and the wall drill (sequential on top, made a bit bigger below.) Dotted line is the 'Piston Line'. Note the 'Piston effect' on the tip of the jo staff (one end only moves toward or away from the wall)

The next step is to allow the bottom tip of the jo to continue rightward along the baseboard as the 'top' tip moves toward the wall along the perpendicular piston line.

This should bring to mind 'the shovel' on the baseboard. You might also sense a 'sweeping' effect on the 'piston line' as you bring the 'top' tip of the jo closer to the wall.

Do not stop until you have the jo covering the right half of the baseboard line.

Advanced Connection Builders

Move back and forth with these actions.

If you do not have sufficient wall space, you could just as easily do this drill with a pen or a pencil to replace the jo and the spine of a book or the edge of a box to replace the wall.

The important aspect is that you find the 'piston like' motion of the 'top' of the broom/shovel created by the jo (or pen).

The 'piston line' is important, for the jo moves purely toward or away from the wall (as appropriate). This will be beneficial to us when we intend to create compression or extension connection with Uke.

Before we explain how we use the piston in Aiki, we must highlight that second important aspect that, like a broom, the shovel 'rolls' the load.

The 'scooping' action of a shovel 'rolls' the load upward and onto itself.

In a standard shoveling scenario, you could liken it to the dirt being a ball and the shovel is the inclined plane that 'moves toward, under, and past the ball'.

The dirt 'rolls up' but does not move in the X and Y horizontal dimensions, solely upward (uprooted!) in the Z. (At least until the dirt moves high enough that the shovel is hitting the dirt horizontally.)

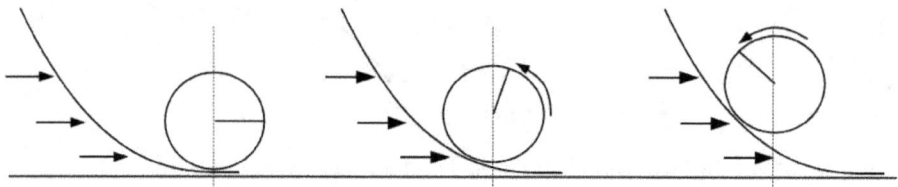

Figure 88: the shovel moving rightward, 'picks up' the dirt, but has the effect of rolling the dirt onto the shovel as the shovel passes by

What we do not want to do is 'strike' the ball, as we would when putting backspin on a cue ball (hitting 'low' in a game of Billiards).

Although using the shovel as a Billiards stick will make the object 'spin' toward us (i.e., rotate in the same direction as it would when 'rolling' up the inclined plane of our shovel), it is a bad thing, for instead of having the object roll onto the shovel blade, it is instead projected outward and we would lose contact/control of the ball. (the backspin overcomes the effect of friction and the cue ball projects forward even though 'rolling backward')

The Broom and Shovel

Remember, we want the shovel to roll the load up the 'ramp' of our shovel blade as the blade wedges underneath the ball. (We then reduce the tilt of the shovel blade to keep the dirt from rolling off, move the shovel to where we desire, and then drop the load. Ala truly snow shoveling!)

These concepts are sometimes mind-benders, but they are integral to Aiki, so I will withhold apologies as I give you yet another (super important) mind-bender drill to reinforce this 'broom and shovel are inclined planes' concept.

You will need a mid-sized ball and a jo staff. I suggest a 65-centimeter inflatable exercise ball that most gyms have.

The drill is very much the same as the vertical 'broom to shovel' exercise we just completed.

The difference is that we 'broom' and 'shovel' the exercise ball leftward and right ward, putting our emphasis on the 'piston' in the top of the jo moving directly upward and downward as the ball is rolled along the length of the jo. (call it 'Fancy Aiki wheel/ball trundling with a jo') See Figure 89

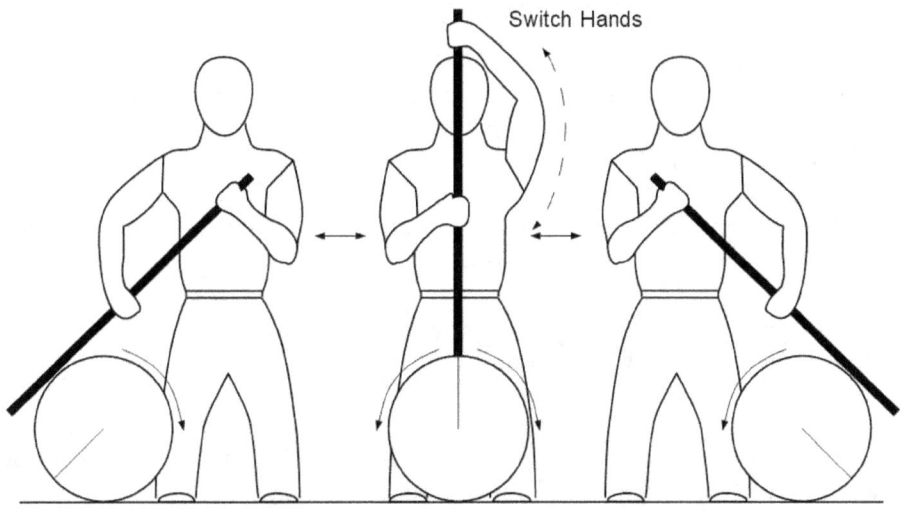

Figure 89: broom and shovel roll the ball drill

The ball will keep our jo staff from touching the floor and the 'zipper ripper'/'rolling a hoop' action of a straight object rolling a ball should be obvious.

Our goal is to come to a full vertical with the jo as it reaches the 'north pole' of the exercise ball.

The 'mind-bender' part is that, in this drill, when shoveling (i.e., when moving the ball outward from the middle), the ball is actually rolled 'underneath' the shovel (as the jo staff passes forward/over the ball).

This 'reverse side' of the shovel action is akin to the dirt passing under the broom bristles, but it is fine, there is utility in 'rolling the ball under the shovel'. (Are you catching on? Can you think of times that you 'roll' Uke under your own arm? Uke wants to roll but cannot… it becomes a cut… try it at dinner tonight, stop sawing at your meal, and instead try to 'Roll your dinner under your knife'. Double up by paying attention to your fork too!)

If you maintain a 'thumbs pointed up' grip the entire time, the shovel portion of this drill may put you in the mind of moving a small punt boat through shallow, standing water by pressing a staff into the water's bed. When imaging this 'moving the punt boat', can you sense a 'piston line' dropping down as the pressure 'squeezes' your boat forward. (the tip of your pole acting as a shovel being stuck in the river bottom, forcing you away, versus pushing the tip of the shovel outward). [Important analogy! Therein is first secret to generating outward force. Another time, another book maybe)

The next step is to replace your jo with your arm and instead of a ball, start rolling Uke, but we need a bit more information about shields before we can go there. Let us talk about that now, so here is the closing comment…

Great things come from simple machines.

In this case, from levers and inclined planes, we derived brooms and shovels; then linked them together with a piston.

Focusing on rolling the load versus crushing or hitting it was important too (and we will learn why Rolling is so important in just a bit).

I guess chores have always been the secret to better self-defense. Sweeping and shoveling (and learning to dodge wooden spoons!) in my youth was not wasted time. I am sure you had your practice at that as well.

I wonder if all parents also secretly know how to perform Ikkyo…

Shield Theory

Go take your Aikido shield off the dojo wall and let us explore some nuance!

What?!?!? You do not have a shield in the dojo?

Shield Theory

How very 'traditional Aikido' of you…

… but you have taken the time to learn how to block, correct?

I thought I had. Especially given all the time I spent in Tae Kwon Do throwing my arms in the air practicing inside, outside, high, low, and X blocks.

The reality is I did not truly learn what it meant to 'block', more aptly 'shield' myself from an attack until many, many years later.

What was this revelation that changed my definition of what it means to 'shield' myself from attacks?

Surprisingly (to me at least), the missing element was learning to assert my transmission. (that pivot we spent so much time talking about)

What we are going to learn is that, when intending to present 'a shield', instead, a great many of us (me especially) are actually presenting 'a rail', a path for Uke to ride along on their way to striking us.

If you are like me, you do not want to get hit, and since creating rails (with our arms) guides Uke directly to us as their target, it practically ensures that we will get hit.

We can fix this!

Let us first learn how to create a shield before we explore how to meet Uke with that shield for the purposes of redirecting Uke and repositioning ourselves in relation to Uke; (i.e., tie it in with broom and shovel!)

A Shield, not a Rail
Let us fix the rail first.

What is a 'rail'?

It is a path along the surface of the shield that eventually leads to a place where Uke can make contact with us (to strike, cut, poke, etc.)

Obviously, a rail is not desirable. The shield should prevent us getting hit.

Here is the test to detect if we are committing this error, and at the same time the drill to cure what ails us (if it does ail us at all)

Advanced Connection Builders

Stand with feet in any position you wish, and with your strong hand in front of you, assume a sword-hand type blocking pose.

Have Uke stand to your 'outside', perpendicular to you, and place the palm of their hand on the underside of your arm, near your wrist.

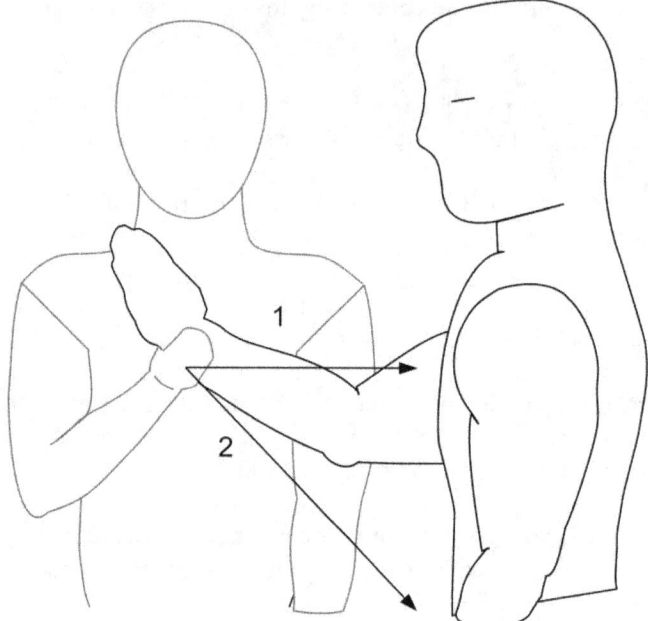

Figure 90: setting up for the 'shield' drills

Uke should test your block by moving their hand through yours and toward your chest (Arrow 1, Figure 90), but, of course, your sword-hand style block should easily be preventing them from touching you.

Excellent block/shield correct? Maybe…

Now is Uke's cue to keep that 'moving toward your body' intent with their hand as they slide down your arm, toward your elbow, with the intent to pass over your elbow and touch your thigh. (Arrow 2 and 1 at the same time, Figure 90) I ask that you have Uke try to hit your thigh because it hurts less than having them 'ride the rail' into your ribs or belly.

If Uke reaches your thigh, you were creating a rail with your forearm.

What happened is Uke rode along your appendage and eventually made contact with their target: You!

Shield Theory

Many are able to stop Uke from touching their thigh on the very first try, but just as many cannot.

If you did not stop Uke, try this drill over and again until you succeed.

The good news is that it typically takes no more than a few attempts before most get it, and even in the worst cases, I have not found anyone that could not teach themselves in just a short bit of time (minutes).

Stopping Uke from touching your body (thigh) is success, but we would rather that Uke stays 'stuck' on your forearm well above your elbow. Performed perfectly, Uke stays above the middle point of your forearm.

Excellent, so now we have a simple drill, one most can already perform, or catch on in just a few minutes, but can they explain what they did? (Likely not, and where is the Mind-Body connection in that?!?)

What is the physical difference between the success and failure of this drill?

Answer: the assertion of your 'transmission' beneath Uke's sliding hand.

To better understand, let us consider a ball and box on a teeter-totter.

With the box on the right side driving the teeter-totter to the ground, the teeter-totter becomes an inclined plane. (Figure 91)

The ball on the higher side starts to roll down the incline of the teeter-totter.

To stop it, we do something physically impossible with a real teeter-totter.

We leave the high-end tip of the teeter-totter in place (i.e., where it is in the air) as we raise the support and, of course, the rest of the teeter-totter, including that box on the bottom, to level with that upper tip we left in place; (i.e., we level the teeter-totter by ripping the support out of the ground and bringing it up into the air. Like I said, impossible in reality.)

(Note: the 'support' is whatever we were using to keep the teeter-totter off of the ground originally.

We cannot call that the fulcrum any longer, because, technically, we are now pivoting at the 'high side tip of the teeter-totter'.

This is a place where second class levers fit into martial arts!)

Advanced Connection Builders

The new (lack of) tilt causes the ball to slow to a stop (at the point we level out the teeter-totter's plank). If we go too far, we start rolling the ball back to where it came from. (See Figure 91)

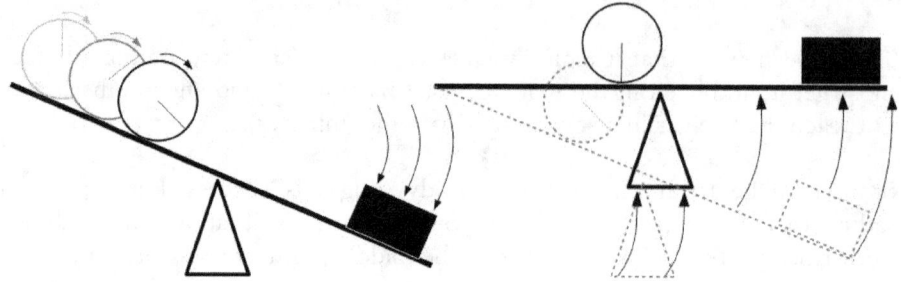

Figure 91: object rolling down after lever is tipped. Moving the pivot (tip of the fulcrum/triangle) upward to level off the incline and stop the object

You might think of this leveling lift as a 'wheel-barrow' (i.e., second-class lever) with the 'higher' end of the teeter-totter acting as the stationary 'wheel' end of the wheel-barrow; and the teeter-totter's support (the middle) being lifted as though it was the 'handle' of the wheel-barrow.

(The low end of the teeter-totter could also be described as acting as the handle, but I prefer that you assert the middle of the teeter-totter arm. We want to stop the ball's roll in the middle of the teeter-totter, not after it has rolled to the bottom.)

Applying this analogy to the drill; i.e., replacing the teeter-totter plank with our arm... we have to flip things around a bit...

our wrist is the 'high-end' (left side) of the teeter totter, our elbow is the low end (right side), our transmission is the support that you move away from you (instead of up), and obviously, Uke's hand is the ball trying to roll to the middle. (See Figure 92)

To stop Uke's hand you have to keep your hand in position as you assert (swing) the middle of your forearm beneath Uke's hand.

I admit that this is a bit of a mind-bender, but one that is super important to comprehend. You may have the ability to succeed at this drill, but not knowing why lacks mind/body connection and limits you in other application where this principle applies, and it applies nearly everywhere!

It is in our best interest to develop the ability to see this 'assertion of your transmission' scenario in any direction/dimension.

Shield Theory

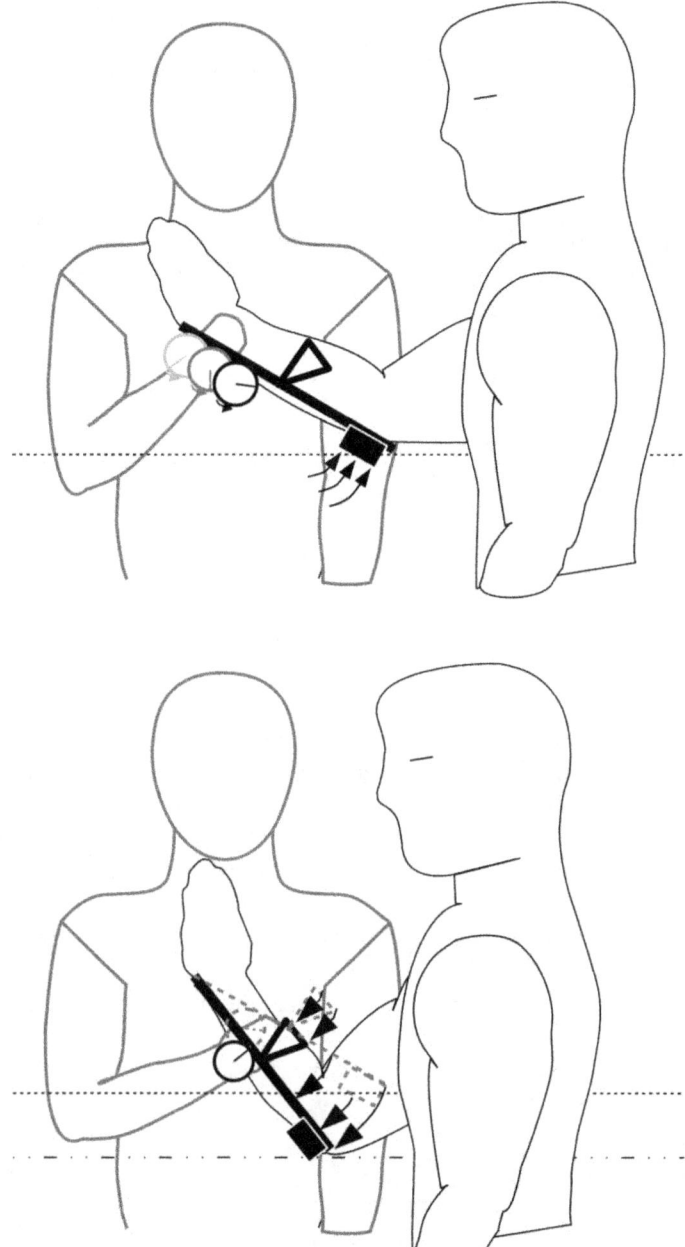

Figure 92: Moving your transmission to stop Uke from 'riding the rail' of your forearm

Let us start toward that goal by 'asserting the transmission' in a bokken.

Advanced Connection Builders

Have you ever thought of your bokken as a shield? You should have.

Yes, we must practice to protect the 'blade' of the bokken as we place the bokken in position to obstruct Uke's attack, but we must also understand that a sword cut is not just an action that passes through Uke as if they were a ghost.

A cut is also a method to move, potentially even throw, Uke. (most importantly, we need to understand that a cut is an attempt to 'roll' an object that cannot roll, and is therefore cut.)

The point is that, if the cut is not clean (i.e., gets stuck in the material it hopes to cut, e.g., because your 'cut' was more like a 'chop', or because you hit hard, non-cut-able material), the cut should still steal stability from its target.

So, what I am going to ask you to do initially is to meet blade to blade, bokken to bokken with Uke.

The action may make you go a bit crazy, but I want you to know the intent here is to learn how best to use your bokken (sword) on an object we truly desire to cut, not one that will ruin our blade.

Uke sliding a bokken along ours is going to be a bigger challenge compared to Uke using their hand. We want to become strong with the ability to present a shield instead of a rail.

Repeat after me: 'This is a drill, not pragmatic sword-on-sword play'…

The short explanation is that we are going to repeat the 'Uke tries to touch your thigh; while we stop Uke by presenting a shield instead of a rail' drill with bokkens instead of hands. (I have to rename that drill someday!)

As always, please be sure to wear protective gear at all times (even with hands. Did I mention 'at all times!'?)

Present your bokken in front of you as you would for most kendo practice.

We want the bokken to be at about a sixty-degree angle from the floor (which is typically a bit higher than you might be if you were pointing the sword at the space between Uke's eyes [to throw off Uke's depth perception of your sword])

Have Uke stand in front of you with your sword almost touching their chin.

Shield Theory

Uke then takes one side-ways step to the left, grabs their bokken with left hand on bottom of the hilt, right hand toward the tip of their bokken, and with right arm outward to their right side, left hand in front of their body, Uke places their bokken parallel to the floor and across your bokken perpendicularly. (much as though they were holding their bokken like the stick of a bull fighter's cape in front of your bokken. See Figure 93)

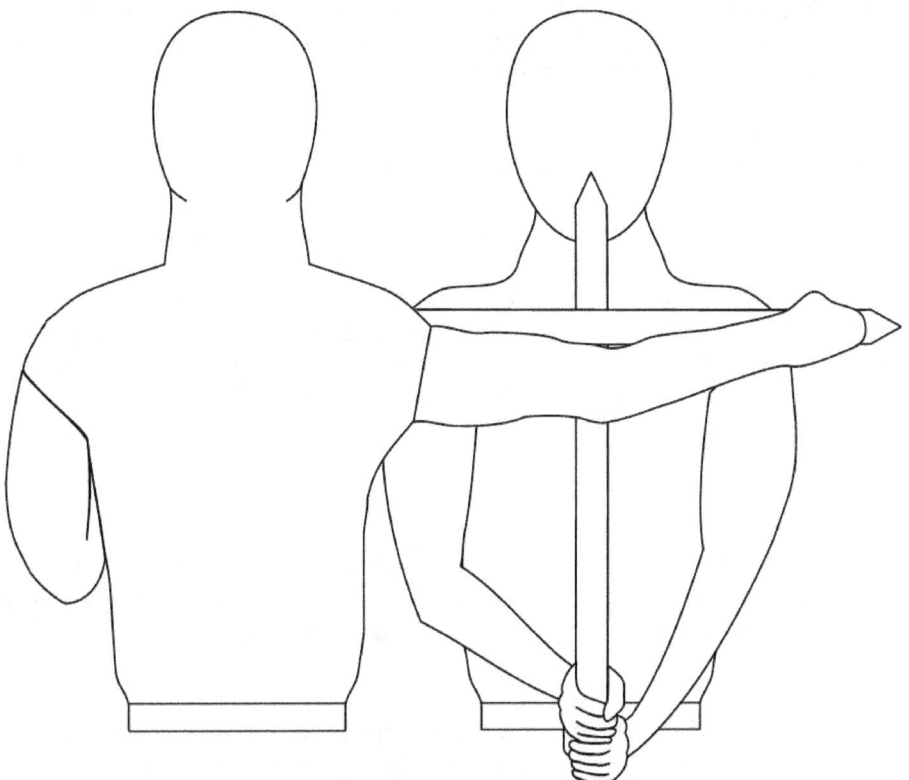

Figure 93: Sword in front of you, Uke to the side with their bokken perpendicular to yours

The point is that Uke's head is not in harm's way of your bokken, but Uke can apply their bokken to similarly as if it were the top of a matador's cape (or the bottom of a window blind).

Just as before with their hand, Uke should try to push their bokken parallel through your bokken and try to 'touch your face'. (Uke must not slip/slide to your left or right, they must keep their blade perpendicular to your bokken and parallel to the floor at all times, not sliding around the sides. Again, this is just a drill. It is Uke's job to keep their sword in position.)

You should find that Uke cannot hit your face. (If they can, stop and ask a senior instructor to get you past this hurdle. The likely mistake is your grip and/or your ability to create unbendable arms [both are best when done using ligamentary grasping/tension! Add your Zhan Zhuang!)

If Uke is sufficiently blocked, ask Uke to keep their bokken parallel to the floor and perpendicular to your bokken as they keep the 'toward you' pressure and lower their bokken in an attempt to touch your thigh; (i.e., as if they were dropping the bottom of a window blind.)

If you do not want Uke to reach you, you will have to 'assert the transmission' of your sword to meet Uke along their path.

It may help to again think of the teeter-totter/wheel-barrow analogy.

You might also describe it as your bokken's transmission heading off Uke's progress. (or 'getting your bokken's transmission lower than and impeding the path of Uke's bokken'. You could also describe it as: 'your bokken's transmission is intercepting Uke's bokken')

Uke must not be able to reach your hands (or tsuba if you have one on your bokken)

When you find success, keep Uke there for a second and then give their bokken a push (for fun) to create some space between you. Using the shield to open up some space between you and your opponent is a common aspect of deploying a shield.

Presenting Uke with a shield, rather than a rail is an important skill. Besides the obvious aspect of no longer giving Uke a path to attack/hit us, it also allows us to way to create and maintain compression or extension connection.

There is no Aiki without connection, thus we need this basic ability in order to effectively action the rest of what we are about to discuss in this book.

How to meet Uke with a shield

We have come a long way already.

Way back in 'Where We Are Strong', we discovered the transmission as a focal point in our forearm that provides us additional leverage as it converts our arm from a third-class to a first-class lever.

Shield Theory

Additionally, we detailed how a broom and shovel are both a hybrid mix of 'lever and inclined plane'.

That broom and shovel explanation also revealed that the pivot/fulcrum of the broom and shovel changes positions (glides) as we use them to move the load in a very particular manner: by using the broom and shovel as inclined planes that roll the load (ala wheel-trundling)

We also learned to use our transmission to 'shield' ourselves and stop creating rails for Uke to travel along and strike us.

These concepts are on their way to coming together.

Tired yet? Come along, we are just about to start into the really good stuff.

Let us explore how to meet Uke with our shield for the purpose of redirecting and reposition ourselves in relation to Uke; (i.e., tie it in with broom and shovel)

What we are going to learn is that the most efficient way to initiate contact with our shield is as an X-block (or, at least, half an X-block; i.e., one arm of it), and once connected, how to guide Uke's attack to our advantage.

When we are done, we will have insight to a technique we call 'the hourglass'

Let us get started

In any combat, especially hand-to-hand combat, the one position that must always be defended is called 'the center-line'.

There are a lot of different places where we refer to a 'center-line', so let us synch on exactly which one I mean.

This 'center-line' is the shortest distance between you and your opponent. It is an imaginary 'direct-line' between your centers (your COB's).

You might think of it as your military 'front-line', the area closest to your enemy.

This center-line gets shorter as either you or Uke approach the other, and obviously, extends as either of you move away, but it has not yet 'moved'.

The center-line does not 'move' unless one of you moves laterally.

Advanced Connection Builders

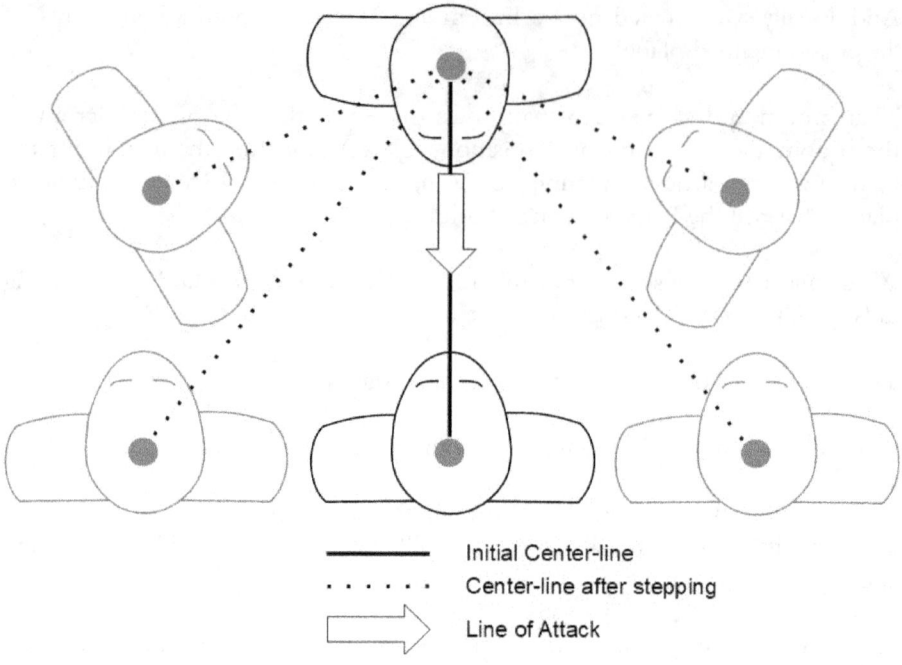

Figure 94: Initial Center-line is congruent with Line of Attack; Center-Line changes as we move laterally (i.e., any direction except directly backward or forward); Side-steps and Turning Steps depicted

This implies that the position of the center-line is always directly between you; in motion, in stillness, always.

It also means that it does not matter whether you, or Uke, are actually facing each other. You could present your side to Uke (or vice versa), you could face each other, face away, you could even spin around in place, but the center-line would not move.

Important concept that many get confused about: The center-line is not the line of attack.

An attack will most generally be initiated along the center-line, but once we move 'off the line of attack', we will have effectively moved the center-line, thus separating the center-line from the line of attack. This is the essence of evading an attack. (Since the center-line is between you, the path to you, any other route will not lead to you. Seems obvious, but for the sake of precision in words, it has to be pointed out.)

What does it mean to protect the center-line?

Shield Theory

It means we have physical obstruction (e.g., shielding), along the center-line.

That means we have to position our shield across the center-line.

Try something for me.

Put your hands straight out in front of you (like doing a push-up in the air.) (left side of Figure 95)

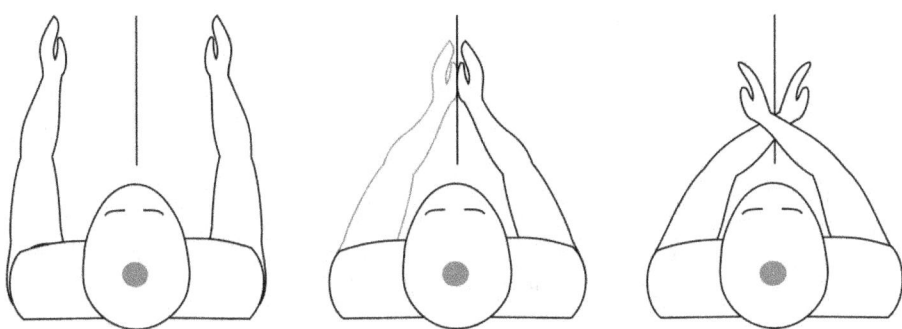

Figure 95: blocking positions: straight, wedge, and x-block

If Uke was standing directly in front of you, facing you, is any part of you obstructing the center-line; can you keep Uke from reaching out and touching your nose?

Epic fail, right? Correct!

Put your hands out again, but this time, with the palms of your hands touching (i.e., creating a 'wedge' or like getting ready to dive into a pool.), fingers pointed at the center of our imaginary Uke's chest (middle image of Figure 95).

Leave your right arm exactly where it is and drop your left to your side. (note the shaded left arm in the middle image of Figure 95)

Go ahead and raise your right hand to a 'sword hand' position.

Have you noticed that your right arm is entirely on the right side of the centerline? (No part of you is crossing the center-line.)

How well does this position impede Uke when they are standing directly in front of us?

Advanced Connection Builders

Answer: not extremely well.

We are likely in a very good position to shield ourselves if Uke uses their left hand, but we will have to move our right hand further across the center-line if we want to obstruct Uke striking with their right.

The reason is because Uke's right arm is coming from their shoulder, not the middle of their chest where the center-line is positioned.

Results so far: leaving a gap between our hands hardly works at all, and one handed 'wedge' position has big weaknesses.

It can get even worse for both of these positions if we swing our arms sideways to get into these positions ala the typical 'inside' or 'outside' block practice.

Timing is everything, if we are late, we get hit. If we are early, we might still get hit or, at best, create a random, ineffective block.

So much for most of my work on Tae Kwon Do katas. (My fault, I did not understand.)

Was there any gold in all that TKD training? There was! The X-block!

Instead of stopping at the point where our hands create a wedge aligned on the center-line between us and Uke, let us cross over with each hand and form the X-block.

The magic in this is obviously that we are fully obstructing the center-line with both hands. Drop an arm and we should still have a decent chance of catching Uke attacking from either side with not just the edge of a shield (i.e., our hand) but with the front face of our shield (i.e., our forearm/wrist).

More than this, what we have done is put our transmission on, or very close to being on, the center-line and this is advantageous, for if you have not figured it out yet, the center-line is also 'the piston line' a.k.a. the 'inflection point' where we convert the action of our arm (in this case) from that of a broom, into a shovel.

Here is how it works…

The transmission is the 'projection' of our COB within our arm. (We can place the transmission elsewhere. For now, we will use it within our arm.)

Shield Theory

As the COB's representative, the transmission offers the maximum physical presence we can offer without physically putting our hips into the mix, thus, it is how we best obstruct the center-line, and it doubles as the best source with which to generate compression or extension connection.

Keeping our transmission focused on moving toward or away from Uke's COB along the center-line creates and maintains our compression or extension connection (respectively) and allows us to tilt our arm (with the transmission as the fulcrum) to defend the center-line without giving up the connection and center-line obstruction. (Please read this and the two previous paragraphs again at least twice before proceeding; i.e., starting at "The transmission is the…")

By attaching ourselves to Uke at parts of our arm other than at the transmission (e.g., our hand), we can produce leverage; i.e., convert our arm from a third-class into a first-class lever. (Worst case, if Uke attaches to us at the point of the transmission, we can re-position our transmission pivot and shift the new location to the center-line to regain our advantage.)

Let us drill keeping our transmission on the center-line, as we use the rest of our body to convert our arm into a broom and then shovel.

This 'basic' drill teaches us how to replace the jo staff in our 'horizontal broom and shovel' discussion (Figure 87) with our arm. We use our hand as the broom and shovel heads, the transmission as our piston point, and by moving the transmission along the 'piston-line'/center-line (using a single-weighted stance and twisting our hips) we move Uke sideways; putting ourselves in a martially advantageous position. (without actually stepping!)

We begin in the standard Tomiki Aikido kata starting pose.

Tori and Uke facing each other, right wrists/back of the hands touching; Both in a 'cat stance': right foot forward and light, ready to kick; left foot back and fully weighted, hips squared to each other.

Hmm… Did you notice that, with wrists touching, you are both in a subtle x-block position?

Open things up by dropping your right hand to your side (yes, stop defending for one second), and letting Uke make that subtle shift to pointing their hand blade directly at your nose.

Bring your right hand up again, this time, in a full 'half an X-block' position.

Advanced Connection Builders

Meet Uke's wrist from underneath with the space between your wrist and middle of your forearm (a.k.a. just to the left of your transmission); your hand will be on the 'outside' of Uke's arm (in other words: keep the focus of your sword hand toward Uke's right side shoulder, your transmission just slightly right of the center-line, and your sword hand should end up almost touching the outside of Uke's arm. Left side image in Figure 96)

It will help to create a bit of extension connection with Uke right now (i.e., take the slack out of your and Uke's arms by moving backward just a bit), but with or without removing the slack…

 here is where all that drilling our hip twists on a door frame at the end of 'Where We are Strong' pays off…

From this position, maintain your weight on your left foot throughout the entire drill, and with a hip turn, rotate your torso rightward (clockwise if referenced from above).

Allow this hip turn to move your transmission point further away from Uke along the center-line (piston-line!), thus creating an extension connection with Uke. (or if you already created extension connection earlier by taking out the slack between you, it will be enhancing the extension connection). Your transmission moving on the center-line is the first part of the 'piston' action.

This hip twist will simultaneously create 'the broom' action in your arm/hand. You should find Uke being 'swept' toward your right. (This action should fully position your transmission over the center-line. Center image of Figure 96)

The focus for now is the hip doing all of the work. Your 'unbendable arm' and extension connection with Uke should keep your arm outward with very little muscular exertion in your right arm.

When you cannot sweep rightward any further, you should find that your entire arm is very close to being aligned the center-line. (see middle image Figure 96 again.)

Congratulations! you have made it to the 'inflection point', time to switch from broom to shovel.

Put all the tension that you created in your left hip to good use by uncoiling yourself counter-clock wise, back toward the starting position (squared hips in relation to Uke).

Shield Theory

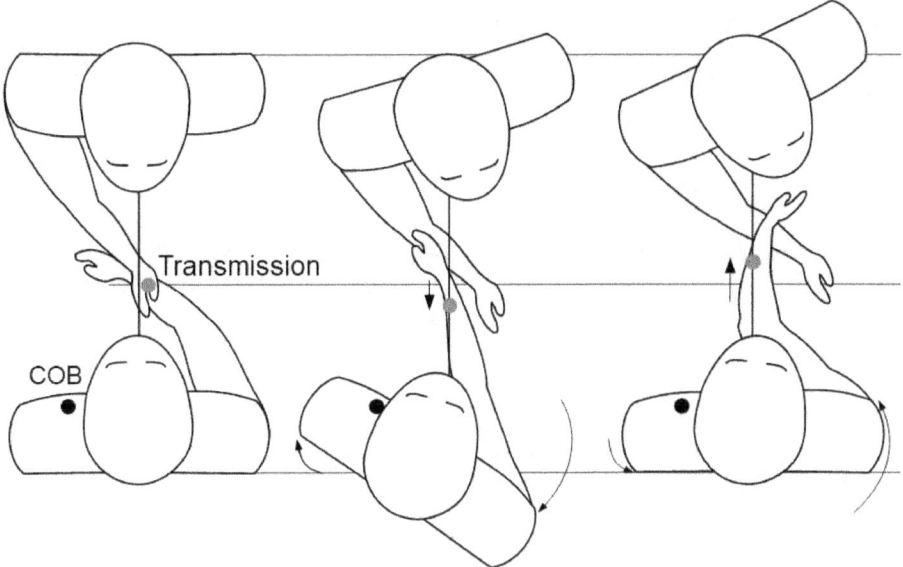

Figure 96: broom and shovel created with a hip turn, moving Uke rightward; keeping the transmission on the center-line

As you uncoil, move your transmission directly towards Uke's COB (i.e., along the center-line) as your right hand 'shovels' Uke's hand further rightward. (right image in Figure 96)

It might be a bit awkward at first, but you will have to allow your right elbow to come across your body some if you hope to keep your transmission on the center-line as your hand shovels Uke rightward.

Depending upon your shoulder flexibility, you might need to rotate your hips a slight bit past square, but no matter, take a look at what else has happened.

Where did you end up in relation to Uke?

You end up outside of Uke's arms, ready to strike. (Probably the best start of an Ushiroate (#5) you have ever done!)

Can you sense that you fully emulated the example/drill/image in Figure 87 with your right hand as the part of the jo against the wall and your transmission as the 'piston' end of the jo?

Take note! You have not even stepped yet, solely wiggled your hips, but you are positioned as if you have done a turning step as we do when sidestepping Uke's attack; (i.e., Tenkan-Ashi to the forward left)

Advanced Connection Builders

You could step in with your right foot at this point, but you could just as easily knee, kick, or stomp on Uke as you enter; never mind what your left hand can do and it took no strength, solely hips and weight.

Let us take additional notice of the ending position your right forearm. It is actually very important!

You originally met Uke's wrist with 'half an X-block'. Your 'shield' met Uke because it was obstructing the center-line with your right hand on the left side of the center-line, your right elbow naturally on the right. (Your transmission subtly on the right side of the center-line)

When performing the back, then forward piston action that drew Uke across you (as you maintained your transmission's place along the center-line), your right hand and elbow switched sides.

Your right forearm assumed (as best it could) the position that your left forearm takes when performing its half of a normal X-block; (i.e., elbow on the left of the center-line; hand on the right)

We call this ending pose a right-handed 'inverted' X-block. (Inverted X-block: when we position our elbow on the opposite side of the center-line from our shoulder and our forearm 'doubles back' across the center-line.) (see right side image of figure 96)

You should find yourself in an inverted X-block position more often than you might think. (It is, after all, a classic Kung Fu pose; and for a reason!)

In fact, I suggest you try it to prove its usefulness. Let us reverse the drill.

Set up in the starting pose as before, then drop your arm; Uke adjusts, but this time, bring your right arm up in an inverted X-block to meet the inside of Uke's wrist with the inside of yours; (i.e., You will be positioned as in the right side of Figure 96, but under Uke's arm; Uke's arm being in the original starting position on the left of Figure 96.)

Use that same rightward/clockwise twist in your hips to draw your transmission away from Uke as your right hand sweeps Uke's right hand leftward. (be sure that your hips do all the work!).

Notice that it is a short trip to the 'inflection point', but when you get there, uncoil your hip counter-clockwise and begin shoveling Uke's arm the rest of the way leftward. (opposite direction of the previous drill)

Shield Theory

Pretty common way to make contact with Uke is it not? (hint: It is! In fact, it is exactly how we should be using our hands as we pass Uke's hand across us, ala when we 'pass' Uke's arm to transition ourselves from the outside to the inside of Uke's stance and vice-versa.

I expect that worked well, so let us find out why that was so valuable by experiencing the contrast.

Let us try the inverted X-block approach again, but make the mistake of allowing our transmission to move rightward and off of the center-line; (i.e., at the start, allowing our right elbow to drift to the right side of the center-line and join our hand on that side.) (left side of Figure 97)

In effect, it means brining our right hand up into the 'half of a wedge' type position depicted in the middle image of Figure 95.

This approach lacks a lot of leverage because your transmission is not in the correct position to engage your hips and create the broom to move Uke.

In fact, your only option for 'shoveling' Uke toward the left is to twist your hips counter-clock wise; (i.e., last half of our successful leftward try).

It can work, but you were not protecting the center-line well at the start.

Worse yet, at the end of the shovel, your hip is twisted counter-clockwise (closed left hip kua; hips not square with Uke; facing left-ward).

The angled, hip kua closed position leaves you with fewer options to launch responses to Uke's attack (you will have to uncoil to hit Uke with a left hand and even to kick efficiently. Unless you take great care, as you uncoil, Uke will have time to follow along and do the same to you!)

This error in setup reveals that, when staying in place, this classic Aikido arm pose (half a wedge, versus the classic Kung Fu, inverted X-block pose [or even the 'half an X-block pose] does not just deprive us of leverage, but also a great deal of martial advantage. (See Figure 97, hips not pointed at Uke)

We must be careful of how our levers are positioned at the start of our interaction with Uke.

Keeping that transmission on the center-line, especially at the start, is important for protection, leverage, and keeping your martial options available to you!

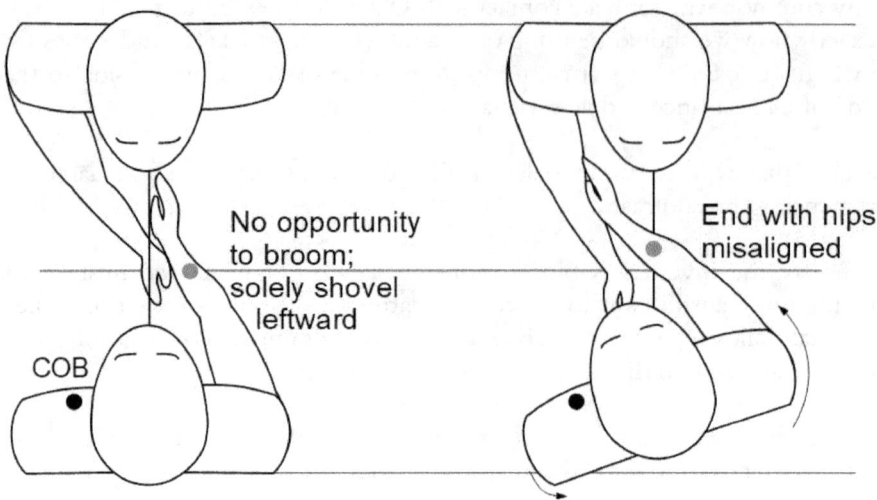

Figure 97: wedge type start to meeting Uke leaves us with solely option to shovel Uke's arm leftward; at completion, we are not in our best position

The saving grace for the classic Aikido 'half-a-wedge' arm is to step to the area 'inside' Uke's arms, ala the Tomiki Shomenate (#1) kata; i.e., the half-a-wedge arm pose requires some stepping to (re)gain advantage.

Stepping instead of staying in place does open a world of martial opportunities, but we do not necessarily have to step with the x-block.

The 'half-a-wedge' arm does also have that 'hole' or initial openness on the left side of the center-line (Tori's perspective).

This invites Uke to charge on into that space. It is a lot easier to react to Uke when you have an idea of where they are headed to.

Now you know! (You likely already knew and were taking advantage of it.)

Before we move on, one last dimension of movement to consider.

We have moved Uke rightward and leftward; can we move Uke upward and downward?

Of course!!! But let us call it 'Paint and Scrape' instead of 'Broom and Shovel' (a paint brush is just another type of broom. A paint scraper is just another type of shovel.)

Shield Theory

Learn to 'Paint and Scrape' Uke downward by getting into the original starting position from Figure 96 ('right hands, wrist to wrist', left-leg weighted cat stances), but instead of dropping your right hand, raise it.

Uke shifts their hand to point at your nose.

Bring your arm down to rest on Uke's.

Be certain you are in the 'half an X-block position' and that you made the drop as if 'painting the wall'. (It may help to remember the Tai Chi sword two-finger hand pose and the 'finger painting' we described in the metaphoric comparison between sword and calligraphy brush.)

You should also check that you have met Uke's wrist with the area somewhere between your wrist and mid-forearm.

Important!! Use the same clockwise left-side hip twist to move your transmission away from Uke as you continue to drop your right hand in that 'finger painting' manner.

Broom/paint downward, and when you (quickly) reach the limit, uncoil your hips and continue moving Uke downward by changing to shoveling and scraping downward.

Be sure your shovel/scraper is driven by the transmission moving toward Uke in that piston fashion and that the force is generated by uncoiling your hip counter-clockwise.

It may at first be awkward to have your hip turn generate this 'broom and shovel in the vertical dimension' (Paint and Scrape), but work at it until you get it. We want all of our hand/arm actions to be powered by our hips.

With the power to move side-to-side, and up-and-down, we can put these actions together sequentially to 'draw a square'/'draw a box'.

It should be obvious how, so practice a 'drawing a box' drill, as you will find yourself using each of these four dimensions separately in lots of other techniques, but note that when you do 'draw the box', it will be extremely difficult, nearly impossible, to have your transmission on the center-line.

Instead, make it your goal to keep your transmission asserted such that Uke never has the ability to move their hand upward toward your elbow; (i.e., always maintain a shield, never create a rail).

Advanced Connection Builders

There is a somewhat fine line for how far off the center-line your transmission can go (and how far the transmission moves/shifts within your forearm), but you will know you have made a mistake if you lose the necessary leverage to move Uke or Uke can slide up your arm to hit you.

Practice drawing the box in both a clock-wise and counter-clockwise fashion, and when you are ready, let us get into the next lesson that makes things much more efficient by knocking off the corners and rounding our 'box' path into a circle, thereby creating 'The Hourglass'.

The Hourglass

I can almost hear you ask, "If we are drawing a circle with our hand, why do we call this the 'Hourglass'?"

When we round off the square to instead draw a circle, our transmission becomes the skinny center between two circles: one traced by your hand, the other traced by your elbow. (a hidden circle!)

If you could do it fast enough, you would have virtual cones on each side of your transmission. It would look a lot more like a bartender's tool used to measure liquid called a 'jigger' than a symmetric 'hourglass' because one cone is bigger than the other; but not everyone is a bartender, so hourglass is more relatable.

The hourglass allows us to keep our transmission firmly planted on the center-line, and that is something we cannot do while drawing a box; it also relates to fencing/staff work in a real-world way…

The 'hourglass' can meet Uke at any starting position on the 'circle' we draw and move Uke along that circle, all the while, never giving up our protection of the center-line as we progress! (and opens opportunities to attack our opponent if they are not adjusting correctly.)

Remember our 'against the door frame' hourglass drill at the end of the 'Where we are strong' section of this book? We are doing nearly the same drill, but with our hand outward and in contact with Uke, hand-side cone open to Uke (instead of, as in the Figure 78 drill with our arm sideways in front of our bodies and against the door frame; where we had both cones open to the sides of our stance.)

The other major difference is that we can now view the pressure we apply to conjunction point (e.g., Uke's grab) as a 'lever powered' broom or shovel and we do not have to get over-explanatory to do it.

Shield Theory

Let us give it a (quick!) try!

The quick path to learning to 'knock off the corners' of our square is to practice the original leftward/rightward broom and shovel drills, but instead or purely horizontal movement, follow an arc as you progress side to side.

Spelled out a bit more… get into the left side of Figure 96, back of the wrists starting pose, both you and Uke in a left-weighted cat stance…

Draw the first quarter of the circle clock-wise by starting on the left with your right hand in the nine o'clock position and broom upward toward twelve o'clock as you rotate your hips clock-wise.

Next, follow through 'high noon' by uncoiling your hips, eventually returning your hips to square with Uke as your hand switches to shoveling and ends in an inverted X-block at the three o'clock position on the right; (i.e., finish the top half of a circle).

Again, all we did was re-perform the drill in Figure 96, but were a bit 'fancy' with an 'up and over' with our hand.

Next, draw the bottom of the circle by letting your right hand dip downward with a broom function from the three to the six o'clock position while using a clock-wise hip twist to motivate the action (might feel awkward).

Lastly move over and upward as a shovel as you return your hand and Uke to the nine o'clock position, your hips moving counter clock-wise, and returning to square with Uke.

When practicing with an Uke, the secret is keeping the connection to Uke's COB via your transmission and in finding how to gently shift between broom and shovel.

You can practice this skill solo by actually tracing a circle upon a wall.

As just described, it took two full 'hip-twist and return' actions to complete this drill.

Next, we complete the circle in just one hip coiling/uncoiling action.

Coil your hip and broom Uke rightward across the entire top of the circle, uncoil to move draw the bottom of the circle as you move Uke completely leftward.

Advanced Connection Builders

It typically means that you will also have to commit to the broom or shovel for the entire pass from one side to the other, but you could play around with using just brooms or just shovels for the entire circle.

Captain obvious adds, "Rotate in the other direction!" Thanks Cap!

Translate this empty hand drill to 'weapons' work...

Replace your forearm with your favorite martial tool and keep Uke from moving upward on the instrument by asserting the device's transmission as appropriate.

We experienced an assertion of our bokken's transmission just a few drills ago, so I suggest, this time, using a jo staff (to mix things up).

All the same sweeping and shoveling actions apply, but we need to take constant account of the position of the jo's transmission as we find ways to coordinate ourselves into making 'drawing the circle' comfortable and driven by the twits in our hip(s).

Could you motivate the jo without the hip twisting action? Sure, but you lose a lot of leverage. Do not cheat yourself.

Finally, one last highlight to make before we move along to our next topic.

So far, we have moved Uke on a singular plane. Up, down, left, and right are still happening on a two-dimensional platform; i.e., on a 'wall'

Personally, instead of focusing on the hand touching and painting a circle on an imaginary wall, I instead like to focus on keeping my transmission positioned under an imaginary waterfall (e.g., a thin sheet of falling water a few inches from the wall I am painting the circle upon when doing solo practice).

How?

I imagine my hand is on the other side of the sheet of water contending with Uke (or the wall); and neither my hand, nor Uke ever get wet.

Why?

There is one last dimension that you must master: Moving Uke toward and away from you. In other words, switching sides of the waterfall.

Shield Theory

I prefer to call the 'waterfall line' the Mason/Dixon line, for it emphasizes a 'my-side/your-side' of the waterfall connotation.

This 'switching sides of the Mason/Dixon line with Uke' is more commonly called a hip-wheel turn. Exactly the same kind we investigated in the second half of the (not so) 'Basic Terms' 'Bounce it off the Rock' section. (the one where we perform the full hip turn)

Flip back to almost the very beginning of this book and check it out again with new eyes.

That drill is very much another broom and shovel situation; i.e., revisit your concept of the 'centripetal arm' as 'finger-painting with Tai Chi Sword Hand' and apply the concepts of broom and shovel (vertically), but follow through the bottom of our vertical 'Paint then Scrape' actions, and raise your hand as you shift your weight and step 'through the waterfall'.

The 'hourglass drill' emphasizes 'keeping our transmission on the center-line'.

Performing hip-wheel does the same, but it means the focus is 'keeping our transmission in synch with our COB', for as we perform the hip-wheel turn, Uke moves across us, and so too does the center-line.

We have to keep our transmission between ours and Uke's COB's; (i.e., the transmission stays on the centerline at all times, even in the hip-wheel turn!)

The point here is that we understand that hip-wheel turns are expressed in our hands as broom and shovel actions (e.g., as we wave our hands like clouds; not everyone feels the 'shovel' near the end of the wave)

Never lose sight of the fact that the transmission is a fulcrum.

You will quickly figure out how to step in various ways and maintain a fulcrum; i.e., the fulcrum can 'move' around but the rest of the lever has to come with it. The 'waterfall' is one way to start to build that coordination and awareness.

The universal way to learn is to feel for the leverage, that it never wanes.

In interest of the size of this book, I leave it to you to go beyond the 'Bounce it off the Rock' drill and explore other hip-wheel turns.

Remember the guidelines above and you will not go wrong.

You have become a lever master, sweeping and shoveling Uke with your 'Shield'. Most importantly, you have connected this with the action in your hips; (i.e., you have conscious, intent driven application of your transmission as a projection of your COB. Your shield = COB + transmission; one or both moving as appropriate to create 'techniques')

We drilled 'hip wiggles', by 'posting' onto one foot and twisting our hip back and forth. We just as easily turned this into full hip-wheel turns (ala the second half of 'Bounce it off the Rock') We even toyed with bending and straightening our knees to make our hand and elbow 'teeter totter' at the transmission.

Even with all of this success, Aiki may not be fully implemented.

You likely have been hit or miss; i.e., one technique good, the next one perfect! Ok, maybe a few 'not so good' ones too, but I have a lot of faith in you!

There is a reason for lack of true Aiki, and it is almost assuredly due to acting as a level surface for Uke to rest upon.

We have to ensure that we are creating an incline for Uke to resist against!

Why?

If brooms and shovels become inclined planes to 'roll' the objects of their intention, so too must we!

Let us learn to constantly 'wheel trundle' Uke as we explore a core secret of physical Aiki: Rolling!

Rolling

Physically, Aiki is the control of stability and balance.

To control these traits, you have to first understand them.

Balance is binary. You either have balance or you are falling, there is no in-between. You know when you have 'broken Uke's balance' when they fall.

To make Uke fall, unless you can outright pick up your Uke and then drop them, you going to first have to control Uke's stability.

Rolling

Controlling stability is a bit more complicated.

An upright person has differing levels of stability, from almost none to 'rooted'. The only time that they have zero stability is when they are falling.

Even more perplexing is that, at any given instant, the measure of stability is different depending upon the angle of approach! A high level of stability in one direction does not ensure stability in another (e.g., strong and weak lines in their stance, muscular versus ligamentary tensions, etc.)

Although complex, controlling stability can be mastered.

In fact, if you have been following along, you know more about stability than you might at first give yourself credit for. Let us prove that statement.

Stability is based on three things that you already know a good deal about:
1) distance of the COB from/to the edge of the stance (the Pin, in 'Pin and Spin'!) [the height of the COB also factors here]
2) the amount of stress on the structure (e.g., how well we have torqued Uke's chain [stressed Uke's QL's]), applied pressure to the weak lines of their stance [no-line/20], bound Uke's hip joints [Three-Legged Dog spots]), and/or ground-up locking (grabbing Uke's foot with the floor and twisted their leg)
3) the final aspect: footing; a.k.a., relying on an unstable foundation

Each of these three 'willows' Uke in their own way, and can be used together in any combination.

You should know a great deal about the first two: Pinning and Structure.

If not, we have an entire other book that explains 'Pin and Spin' <wink> and, so far, in this book, have detailed the effects of Teeters and Boxes upon Uke's QL's, perfected the 20, explained Three-Legged Dog Spots (TLD's), and more!

What we have not yet done is complete the curriculum by inspecting the Aikido equivalent of 'giving Uke unstable footing'.

The word 'Footing' is a bit misleading. In Aikido, instead of making Uke step on banana peels or walk on ice, most often, we are engaging Uke's upper body.

That should beg the question, "Does the upper body have 'footing'?"

Advanced Connection Builders

Yes! It does! (or, more accurately, it does in certain situations!)

Imagine getting into a plank position for performing a pushup.

For all intents and purposes, you have converted your arms into 'legs', ala the legs of a table, and your hands have now become 'feet'.

While still in your imaginary planking pose, imagine that floor under your left hand suddenly drops on one side and becomes a ramp.

You are not such a stable table anymore are you? (rhymes!)

Initially, this scenario can sound awkward and unusable for direct correlation to Aiki(do), but planking to do a pushup is very much the same as pushing on a wall. (depending upon how many hands used, pushing on the wall is akin to a one, or even two-handed 'standing plank' pushup, with 'feet on the floor, hands on the wall')

If it is the same as pushing on a wall, it is also the same as when we push on Uke or when Uke is pushing on us.

It is our job to be sure that the Conjunction Point ('CP'), the point at which we meet Uke, presents an incline that 'rolls' Uke and not a stable, flat floor.

How?

You have been preparing for this moment throughout your reading (both 'Six Precepts' and 'Codex')

Your most direct brush with this concept was very early on when you learned about the elephant arms drills... when applying the 'albatross' and 'cheat', you created 'the void for Uke to roll into'.

Uke sensing a desire to 'roll' off of your hands was/is the secret.

Let us dig a bit deeper on that concept.

To inspect this phenomenon, it sometimes helps to equate the conjunction point to the keystone of an arch.

One simple way to create the situation is yours and Uke's right hands coming together, palm to palm, as if your hands got stuck together when making contact for a 'high five'.

Rolling

This 'coming together of hands', along with a slight bit of moving toward and resisting each other, sets up compression connection.

In this case, both of you are mimicking the 'standing plank', the 'standing one-arm push-up' as you together create an arch.

If both sides present their hands as a flat, stable surface (in relation to the synthetic gravity of compression connection) Uke and we will both be able to find support (balance), and thus, also find stability.

What if, instead of acting as one person doing pushups against the other, we cheat, change our role, and become the 'vertical floor' (ala a wall) that Uke is pushing on?

We could then de-stabilize Uke by creating 'unsure footing' for their hand by 'tilting the vertical floor' (e.g., a vertical 'Albatross' and/or 'cheat')!

Presenting Uke with 'unsure upper body footing' is the root concept/goal.

Just be careful not to confuse this goal with one major misconception that comes up as soon as we use this wall analogy: Uke pushing against a door.

'Unsure upper body footing' is radically different than when 'Uke tries to open a door, but someone opens the door for them from the other side'.

Why is it different?

Well, for one, most persons do not actually fall or flip when they try to push (or step) upon a surface that is not there. They might get embarrassed or startled, but most catch themselves quickly.

We want to lead Uke's movement in a continuous fashion. The 'door not being there' is a singular 'point in time' event.

We cannot sequentially lead Uke in the scenario where we are not actually part of Uke's 'synthetic gravity' situation and have zero influence on their motion; (i.e., no connection, no Aiki!)

Since, instead, we would rather 'tilt the floor under Uke', and give Uke the continuous 'rolling' sensation, it is very handy to understand why a ball rolls down a slide (i.e., down an inclined plane) so we can duplicate the intention in Uke.

Advanced Connection Builders

What I offer here is a 'custom' description of the mechanics.

It will likely make a few physicists crazy as I ignore normal force, thus, do not use this explanation in school or at work, but it works fine in the dojo.

Here we go…

A ball rolls down a slide because the slide (inclined plane) holds its shape and delivers indirect pressure (upward, in opposition to gravity) to the ball's center of mass (or as we named it, the Center of Balance, 'COB'). (See Figure 98)

As we learned in 'Six Precepts', pushing through the Center Of Mass/COB of an object is called 'direct pressure' and direct pressure will attempt to move an object directly away from the source of the pressure.

A floor does this well, and that is why a ball bounces when it falls and hits the floor. Gravity pulls, the direct pressure from the floor (pressure through the ball's COB) stops the ball, and gives the ball something to rebound against.

It is very much the same when you bounce the ball off of a wall, but you have to ask, "Where is the gravity in bouncing a ball off of a wall?"

Earth's gravity becomes negligent compared to the force you use to throw the ball. The 'wall and ball' scenario is instead, an example of 'synthetic gravity'; i.e., 'horizontal gravity', specifically, in this situation, compression connection! Same direct pressure, same 'bounce', just horizontal.

Conversely, pushing on an object in any dimension other than 'directly through the COB' will create indirect pressure, and begin spinning the object, and that is exactly what the slide is doing to the ball!

Here is the trick… in our 'ball and slide' scenario, the ball's COB is following gravity's influence and attempting to move directly downward toward the Earth.

In order for the ball to be stopped/supported, pressure must be delivered through the 'South Pole' of the ball and directed through the ball's COB (and effectively out through the 'North Pole'; i.e., pressure passes through/along the ball's 'Axis'. Just ask your local rock balancing enthusiast how it works, for what are rocks but misshapen spheres/balls?)

When we bounce the ball on a floor, gravity conveniently aligns the contact with the level floor to the 'southern' point of the ball.

Rolling

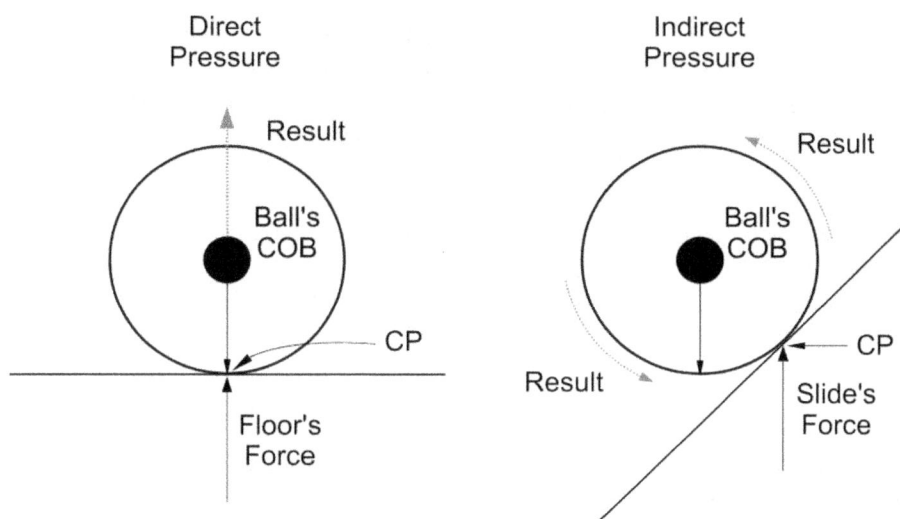

Figure 98: Ball follows Earth's gravity, (left) meets the floor, direct pressure is created, and it results in a 'bounce'; (right) ball meets a slide, indirect pressure results and the ball spins

Not so with the slide! The inclined plane is 'never level'; i.e., it is 'uneven'!

The slide is not active, it is inanimate. The best it can do is hold its shape in the face of an object trying to follow the force of gravity downward. (do not scoff at this subtle but incredible talent, you will be using it very soon!)

The magic in the slide is that it is slanted/uneven. The slide cannot physically make contact with and push upward upon the 'South Pole' of the ball. At best, all the slide can do is 'partially support' the ball, by making contact (Conjunction Point! CP) with the ball somewhere closer to the 'Antarctic Circle' of the ball, but certainly never at the 'South Pole'.

This means that the slide's upward resistance against the ball will never be direct; it will always be an 'indirect touch' to the ball's COB.

That is an advantageous constant, for the ball has a constant of its own (one we all share): the ball continues to obey gravity and is drawn directly downward through its COB toward the Earth.

The ball's continuous obedience to the force of gravity influences the ball's 'South Pole' to 'fall into' the vacancy located just to the side of the CP and the whole scenario repeats itself with an ever-changing CP.

We call the interaction between these particular constants: 'Rolling'.

Rolling ensues for as long as the ball has a length of non-level, uneven surface to roll along. In this scenario, the level plane is met when the ball reaches the end of the slide and lands on level ground, where the Earth provides the resistance in direct opposition to gravity, projecting that resistance upward though the COB of the ball, ala 'direct pressure'

Thus, we should recognize Rolling as a continuous action (condition), not just a point in time event…

You might be wondering why I am so excited about such a subtle profundity.

Entertain this…

What if the ball was flat on one side? See Figure 99 to envision how that scenario plays out.

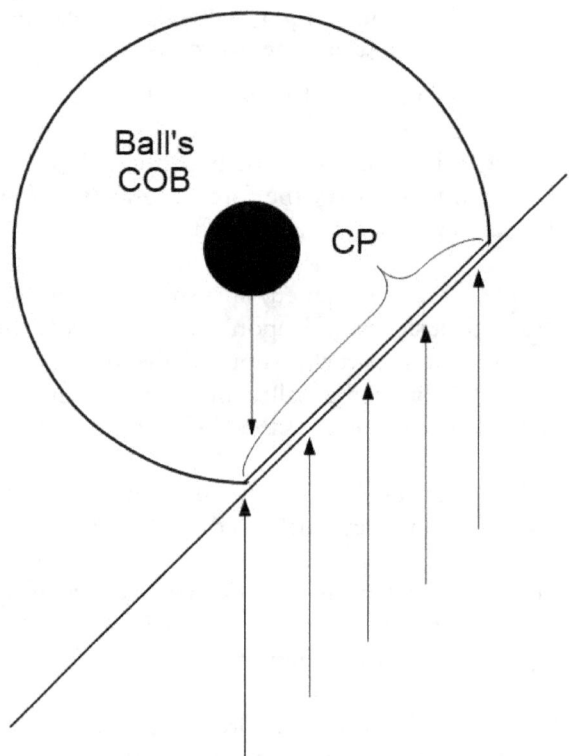

Figure 99: Ball with a flat side, stable and stationary, but precariously positioned as any tipping to the left will cause a roll

Rolling

The ball is stopped, for the CP is a lot larger and is underneath (supports) the downward force of the balls' COB.

The incline is positioned such that the downward force of the COB is precariously near the edge of the CP 'footing'/'base'.

It would take the slightest counter clock-wise rotation to initiate a roll.

Just before we get to that roll, there would be just enough room in that rotation to lift the flat side of the ball from the slide so that the ball was balanced and pivoting on the corner of the ball's flat edge.

The ball could rotate back to resting its flat edge on the slide, or it could go into a fall/roll along the slide.

What we have here is a treacherous stability (instability!), and it is a condition we can use, let us call this 'Almost Rolling'

True rolling has universal, real work-a-day utility when we want to move something that we do not want to pick it up.

Almost rolling has martial utility when an object (e.g., ball, soon to be Uke) meets an obstacle that puts the object in the condition of 'I would be rolling if not for this little but of resistance stopping me'.

The ball is inanimate. It cannot sense the precariousness of its situation. If the ball could sense, it might want to correct the situation due to an innate sense of fear of falling (or more accurately, fear of a sudden stop!)

It is in 'almost rolling' that we find the spirit and essence of physical Aiki.

What do we do with this knowledge?

To be blunt, do not be blunt!

We want to avoid creating, in reference to synthetic gravity (i.e., compression/extension connection), a level surface for Uke to rest upon.

How?

You already know how! You proved yourself capable when you practiced the elephant arms drills. Do you remember how it feels when Uke is rolling into the void? (When Uke is truly 'willowed')

You positioned your hands as if they were a 'slide', an inclined plane in reference to the synthetic gravity between you both (called 'connection').

In the elephant arms drills that compression connection is pretty much aligned with Earthly gravity, but that is a convenience we use to make learning easier.

Oh, you mean you want to know how to perform in other contexts?

Ok, but in order to learn to present Uke with an inclined plane instead of a flat surface (and have the awareness of what you are doing…) we are going to have to synch on what it means to 'cut', because 'cutting' is one answer.

To really understand what 'cutting' means requires more explanation before we get to a drill, but trust me, it is worth it. [Mind-Body Connection!])

Scoring, Chopping, Slicing, and Cutting

So, you think you already know how to cut, huh? I am sure you do, but let us be particular and synch on what we mean when discussing the words: Score, Chop, Slice, and Cut in the dojo…

Scoring is the easiest. Ever score an orange so you could peel it?

Your cut is superficial, and in most common methods, you are performing a 'broom' action with the blade, allowing the blade to 'bite' just a little as you make your way around the orange.

This action is very similar to the drill in Figure 89 where we 'roll the ball with our Jo staff', the big difference is we typically rotate the orange on the blade instead of 'wheel trundling' the orange with the blade. Scoring has its utility and is a quick and easy way to start into this discussion.

Next is 'Chopping'.

To reveal it, let us first align with true, Earthly gravity. Imagine this.

Place a straight backed, squared knife (e.g., a cleaver) blade-upward with the spine on the ground.

In this scenario, the blade is an extension of the ground, albeit a very thin, wedge shaped extension, but from a side angle, it looks like just another flat, level surface upon which to rest.

Rolling

Now drop a perfectly round orange on the blade.

From the side, you might expect the orange to bounce on the blade.

After all, the blade just looks like a flat surface. (ala left side of Figure 98)

Instead, if the orange falls such that the 'south pole' center of the orange touches the blade, the unsupported sides of the orange will progress and cause the blade to 'wedge' itself into the orange as best as can.

This is a 'direct touch' to the orange's COB, and most importantly, this is <u>not</u> cutting. This is chopping! Otherwise known as 'Cleaving'

Let us remember that motion is relative.

In the kitchen, instead of the object coming to the blade, chefs take their kitchen cleaver and position the blade perpendicular to the counter and 'drop' their blade directly downward as they 'chop' through their ingredients.

They are not 'cutting', they are chopping. This difference between chopping and cutting is tremendous.

Chopping = blade perpendicular to the line of force, line of motion, line of synthetic gravity. Line of force passes (directly) through the objects COB.

Ask yourself, "Would you chop a steak with a cleaver at the dinner table?"

Never! (unless you are overcompensating for something?)

You cut a steak, and cutting is performed with a blade at incline, but…

You might think that is all there is to it. 'Slant the knife and drop it directly downward', but although extremely useful, in my kitchen (and in the dojo), that is called 'slicing', not 'chopping', and certainly not 'cutting'!

What is the difference between 'slicing' and 'chopping'?

Indirect/angled pressure versus direct/straight-through! (respectively, think 'orange dropping onto a slanted blade and the blade passing through')

This is a gruesome analogy, but to understand slicing, let us consider a guillotine cutting a carrot (I know, gruesome, but the carrot was sentenced to soup and a salad for the crime of being wantonly and publicly tasty!)

A level blade dropping on the carrot will 'chop' directly through the carrot.

Dropping a slanted blade will, at first, act just like the 'ball and slide' on the right of Figure 98, but it does it in reverse; or maybe better said, as if up-side down, as if gravity was pulling the carrot upward instead of down.

This is due to the blade initiating contact with the carrot off-center, not the top (Arctic Circle instead of the carrot's North Pole).

The effect? The blade will, at first, attempt to 'squeeze the carrot out from under the blade'. (I seem to remember watching someone try to cut a bottle by driving a cleaver angularly down on it. If I remember correctly, it is all in the reflexes. [Do not try it! I also remember it starting big trouble])

Although the carrot will be influenced to 'roll' to the high side of the blade, unfortunately for the carrot, friction against the bottom of the guillotine will inhibit the carrot from rolling and the blade will begin to separate the carrot. (ever see the depression on the bottom of the guillotine, now you know what it was for, to prevent that 'squeezing out' movement.)

There is a point where the slanted blade passes through the very center (COB) of the carrot and the blade will be making 'direct' touch to the carrot (technically, chopping), but only at that instant.

At all other times, there is indirect touch between blade and carrot. That is good, because a true cut happens when an object is 'rolled' along the blade.

The object being cut will not be able to actually roll, but the intention to roll must be there if it is to be an efficient cut. (Read that again!)

This is a pure expression of Aiki, that Uke senses the intention to roll even when actual rolling is not happening. That sensation of 'almost rolling' is a powerful de-stabilization (willowing); the effect of (almost) rolling is often stronger than that of 'Pinning' or 'de-structuring' Uke.

So, slicing = blade slanted in relation to the line of force (a.k.a. line of motion, line of synthetic gravity) and passes (directly) through the object's COB.

That particular slanted blade scenario is good, and in most cases, better than 'chopping', but it still is not creating the maximum amount of 'rolling' possible.

That is why 'slanted but direct touch of a blade' is 'slicing' and not 'cutting'.

Rolling

To truly 'cut', we want to avoid that little 'instant of direct pressure' we highlighted in the slicing action.

In order to do so, we need to switch our analogy from 'ball and slide'/'guillotine and carrot' to our new friend 'wheel trundling' and two old friends from 'Six Precepts': 'Tomoe to the Front' and 'Tomoe to the Rear'.

In wheel trundling, we do not want the stick to penetrate the wheel, we simply want to roll the wheel.

We achieve this by gliding the stick along the surface of the wheel; (again, ala Figure 89; broom and shovel roll the ball).

In 'cutting' our carrot (this time with a knife), we perform almost the same action. We attempt to 'roll' the carrot along the cutting board with the knife, but the blade 'bites' as it moves and, instead of rolling the carrot, it begins separating the carrot.

That 'bite' is the assertion of the transmission within the knife.

Yes, the very same kind of transmission you used to create 'the shield' with your arm, bokken, and jo.

The direction we send the 'bite' toward in relation to the COB of the object we are cutting is important, as it defines one of two cuts: 'Front' or 'Rear'.

The first cut that we can create, the 'Front cut', directs the blade's transmission 'Tomoe to the Front' or 'passing over' the carrot's center.

In context of the carrot, the blade, and cutting board, the carrot lays across the board at a 45-degree angle, near tip in your left hand and far tip is 'away' on the right, so that it will roll away from us as we push the blade outward/forward to cut.

Think 'wheel trundling' standing directly behind an auto tire (you cannot see the rim) and pushing the wheel forward.

'Tomoe to the front' of the carrot's COB means driving the blade's transmission to a point above the center of the carrot. (Figure 100)

Take a moment to let that settle in by relating this cut to 'Tomoe to the Front' of a standing Uke, which is easiest to be thought of as a 'hook punch to Uke's nose'.

Should Uke lay flat sideways across you (with their head to your right, feet to your left) you could imagine rolling Uke by pushing them with a hand on their shoulder and hip; making Uke face away from you in the initial movement; (i.e., 'Rolling Uke over')

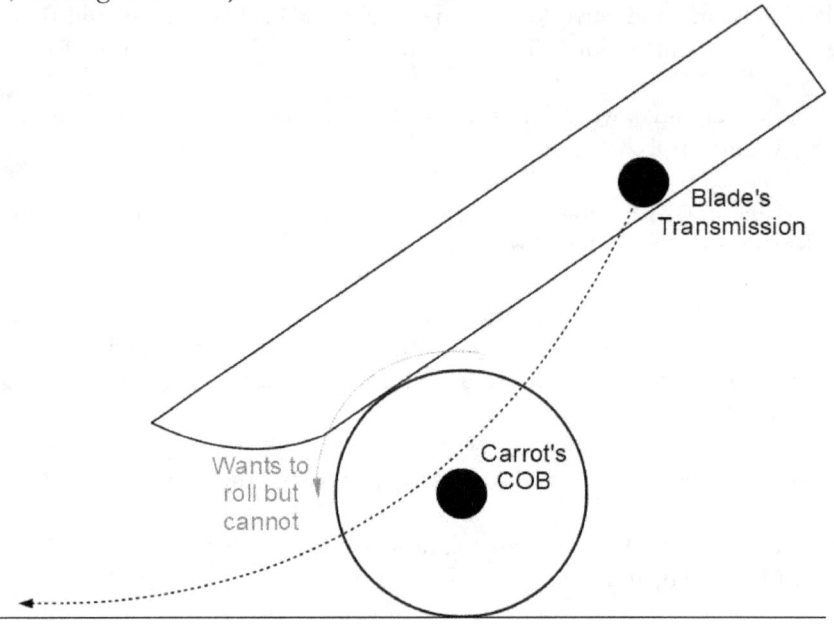

Figure 100: 'Top Cut'. Akin to 'Tomoe to the Front' pressure on Uke. The blade is trying to roll the stationary object, but cuts instead

The 'Front (top) cut' will attempt to roll the carrot almost as if we were wheel trundling. The carrot will try to roll naturally, top side moving in the same direction as the blade (away from us), but the 'bite' of the blade's transmission will compress the carrot, thereby increasing the friction, and again, although intending to roll, the carrot will instead be severed.

If we do not truly 'bite down' with the blade's transmission, all we will accomplish is rolling the carrot across the board, just as we would roll the wheel in wheel trundling. No cutting happens! I call this 'gliding' the blade.

If we are lucky, we might score the object, but even that is not for certain.

Practice this cut on some salad ingredients that need to be cut (and eaten).

Use your blade to roll a carrot or cucumber toward and away from you a couple of times before you 'leverage' the transmission of the blade to bite into the vegetable before it is allowed to roll. (next step: have a snack!)

Rolling

The second, and the most efficient cutting action, requires moving the blade's transmission to bite 'Tomoe to the Rear', or in this case 'under' the carrot's COB. (See Figure 101)

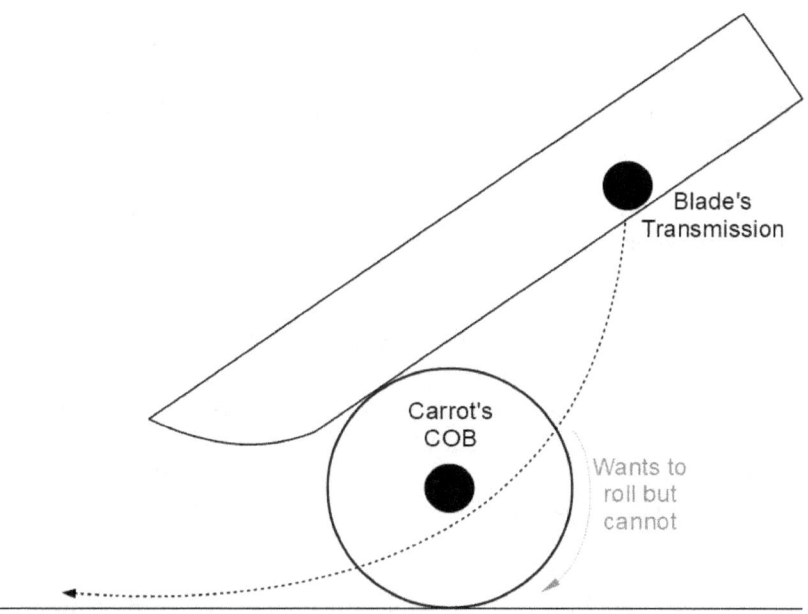

Figure 101: 'Bottom/Rear Cut'. Akin to 'Tomoe to the Rear' pressure on Uke. The blade tries to roll the object with 'backspin' but cuts instead

In this 'Rear cut', the 'roll' that you induce is akin to 'putting backspin on a cue ball' (in a game of billiards), but with the pool stick in the wrong position.

It is like 'mind-bender' shovel the ball move, but rolling unnaturally, against the direction the ball is rolling.

If it were possible, it would be like trying to push/fold our lying down Uke's shoulder and hip under them as we rolled them across the floor. (have Uke's face initially turn to face us we push them away i.e., roll them under, instead of over). This is almost impossible with Uke lying on the floor, but wait! It is so much easier, and useful, when they are standing up!

Your carrot is much more rounded than Uke, so it will have a greater ability to 'roll under itself' as you lower the path of your blade's transmission to drive the lower half of the carrot away from you, but, due to the friction of the cutting board, the carrot will be held static as we 'bite through' with the blade.

Advanced Connection Builders

This is a very convenient situation in the kitchen.

The intent to roll is established and the friction acts as a 'fork' as it holds the object in place as we drive the blade's transmission.

Another key feature of this cut is that the carrot/object is making contact on the top half of the blade. In effect, you are 'creating a shield' that would have kept the carrot from reaching the hilt of your blade (if the blade was not passing through the carrot that is.)

The 'roll' trying to move in the opposite direction of the motion of the blade is the most efficient manner of cutting we can create. This is a true 'cut'. The truest of all cuts.

Please do not, but if you doubt that the carrot is trying to roll in any of these scenarios, you will be risking cutting yourself by performing these cuts again with a wet, slippery carrot and a cutting board coated with oil.

Again, please do not try that! Trust that you will find that the 'Tomoe to the Rear Cut' is the easiest, especially when friction between carrot and board is greatly reduced.

So, we have learned how to cut while projecting our blade upward and outward.

Excellent, because it ties it together with our 'shield' lesson, and, if you did not catch it, let me add that this 'healthy snack time' kitchen drill was also a study in 'shoveling'. Think about it. (Bosu ball rolled under the jo staff…)

Now, before we move on to other nuance hidden in cutting, we should have a true Aiki(do) drill to bring this cut home.

There are many ways to assess someone's skill at cutting once you know what to look for. The 'Do' (sounds like 'dough'), or 'sideways' cut is a dead giveaway to find out if someone knows how to cut or is just gliding their blade across surfaces.

The tip-off is if Tori is actually 'nudging' Uke to move as the bokken is brought toward and then across Uke's belly.

Bokken are not true swords, so of course, and thankfully, there will be no penetration through Uke's belly, but the 'bite' of the wooden blade's transmission should be influencing (reducing) Uke's ability to stand in place.

Rolling

All too often I see persons perform this sideways cut in repetition and, as I watch the practice weapon 'slide' along Uke, I oft wonder if Tori would know how to deliver more than a 'scoring' of Uke's gi with a real sword.

Your cuts should constantly influence (roll!) the material they intend to separate, especially if you get stuck in the material; (i.e., always generate instability).

Therein lies the drill, and likely the hardest of the sword cuts to learn at first: the horizontal Do (doh!) cut!

Let us try 'cutting', thereby 'rolling' Uke with a bokken.

We will start with a chop, then slice, switch to a 'Front Cut', compare the experience with the 'Rear Cut', and finish on a sour note with a Glide…

Uke stands in front of you with a right foot forward stance, hands in the air (or behind their head).

Our only request of Uke is to expose their oblique and belly muscles and stand a bit strong. (In just a moment, they should/could protect their ribs by helping place your bokken where their ribs are not or, by putting their hand between your bokken and their body. Safest if you include protective gear!)

Assume a right foot forward stance, your bokken parallel to the floor, with the tip to your left (if holding the bokken correctly, your right arm is over your left in a somewhat modified X-block, hint, hint), and place the middle of the edge of your wooden blade on Uke's right side oblique ('love handle'). (See top left of Figure 102)

'Blade cleaves flesh; Stick breaks bone'. Do not perform this drill on Uke's lower ribs. We do not cut bone, it dulls the blade, and increases the chances of the blade getting stuck. In this drill, it hurts Uke.

We have an opportunity to practice using a blade correctly, let us do just that.

Start this drill using a direct push to Uke's COB. (Top left of Figure 102)

Accomplish this by putting your blade perpendicular to Uke's '20' lines and stepping parallel to those '20' lines.

Keep your blade positioned perpendicular to Uke's '20' lines and you will be 'chopping' into Uke.

Advanced Connection Builders

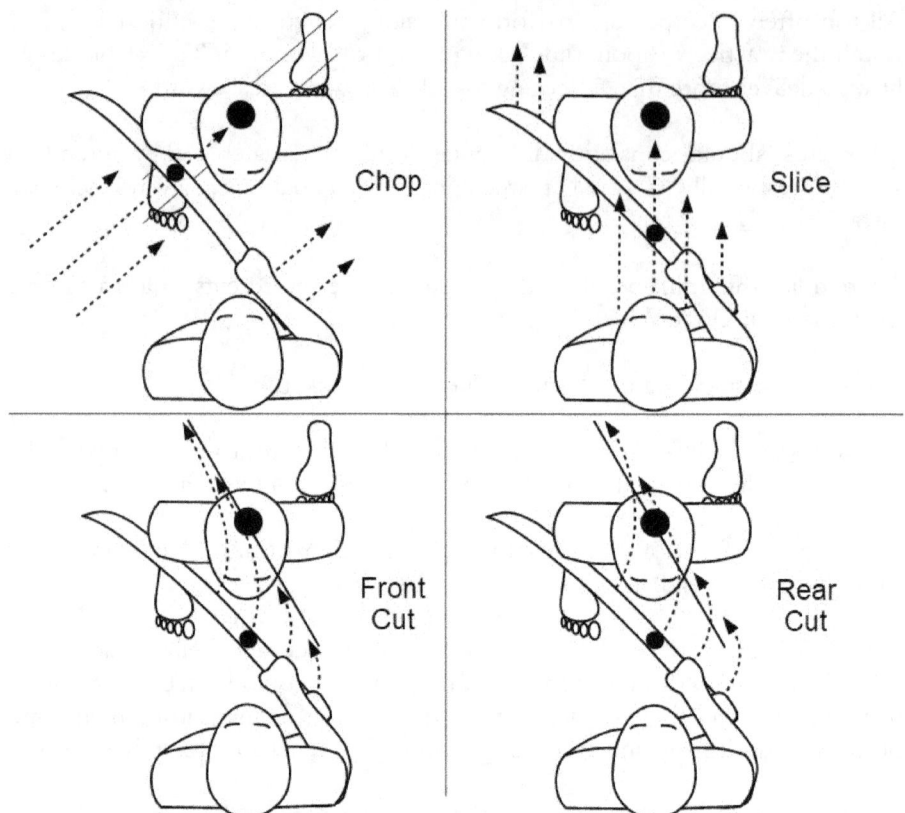

Figure 102: Four horizontal 'Do' (Dough) cuts; Uke's COB and the transmission within the blade depicted as circles

Your bokken was already against Uke, there should have been no striking!

This chop should be very difficult if at all productive, but success is that Uke is feeling solid compression connection. (For a little fun, and to find greater success in moving Uke, use your bokken to move Uke's COB to either of their '20"s before pushing through! It is the 'Secret of the 20' drill with bokken!)

Stay in there, do not back off yet.

This is a 'blunt force'. One that would have some penetration if you were using a three-foot razor blade, but it would only be moderately effective (compared to the full potential). Let us progress.

Bring your bokken a bit more across Uke's front, but still angled, and move your bokken as if passing directly through from Uke's front, to their back.

Rolling

This is the 'Slice' (Top right of Figure 102)

Again, stay there while I point out a few things.

Done correctly, you never lose the feel of pushing your blade forward/outward and Uke never feels any less (or more) compression connection, but you should both be able to sense a bit of instability being created in Uke.

There is an element of 'rolling' being formed. Because the tip of your blade is to the left of Uke, Uke may sense desire to 'roll' up to the blade to your hands. (Uke wants to turn their face rightward from their perspective [CW from top down view] and 'roll' into the space nearer your hands)

What is more is that you should both sense the increased propensity for a true blade to 'slice' and penetrate much more deeply than the 'chop' allowed.

This time, let us cut! A 'Front-Cut' to be exact!

To do so, maintain the bokken in its current position and move the line of force from direct to indirect by directing the force of your bokken's transmission toward the front of Uke's COB. (bottom-left of Figure 102)

Take care not to lose the intention of pushing forward/outward with the blade. (It should feel a good bit like you are rolling/cutting Uke 'under the shovel', ala the Figure 89 Broom and Shovel the Ball drill.)

Notice that I did not say 'way out in front'. I would prefer you go slightly in front of Uke's COB and create a 'mostly direct, partially indirect touch'.

Although Uke should feel a lot of temptation to 'turn to face the right', they are likely still standing there able to absorb your new 'touch' to their center.

Uke would never try that with your true steel blade, but if they had armor on, maybe they would.

Ruin their 'potentially armored' day as we did before by leading Uke's COB to their No-Line and align the pressure of your 'Front Cut' along Uke's No-Line.

This should result in Uke turning away from you somewhat as you willow them. Do not rule out making them fall, but at least push them backward!

We have one final cut to attempt, so reset.

Advanced Connection Builders

When ready, do the obvious and focus the direction of your bokken's transmission behind Uke's COB. (Bottom right of Figure 102)

Your bokken will broom Uke just a bit at the start.

Be sure you are creating that 'mostly direct, partially indirect' touch that keeps you influencing Uke's COB into a spin. Uke should be sensing the desire to turn and face you.

Maintaining the pushing/shovel motion of the bokken should be easy.

Again, have fun of moving Uke's COB to their No-line and give them a nudge to make them fall or at least move away from you.

Consider how deep would a blade go? It would 'roll' Uke all the way in!

Becoming aware of this bit of subtlety is the path toward mastery, but so too is recognizing failure.

Set up with Uke as before and 'glide' your blade along Uke.

That is right! Just 'score' along Uke's surface.

It feels awful does it not? Let us not do that again!

Practice these cuts with three persons. Ask the observer what each phase of the cutting progression looked like, both in the effects upon you and in Uke.

What is their opinion of 'gliding' versus 'chopping', 'slicing', and 'cutting' to both the front and rear of Uke?

Switch places and observe and feel it for yourself.

Remember when I stated the Do cut is one of the harder cuts to learn?

Some find this drill frustrating, mostly due to the X-block position of their hands. It can be awkward, especially if you are not practiced with a bokken.

If you must, feel free to alter your grip such that your left hand is near the tsuba and your right hand down on the end of the bokken handle.

This is traditional swordplay heresy, but it removes a great deal of hardship for many. My only gripe is the lack of an x-block position in your hands.

Rolling

If necessary, move your left hand to sit on the Transmission, but promise to practice until you can perform the drill with the correct sword grip.

We have so much more to learn than just horizontal cuts, and more than just shoveling.

So, to speed things up…

…we will leave this drill with asking you to figure out on your own how to perform these horizontal cuts on Uke's other side; i.e., with your sword positioned with the tip on the opposite side of Uke. (Your right, Uke's left. This time use the No-Line with the Chop, and '20's when you cut!)

I would then ask that you try to perform 'chopping', 'slicing', and 'cutting' with your sword vertical, coming at Uke from the (out)side with your blade against Uke's outstretched (ligamentary tensioned!) unbendable arm; (i.e., Standing perpendicular at Uke's side, as if ready to cut Uke's arm off).

Remember that you should start with being in contact with Uke, the bokken simply rises and falls! I.e., just pushing, no striking! Constant contact!

Be sure that you are driving your blade across the front of Uke, (i.e., avoid Uke's face), Uke's Unbendable Arm should help keep their arm from folding across them as you direct your bokken's transmission both under and over Uke's upper arm bone (Humerus); (i.e., Uke should be practicing the Figure 71 Zhan Zhuang 'side-ways unbendable arm' drill!)

This includes using both a rising and also falling motion; (i.e., Cut up and cut down) The point is that you learn to cut vertically outward, away from you and it 'rolls' Uke's arm.

Lastly, ditch the bokken, use your forearm 'shield' (i.e., your 'sword hand'), and repeat everything we just did with the sword; horizontal and vertical.

To practice horizontally with your hand(s)… have Uke assume a right foot forward stance. Next, use your judgement on where to start/stand in relation to Uke, decide which direction you will step toward, and begin to perform horizontal 'Do cuts' at Uke's chest in lieu of oblique. If you follow along in earnest, (e.g., you remember to use the '20'), you will likely develop a graphic understanding of Gyakugamaeate (#3)!

Next, use your arm vertically against the back/outside of Uke's upper arm and experiment with nearly every technique you have ever practiced…

Advanced Connection Builders

I could go on about putting the broom and shovel into your hand instead of forearm, give you a bunch of drills, and some of you might even thank me.

Instead, I urge you to look inward to your own imagination and use the concepts to put the transmission into not just your hand, but any length of lever arm you can muster (e.g., a foot, shin, tanto, rattan, anything rigid that you do not actually throw at Uke can be part of the study.)

Most importantly, if you are already happy with your results in class, simply keep doing what you are doing in your practice, change absolutely nothing about it except your awareness as you figure out where and how you are already using these levers.

Here is a method to get there.

Ask yourself, "How would I explain what I just did using a broom and/or shovel?", "Where is the Piston Line?", "What kind of 'cut' did I use?", and/or "Where is my fulcrum/transmission in that action?"

The next step after that is to share your experience!

"Wait! What about using the sword like a paper cutter?!?"

We call that cut the 'Leaf Cut', for it mimics the way a leaf might fall.

In our analogy, the leaf drifts from side to side as it falls. In each pass it reaches a limit where it finds a pivot on the lower forward edge; and then the higher rear-side drops. The now lower rear edge becomes the forward edge as leaf then cuts through the air in the reverse direction to find a new stop, a new pivot is formed on this lowered frontward point, and the pattern alternates again and again, side-to-side, until the leaf reaches the ground.

The important points to take from this is that 'the leaf never falls upward'.

As the leaf zig-zags, the lead point stops and becomes the fulcrum for the rest of the leaf to 'wipe through', much as a windshield wiper might.

The 'lack of retreat' seems obvious, but sometimes it is not so obvious when mimicking the action within a technique or within your arm, bokken, jo, etc.

This 'Leaf Cut' is an example of using your martial device as a second-class lever. Our most common use is as the impetus to create our 'shield' (Bring to mind how we 'assert our transmission' to stop Uke from using our forearm as a rail.) Thus, the second-class lever is often how we first assert the 'bite' of our blade, but the action quickly becomes a 'Front' or 'Rear' cut.

I personally like to think of the leaf cut as 'half a teeter-totter' but that is a much harder analogy to explain (and I admit there is not much advantage in the understanding, but it was how I first perceived the action.)

Technically, the Leaf Cut is not an example of a broom or shovel, so it was not a focus of this section. That is the primary reason it has not received the same kind of focus; that and the fact that you will be working with the 'Leaf Cutting/second-class lever action' plenty. It does not need much more explanation than it needed in 'creating your shield'.

Our next step in this book is to cover something new! Let us cover one last secret to becoming the inclined plane and highlight a core simple machine used in all of Aiki: Wheels (pulleys), Cams, and Tomoes.

Wheels, Cams, and the Tomoe

Lots of people become afire when they apply circles, squares, and triangles to Aikido, but pound for pound, the two best shapes to describe the mechanics of Aikido are the Tomoe (yin or yang shape) and the 'tear drop'/'Egg' (or from here on out, called the 'Cam'; ala a Cam as used in your car's Cam Shaft or, more advanced, like the Cam in a compound bow).

Before we can describe the action of these marvels, we need to recognize them as extensions of their base component: the wheel.

A wheel can be used as a lever. When we use a wheel as a lever, we typically call it a pulley, other times we call it a 'roller' (an apropos name!)

A pulley can be viewed as an infinite number of teeter totters (think of having so many wheel spokes that the rim becomes solid), each sharing a common center, and each teeter-totter stepping up when it is their turn to add a little lift and carry the object it is touching around the pulley's circumference.

A single pulley does not give anything in the way of mechanical advantage (or disadvantage!), but it is a great way to change the line/direction of force.

The change in direction is realized as 'pulling downward on one side of the pulley lifts the other side'. (Call to mind 'The Belts' from 'Six Precepts'!)

Aikido uses the modified wheels called Cams and Tomoes to move Uke.

Advanced Connection Builders

Some will insist that Aikido mechanics are like gears.

That is close, but the reality leans more heavily toward augmented wheels/pulleys.

I was in their 'gear' boat, and it took me a while to get past that 'gears' notion, but once you comprehend how a serpentine belt works in a car, you come to realize how extension connection works (and why too much or too little tension ruins your car and or your technique!).

Could you argue a bike chain? Sure, but show me the continuous 'teeth' that interlock between you and Uke. Gears without teeth = wheels.

If you could guarantee the friction, you would not need the teeth, but alas, wheels tend to slip when slippery. Which is why grappling does not work well with sweaty people!

Disgusting, but Aikido is a grappling art, and the flaw argues for wheel/pulley vs. gears.

You also begin to understand the importance of compression connection when you wrap your head around how a capstan and pinch roller are used in a cassette player. (There is a legacy reference! Look it up youngster!)

I guess you could contemplate a baseball pitching machine, in particular, the version that has only a single wheel. The wheel 'rolls' the baseball against a flat metal 'plate' (which in turn throws the ball). And that is the point, one object rolling the other due to frictional contact. Again, too much, or too little pressure and things go awry; sweaty people again need not apply!

Where the wheel is insufficient for our Aiki(do) needs is that it presents a consistent force for Uke to resist.

For all intents, walking along the top of a ball, is not much different than walking on a treadmill. A ball large enough to walk on is not quite as flat as a treadmill, but a consistently positioned, somewhat rounded, virtually level surface is presented to the pedestrian in each step. (Top of Figure 103)

We prefer to make Uke contend with inclinations and declinations.

So enters the 'Tear Drop'; i.e., 'the Egg', or otherwise known as a Cam.

It is not much more than a wheel with a lobe on one side.

Rolling

The utility of the Cam rotating is that instead of keeping the even plane as our perfectly round wheel does, the Cam's 'lobe' area 'elevates' and 'lowers' an object resting on top of the Cam as the Cam rotates.

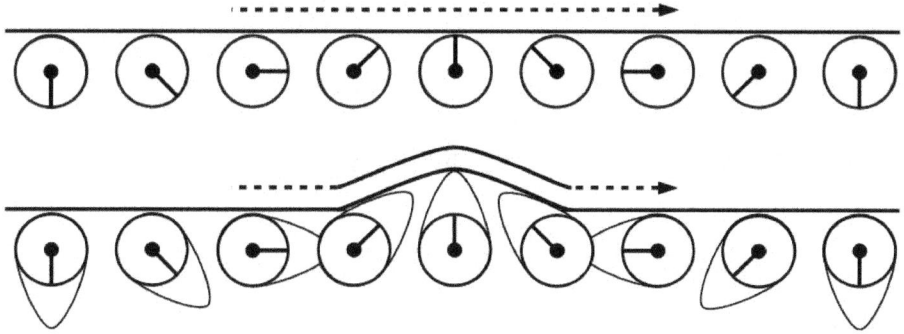

Figure 103: How a wheel (top) and Cam (bottom) affect the path of an object it supports as they spin (not to perfect scale)

A person walking along the top of our Cam would experience a period of 'flat' ground while walking along the hub/wheel of the Cam, but would then have to walk 'uphill', then 'downhill' respectively has the Cam's lobe came around. (See bottom of Figure 103)

What are 'uphill' and 'downhill'? Inclined planes of course!

Auto engineers have known this for a long time, and is why they have been using Cams as part of the mechanism to lift and lower the 'lifters' in an engine which ultimately open and close the valves in the car's cylinders. (among other uses for Cams in your car)

Go ahead and look up how your car's Cam 'lifts' (and lowers) the valves of your car, or you can use this analogy instead…

Get a hula hoop, place it on your forearm, place your fully extended arm horizontally, directly out to your side, and then spin the hula hoop around.

Notice how your arm is the true 'hub'; i.e., the 'wheel' portion of the action, and the entirety of your hula hoop is now the lobe. How so?

A person standing in front of you never gets any closer to your arm, but the hula hoop vacillates from closer to, then further from that person.

Careful or you might hit them!   Hmm… That is interesting…

We could use the Cam action to strike Uke, but for the sake of Aiki, would we not rather have our Cam (or Tomoe) 'roll' Uke?

How do we become a Cam that rolls instead of strikes Uke? We take a lesson from Pinball! (and steal a bit of wisdom from bowling too!)

Ever since I was a young boy, I have played the silver ball.

I am no 'Wizard' at it, but I learned a thing or two.

One of them was the difference between 'hitting' the ball with the egg-shaped flipper and 'slinging' the ball as it rolled along the flipper.

Striking the ball is percussive.

The rubber on the flipper helps to drive the ball away with a bounce.

The action is very akin to a 'chop' when it sends the ball directly backward from whence it came, and more like a 'slice' if the action was not purely perpendicular.

The 'slice' would drive the ball away at a bit of an angle and even give the ball a bit of spin.

Both of those actions have their utility, but they are not our focus.

The special case was when the ball would ride along the side channels (the in-lane) and was directed to roll down across the flipper (toward the exit, otherwise known as 'the drain').

The ball would roll past the 'hub'/'wheel'/'pivot' part of the flipper and onto the rather pronounced 'lobe'/'arm' of the flipper.

Since the ball was already in contact with the flipper (<-- important!) when the flipper was engaged (rotated at its hub/base), instead of 'hit' the pinball, it could 'sling' the ball toward the targets; i.e., the flipper added to the ball's spin as it sent the ball outward by working in concert with (adding to) the energy and direction that the ball was already 'rolling'/'spinning' toward.

It had to be 'just right'.

It could put you in mind of how a bowler releases the ball toward the pins.

Rolling

A bowler's hand puts enormous rotation into the ball as it is 'slung' directionally toward the pins.

We see this effect in Aikido when we assume the shape of the Cam and 'take the space' between ourselves and Uke. If we are good, instead of striking Uke and 'knocking' them sideways, we 'sling' Uke away from us.

This action gives Uke a strong desire to 'roll', and in many cases, we actually turn Uke with this action as we project them away.

That turning 'turns' into (pun!) tension in Uke's body! (i.e., structural attack)

So how do we 'sling' Uke?

Let us start in a familiar position: face to face with Uke, right foot forward cat-stance, but instead of 'wrist to wrist', let Uke point directly at your nose as you bring your right arm underneath theirs in a 'half an X-block' position. (left side of Figure 96 again! Or left side of Figure 104!)

We could act 'percussively' and 'block'/'strike' our arm against Uke in attempt to bounce Uke's arm sideways.

Yes, we could just as easily perform the broom action again, and that would be fine as long as we include a bit more focus on not just moving Uke rightward, but also rolling Uke in that direction (a.k.a. repeat the drill in Figure 96 with a touch more focus on Figure 85)

Let us do something even more subtle and interesting. Let us, instead, keep the focus on 'slinging' Uke backward as best as we can.

From this starting position, keep your unbendable right arm in place (optionally, create a 'lobe' with both arms, with your left arm helping keep your right arm in relative position with your body; i.e., turn yourself into a pinball flipper!) and prepare to rotate your hips rightward, clock-wise.

Create a shield and connect to Uke with compression connection.

This means assert your right arm's transmission toward Uke's COB just a little right of the center-line, and have the Conjunction Point (CP) just a bit to the left). Be sure that the CP is on the lower or 'hand side' of your transmission; i.e., your transmission is between CP and your elbow. CP and transmission cannot be on the same spot on your arm. See Figure 104.

Advanced Connection Builders

'Sling' Uke, and to help you, here is a visualization that seems to help create the proper 'one wheel rotating the other wheel'/'You rotating Uke' feeling.

Imagine, just as in pinball, a ball rolling along the back of your right arm, rolling down from your shoulder, over your elbow and then toward the CP.

When your imaginary pinball reaches and aligns with the CP, begin your clock-wise hip rotation and allow the CP to roll toward your hand in synch with your imaginary pinball.

Success is when we drive Uke back deeper into 'their side' of the aforementioned 'Mason-Dixon Line'/Waterfall between us. Extra points if you can actually get Uke turned a bit.

The measure of success is that Uke is rolled back and away from you more than they are sent sideways.

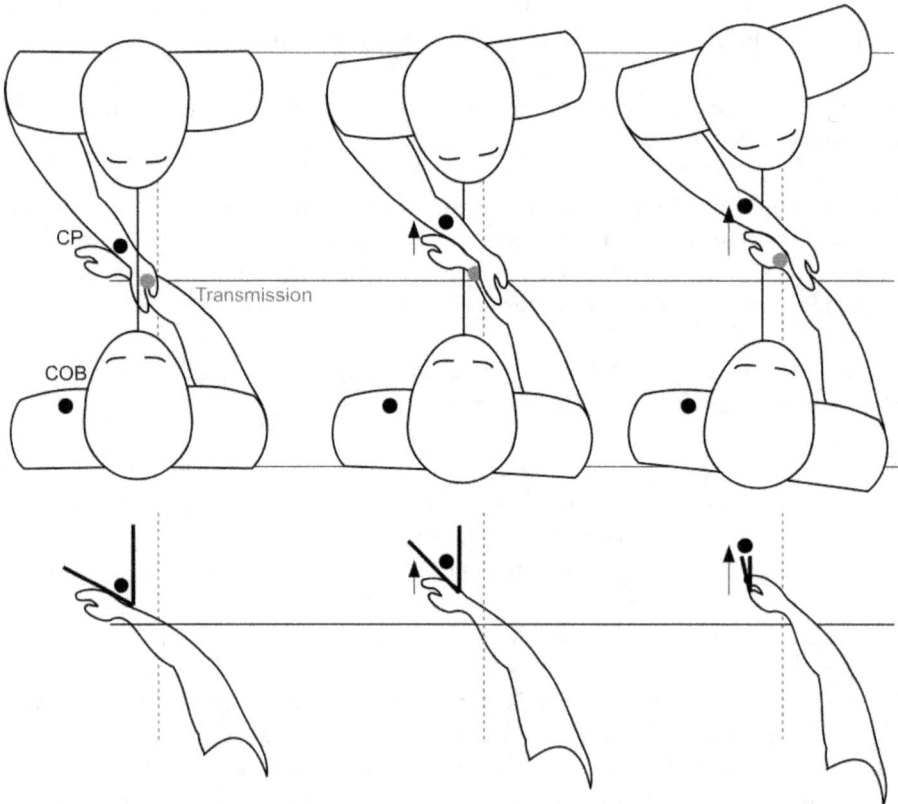

Figure 104: (Top) 'Slinging' Uke back, (b) closing the 'Maw'

Rolling

This is a very subtle way to impart force upon an object.

You are leveraging the effect/flaw in a 'slicing' action. We are applying an angled/inclined blade and the CP is squeezed/rolled out of the situation; much as a bottle would be projected out from under an angled blade brought 'slicing' down upon it. (ala in the earlier discussion about how the slice first attempts to 'roll' the object, but in this case, we desire the lack of friction, no depression in our guillotine, no fork, we want the roll.)

After all, we are not actually cutting Uke, so the focus is on creating the rolling sensation in Uke as we project them.

This action is called 'Closing the Maw', because, from a top down view, it looks a bit like you are closing the mouth of a scissor by moving only one side. (see bottom of Figure 104.)

Can this be done vertically? Of course! (any direction for that matter!)

Great! Now, let us broaden our horizons even more by 'slinging' Uke toward us (before sending them back the way they came).

We call this action 'catching' Uke, and I suggest that, done correctly, you may just experience Aiki-Age in a new way.

Using a pinball analogy, 'catching Uke' is like when we 'catch' the pinball on a flipper by rolling the ball down the flipper so that it rests in the crook between the base of the flipper and the in-lane wall.

We learn to 'catch' Uke using a typical Aiki-Age demonstration, so allow Uke to grab your forearm just above your wrist (pick a hand, left or right).

Now move your transmission toward Uke's COB to create some compression connection, and then feel for the 'true pressure' in their grab.

What I mean by that is that not all Uke's grab with focus of being 'on-top' of your arm. Some put their effort under your arm and feel 'heavy' in their fingers. Still others feel heavy through their palm. Depending upon the position of their hand, this can mean their weight is applied above, below, or on the side of your forearm. This is the true conjunction point! (CP!)

Find the pressure, then, based on the position, allow that imaginary silver ball to 'roll' up your hand/arm starting from your fingers, up your hand, up your wrist, and on toward that conjunction point.

Advanced Connection Builders

When the ball reaches the true pressure of Uke's grasp (i.e., the CP), allow your forearm to pivot so that the CP 'rolls' in synch with the imaginary ball.

As you do so, drive your transmission around the CP and further toward Uke's COB/Sagittal Plane. (You will know this action later as 'drifting'.)

This can also be explained as creating the intent to allow the pinball/CP to attempt to roll up your arm, much like it was directed down one side of a skateboard half-pipe ('U' shaped ramp.)

It might be confusing to hear 'up your arm' is 'down one-half of a half-pipe' but the reason to explain it that way is that you want to reach a point where the 'up your forearm' stops and you then 'feed Uke's roll back to them'; i.e., as if rolling back toward Uke on the imaginary other half of the half-pipe.

Uke never leaves your arm, but the intention here is the trick. There is an 'inflection point' that you have to establish to create the desired effect.

Recalling the pinball analogy, this is like 'receiving' the ball and allowing it to roll down the extended flipper and eventually over the crook of where the extended flipper and the in-lane wall meet, then up the in-lane. (The flipper is analogous to the initial 'down-side' half of the half-pipe, in-lane area equates to the 'upside' of the imaginary half of the half-pipe)

This action can also be practiced solo, two handedly, by resting a jo across your two wrists, allowing the jo to roll toward your transmissions (created in mid-forearm), and then made to do hair-pin through the air, all the while spinning in the same direction it did as it rolled down your wrists. The jo should land on your wrist mid-spin and continue spinning, rolling down toward your transmissions. Repeat with jo always spinning in one direction.

'Catching the pinball' can be done on Uke even without a grab.

I.e., we can 'roll Uke to stop them' at the transmission point of our arm regardless of the type of contact. Uke senses the desire to roll, but cannot.

We can then maintain the compression connection and move our arm as desired; taking Uke with us. This is a lesson in 'sticky hands', but of course, done with your forearm.

I suggest Uke senses the effect described in Figure 99; i.e., To Uke, it is as if they are the ball with a flattened side made to hover between rolling over or returning to stable.

Rolling

These 'slinging' and 'catching' Uke actions can be likened to how a ball is caught in a lacrosse stick's net, so before you doubt you can do this at speed or at the moment of contact…

I am not saying it does not take practice, but it is real physics…

Now, there is a complement to these 'slinging' and 'rolling catch' actions.

'Slinging' and 'catching' are 'projections' in that they move Uke outward, away from us. We 'receive' Uke's force, but 'flip it' at the inflection point to keep Uke's force moving along, but ultimately away from our center.

The 'sling' is obviously a projection, but if you are having issues seeing the 'rolling catch' as a projection, you can think of it as if you were going to continue the 'rolling catch' to the point that you 'sling' Uke's force toward you and past your head instead of feeding them back to themselves.

Here is why…

The complement to 'projections' is acting like a 'vortex' which brings Uke closer to your COB.

In order to become a 'vortex', especially when creating Orbits, we need to switch from using our arms as a Cam/Egg, and emulate, instead, a Tomoe.

The Tomoe (half a Yin/Yang symbol) is a 'broken egg', an egg missing a piece of the shell. More accurately, an egg missing a piece of the 'lobe'.

We want to use the 'inside of the shell', the 'inside of the tail', the 'inside of the lobe' to act as a wave.

Picture in your mind a wave and how it moves a surfer.

The surfer cannot stand on water, so the board is there to displace and spread out the surfer's weight. A wave would roll the surfer, but the board additionally presents a shape that cannot easily be toppled (rolled end-over-end), but regardless, the pressure to roll end-over-end is ever present. There is not much friction for the board as it cuts the surface tension of the water, so instead of toppling, the board slides. The surfer puts a transmission of sorts into the board for both the wave and the surfer to put pressure on…

and the rest looks like fun!

Advanced Connection Builders

The relevant action to take note of is that the surfer is being moved horizontally the entire time by an ever progressing, concave inclined plane.

Yes, very much like a snow plow or when we move Uke with a forward 'Forward-C CAM' in our hips/torso.

For all intents, it is a shoveling action.

If the surfer analogy does not make sense, bring a ball to your bed and place it on your sheet. Lift your sheet and make the ball roll away from you. That is the base mechanic of the wave/surfer scenario.

To create this effect in our arm, we put Uke on the inside of our broken egg shape, a.k.a. the Tomoe! and again, twist our hips.

Try it for yourself by setting up as before: face to face with Uke, right foot forward cat-stance, but instead of 'wrist to wrist', let Uke point directly at your nose as you bring your right arm on-top of theirs in a 'half an X-block' position. (almost the same as the left side of Figure 96 again! This time with your arm on top of Uke's.)

From this starting position, keep your unbendable arm in place (you could again create a 'lobe' with both arms if you would like, it might help with keeping your right arm in relative position with your body) and prepare to rotate your hips leftward, counter clock-wise (from a top-down view).

Create extension connection with Uke this time by creating the transmission in your right forearm just a little to the right of the CP and then shift your weight away from Uke; (i.e., transmission is in your forearm in the space between the CP and you elbow; i.e., the thicker part of your forearm, closer to you than Uke. Once created, shift your weight away from Uke.)

You will want to use a 'catching the ball' action (Uke rolls up your arm, toward your elbow, down half of a half-pipe) as you turn your body CCW.

It may help to imagine a pinball starting to roll from your hand (top of the wave), along the inside of your arm, and moving toward your transmission (lower on the wave).

When the imaginary ball reaches the CP, turn your hips CCW and, although Uke will not physically be rolling, they should feel the desire to. (They should be stuck at this 'inflection point' that emulates the 'bottom of the U')

Rolling

By rotating your body as the 'hub' (in Tomoe terms, your body is the 'hub' area with the small dot of either Yin or Yang in it), your arm (the tapered 'tail' of the Tomoe, the 'wave') will catch and move Uke toward your COB.

I suggest that you allow your right hand to drop to your left hip as you perform the twist as it will generate more draw in less turn, and…

 if you truly have created the 'rolling' sensation for Uke, the twist should seem effortless considering you are dragging Uke along with you.

I deeply appreciate this last drill, as it draws Uke in an extension connection but is radically different than 'The Belts' we learned in 'Six Precepts'.

That is because the 'Belts' and this 'Vortex' are different 'simple machines'.

Moving Uke (the CP) 'as a wave would' is 'shoveling'. 'The Belts' are more akin to 'spooling' or 'wrapping rope around a pulley'. (That pulley is your COB and the ropes are the arms getting stretched around a COB. Like rotating back and forth with loose arms wrapping around you.)

Due to the fact that this Tomoe vortex (a.k.a. the 'rolling catch' along the inside of the Tomoe/broken egg) is a shoveling action, the 'rolling' sensation for Uke is inherent to its nature.

The Belts lack this rolling action. The Belts are inherently a 'tug-o-war'; i.e., linear tugging.

You should be getting the pattern down by now.

How about a drill to put 'rolling' into our hands?

Create a shield, create some connection, imagine the silver ball rolling along toward the CP, allow the CP to join the ball as it rolls by, create the inflection point (the bottom of the U-shaped half-pipe), and continue toward or away from Uke as appropriate to maintain the compression or extension connection.

I will bet you are expecting me to suggest doing a turning step to position yourself on Uke's outside as you placed your palm on Uke's arm to create the shield, create some connection, imagine the silver ball…

Great idea. Try it! You should be able to put together your own drill, but I have something else for you… a solo drill!

Advanced Connection Builders

Here is a quick solo drill that you can use when you are bored (and no one is looking. Please wait till no one is looking. You will look weird!)

Place your left forearm in 'half an x-block' across the front of your body; (i.e., as if Uke was in front of you, you protecting your center-line).

The important aspect here is that your forearm bones are not crossed. You will know you are in the correct position if your thumb is above your pinky. (As you would when sticking your hand between elevator doors to keep the doors open)

Start by placing your right-hand palm against the 'inside' of your left wrist (middle position of Figure 105) and prepare to push your radius bone away and eventually 'roll' your ulnar bone; (i.e., create a 'pinky-pivot')

Cup your right hand a bit and create your shield (transmission should be just above your palm heel), then use the pinball imagery to find that ball rolling down from your right-hand fingertips, over the finger pads, down over your palm, until it reaches the junction of your left arm ulnar bone and your right-hand palm heel.

As the ball finds its way to that point, push your right hand forward such that the left-arm ulnar bone is 'rolled' away from you (your left thumb moves away from you. Much like rolling dough with a rolling pin.)

Your right-hand fingers might still be in contact with your left-arm radius bone, and if they are, should be pushing your left radius bone (index-side bone) downward (depicted in the top of Figure 105), but it is also possible that your right-hand fingers lose contact with your left arm radius bone (not depicted). Either way, you should be creating a pinky-pivot (Kotegaeshi) type lock in your left forearm primarily with your right-hand palm heel. (again, see the top of Figure 105)

The benefit of this drill is that you can feel the effect of the pressure from two perspectives.

You should be able to sense if you are truly giving your left forearm a reason to roll, and at the same time, become aware of how it should feel in your right hand.

Worst case, you learn to shield with your right hand. (That is great too!)

Reverse the situation from this point.

Rolling

Begin again as we started before, (with your left forearm upright, left thumb over left pinky; as if keeping an elevator door from closing) and, with the shield in effect, allow your right hand fingers to drape over your left wrist a bit, then 'roll' your left-arm radius bone toward you; your left thumb comes closer as well. (bottom of Figure 105)

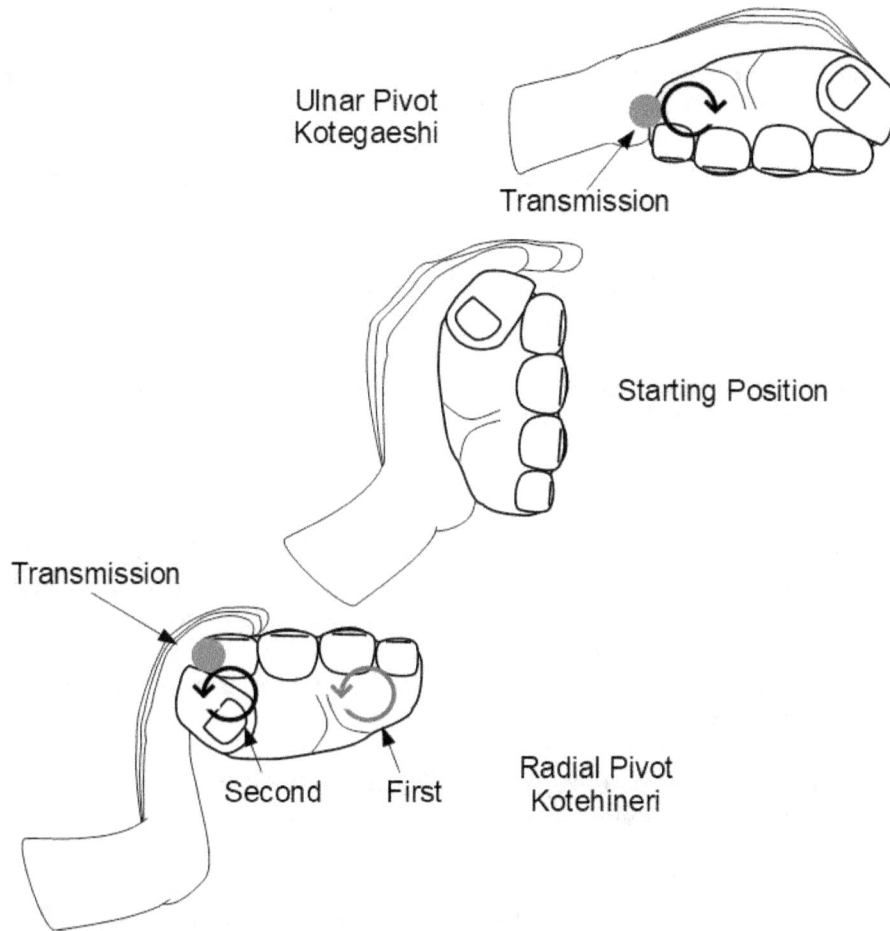

Figure 105: roll Uke in our hands solo drill. Starting position in middle, roll your thumb away from you, emphasize the ulnar pivot (Kotegaeshi), or roll the thumb toward you, then switch to emphasizing the radial pivot (Kotehineri)

It may help to use imagery of a 'pinball rolling up from your palm heel toward your fingers'.

Although this turning begins with a left-arm ulnar pivot, once your radius reaches the point it can roll no further, begin to concentrate on spinning the radius bone (index pivot) as you do in Kotehineri; (i.e., sharpen the pencil).

Practice liberally, but discretely, preferably without onlookers; then feel for the rolling when in class, when practicing with a real Uke.

Using your hand as the broom and shovel might not, on the surface, seem like a 'projection' or 'vortex' move, but once you have created Aiki (sent Uke 'Rolling') you will have plenty of opportunity to move Uke as you will.

Hold onto the 'slinging' (projecting) and 'creating a vortex for' actions of the Egg/Cam and the Tomoe for just a bit, for we cannot go to 'Orbits/walking the circle' (where you can even more creatively apply these concepts) until we put a close on Rolling and there is one more 'Rolling' concept we have to cover in this book.

Extended Shield theory/Rolling, use no effort
'Slinging' and 'Creating a Vortex' are subtle actions.

Let us instead try something very overt: rolling Uke can literally be 'rolling' Uke.

The solo drill with your wrist/hand that you just tried is an instance of this.

The solo drill teaches us the rudiments of creating the de-stabilizing effects of 'wanting to roll' used as a part of an effort to 'collect Uke's joints', also known as 'Torque Uke's Chain' or 'Joint collection'; (i.e., while performing the solo drill, did you sense the index and pinky pivots that you were creating in your left forearm. Are you mentally visualizing how that pressure can be used to apply Kotehineri or Kotegaeshi on Uke?)

How about instead of 'rolling Uke against our hand', we get a bit coarser grained and roll Uke's ulna against our 'forward C' pose?!

Grab Uke's wrist/hand for a typical Kotegaeshi technique. As best as you can, place Uke's forearm level across your solar plexus region.

Now, literally roll Uke's forearm down the front of you (your forward C) as if rolling dough. (See right side of Figure 106; or for a top-down view, go back a few pages to look at Figure 39, but envision Uke's hand against your body)

Take note that your body should be part of the 'partial obstruction', the 'concave inclined plane'.

The goal is to 'leverage' this 'mostly direct, partially indirect' pressure between Uke's wrist and your body; use it to 'partially support' Uke.

Rolling

The partial support will be part of the reason that Uke finds awkward, unstable balance, and best yet, you will find that you require only minimal exertion to provide such rickety support.

Therein lies the true secret to 'soft style' martial arts. It only feels soft to Uke because there is no direct pressure contact.

It feels 'soft' to you, but you are not 'relaxing'. 'Relaxing' is extremely bad advice in Aiki practice. Most interpret relaxing as 'mushiness' and lack of structure.

You became 'softer' by 'resting' (ok, maybe 'posing') in ways that partially support and bind Uke's body, but you lack extreme exertion. Your properly placed, rested, structure requires little effort to maintain.

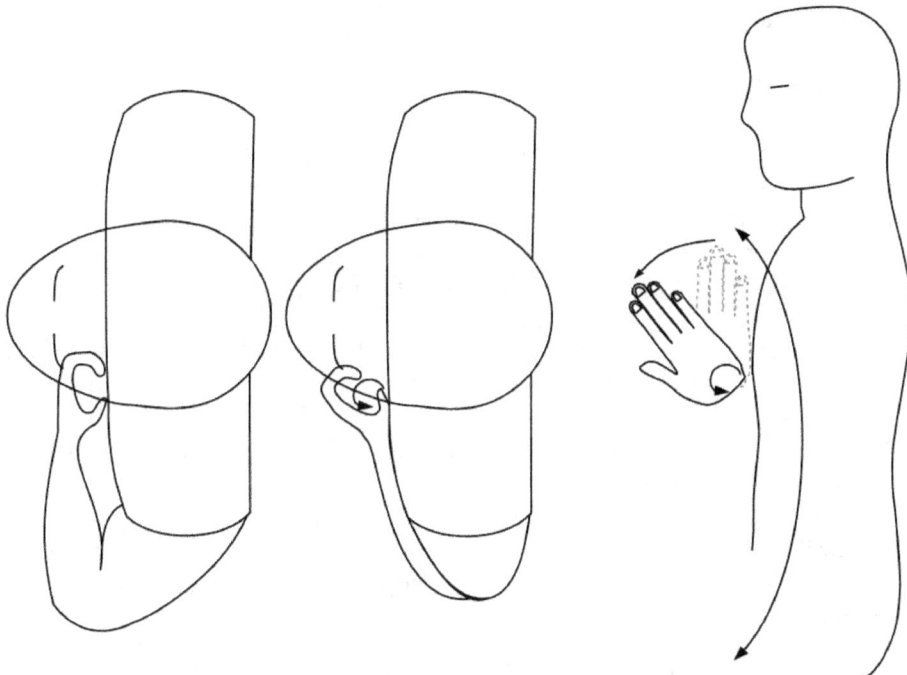

Figure 106: Rolling Uke along our bodies. On left, across our chest to create a Kotehineri, on right, Uke rolled downward along our forward C to create Kotegaeshi

That leaves us with a very important, but easy to forget maxim for binding Uke: Once you start binding Uke in a particular direction, keep going in that same direction or you lessen the binding; i.e., 'keep Uke's Chain torqued!'

Advanced Connection Builders

Starting to torque Uke's chain in one direction, and then reversing that action, typically unbinds Uke and gives them freedom of motion which often results in the loss of connection (in the same kind of way that shifting from compression to extension connection except at the North and South poles ruins the connection and we lose influence over Uke's actions)

Let us try tangibly rolling Uke's hand across our chest area to create Kotehineri before 'extending' the concept against an imaginary line; one that, although does not stay against our body, still keeps Uke rolling in the correct manner.

Practice this 'against us' Kotehineri by standing side by side with Uke, facing the same direction (No care on which of you is on either side of the other).

Have Uke place their near-arm elbow on your near shoulder with their hand hanging down over the middle of your chest.

Go slowly! Let Uke tap out.

Using either of your hands, (using your ligaments!) grasp Uke's wrist as you would for Kotehineri and roll Uke's index finger across your chest. (Much the same as when we 'rolled' our own hands on a table to learn Kotehineri, only this time, you are a vertical table, and you spin Uke's hand. You might have to tuck Uke's thumb to get theirs to roll nicely.)

This is depicted on the left side of Figure 106, Uke (not depicted) is to the left of you, facing same direction, with their right hand draped on your shoulder.)

Practice rolling Uke with one of your hands, and then the other. Then have Uke switch sides they are standing on.

The drill can be performed very subtly and can make you look like one of those 'moving without moving' practitioners; i.e., you do not have to move around a lot to get results.

Here is the 'extended' concept in this advanced rolling arena: you do not necessarily need a physical object to roll Uke against.

An imaginary line is all that is necessary. It is your ability to follow an intent that matters.

This imaginary line is very important, for it need not be a straight line (ala your chest in the last drill), it can just as easily be a 'hairpin'. (just like the

Rolling

hairpin we leverage when rolling the imaginary pinball along the skateboarder's half-pipe 'U' to feed it back to Uke.)

The 'hairpin' is especially useful when Uke is much taller than us.

Shorter Tori seem to find great pleasure in discovering that, if they cannot continue in a certain path due to physical constraints (e.g., working with much larger Uke's, ones that have larger ranges of motion than Tori can produce), they can simply double-back from whence they came.

The only stipulation is that they need to continue the 'roll' in the same direction.

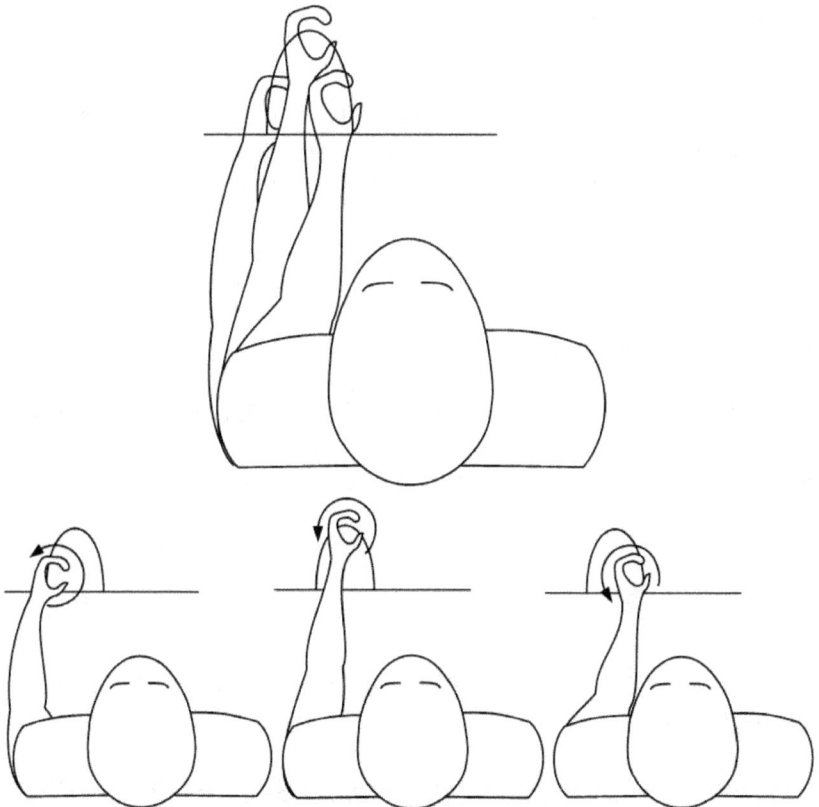

Figure 107: continually rolling Uke along an imaginary hairpin

To help wrap your head around how this works, find a video displaying (or buy!) a toy called a 'rail twirler' or 'whirly wheel'.

The wheel's axle is magnetic and clings to the hairpin shaped metal rails.

The wheel moves forward and backward as it 'rolls' always in the same direction. It is similar to a Mobius Strip, just without the twist.

Substitute Uke's appendage for the wheel's axle and keep rolling Uke in the same direction even when you 'double-back' against the hair pin path.

It will do wonders for throwing taller Uke's in Ikkyo/Oshitaoshi (#5).

See Figure 107 to see what is being asked of you (in a Kotehineri practice).

Let us put a close on this section and move on to our final section of Aiki…

Having Uke experience the 'impetus to roll' is without a doubt the ultimate secret of Aiki (control of stability and balance!)

'Rolling' Uke is the essence of 'unsure footing', and together with the complementary actions of 'reduced base' (aka the Pin) and 'weakened frame' (aka de-structuring), completes our 'Trinity of In-Stability'.

It is easier to throw/'break the balance' of an unstable Uke, so make instability (willowing) the base of all of your Aiki study.

That much being said (and that is truly a lot being said) …

So far, we have investigated the rotational/circular Aiki of our upper body.

How about we take this just a bit further before we end this book?

Let us 'round' ourselves out, and get circular from the ground-up!

Orbits

Let us agree that, as a general rule, it is typically safer to avoid having Uke's momentum traveling directly toward us.

Do not agree? Well then stop moving the centerline off of the line of the attack! (See! You do agree!)

What if instead of Uke moving toward or away from us (linearly), they began to run around us, to orbit us? (circularly)

Would something that was orbiting you ever run into you? Of course not! An object orbiting something goes around it, not through it.

Orbits

Toppling an Uke that is orbiting us; i.e., 'running past us', would be easy!

We could trip, clothesline, and/or twist one of Uke's body parts to have the action result in their toppling. (or roll them against our body! etc.)

We can make Uke run around us! It is not due to Uke-do (i.e., Uke making us look good), it is due to something we call 'Orbits'.

Orbits are a key element of a Chinese martial art called Ba Gua, and is one of the secrets to 'circular Aiki' (if it is not part of Ba Gua, then I have discovered something new, or more likely, the credit is in the wrong place, but this is what I sense when I watch the Ba Gua practitioners I seek to emulate and this is definitely 'circular Aiki')

Orbits are displayed in the motion of the planets, in high-speed vehicle races that involve turning, and in siege weapons (to name just a few).

Ultimately, orbits perfect the motions of both your upper and lower body. Below, orbits are the secret to perfecting your turning steps, and above, they amplify your Aiki Age/Sage and can save you when the 'rolling' does not seem to be working.

I have yet another toy/prop to explain the phenomenon, so you know this conversation is going to be fun as well as informative!

Normally, I would give you a bit of something to think about before getting into a drill, but this time I want you to feel it before we try to understand it.

The Sun, the Moon, and Tori

Draw an approximately two-and-a-half-foot diameter circle on the floor (e.g., trace around the edges of a Bosu Ball or a typical trash can lid).

Stand next to the circle and prepare to walk around it. Invite Uke over.

Grab Uke by the hands (using ligamentary grasping preferably), completely relax your arms (I do mean relax, not resting this time! Mushy arms!) and walk backward around the circle.

In shorter than three steps, you should start to feel Uke begin to launch outward.

It is incredibly powerful and requires no more effort from you than to be able to walk backwards just a bit faster than Uke can move forward.

That 'launch' is Uke starting to 'orbit' you, and you can make it happen in just a weight shift once you understand it. (You will not need three steps.)

Drill over, but if the sky did not open up, beams of light shine down, and a choir of angels start singing, you may have missed it.

Let me explain what is happening without any mathematical equations (for admittedly, I do not fully grasp the detailed physics, but luckily, the true math is not necessary for learning how to use the 'orbit effect' in Aiki.)

You have heard this before, but the same caveat applies…

Pay some attention to the 'folk science' you are about to read in effort to try and understand the mechanics of what you are doing, but if you 'get it', do not try to use these explanations to pass any physics exams!

Remember! I am saying, 'No pressure' to fully get this…

Do not worry if I fail at getting the cognitive aspect across to you. All you really need to know is what you felt in that drill.

Begin…

Let us get started by relating this to the heavens: the Sun (circle you drew on the floor), the moon (Uke), and Tori; (i.e., the Earth; i.e., you)!

For the sake of analogy, what you just achieved is influencing Uke to 'orbit' you, as if they were your 'moon'.

The basics are thus: as you accelerated into your own orbit around 'the Sun' (i.e., increasingly moved a bit faster each instant as you progressed around the circle), Uke's mass was compelled to assume the role of 'the moon'.

Why?

Think of your walking path around the 'Sun/circle' as a race track.

At the start, together, you and Uke represent a singular car, with you as the front axle/wheels and Uke as the back axle/wheels.

Initially, your tugging on Uke's arms pulls Uke along with you and you both share the same lane. (You are the front of a front-wheel driven car! Even though you yourself are walking backwards.)

Orbits

You have created a synthetic gravity between yourself and Uke. We call that synthetic gravity 'extension connection', for, if not for your arms, Uke would have flown away. (just as the moon might fly away if not for the gravity between the Earth and Moon. There is a balance to be found here.)

What happens next is that, since your COB and Uke's COB are not actually the same object, and since your COB is accelerating and Uke is not keeping up, Uke's COB is 'forced' to shift into an outer track. (As I understand it, it is due to a 'conservation of energy' thing, but enough said on that).

That 'forced shift into the outer track' is the 'lightness' you felt as Uke 'launched'. Mission accomplished!

Uke is no longer running in your lane, and thus, not directly toward you! (Racers/drivers in different tracks/lanes never collide!)

Uke, as the backside of your car, has begun to 'spin out' beyond and around you.

For auto racing fans, Uke's launch is called 'drift' typically resultant from over-steering.

From a top down view of the situation, you can see the Uke/Tori car 'drifting' as it continues in an arc around the track.

That drift is ultra, super, hyper, turbo, mega, meta, uber, alpha important and we are coming back to it, but let us first finish the story before focusing on the drift.

What happens as Uke 'floats' into that adjacent track is nothing short of a miracle of the natural world.

As Uke's COB moves into an outer lane of the "race track around the 'Sun'", the extension connection between us begins to increase in influence and Uke is instead launched into a 'race track around us'.

When this happens in a celestial setting, there is no friction in space, so the moon can make complete orbits around the earth for nearly ever. (Maybe, at one time, the Earth and Moon were in a race around the Sun and this is how they got in their situation? The one with more mass took over as the center of a unit? i.e., Earth as nucleus, Moon as electron, gave up the race and are now moving as one atom around the Sun? Was the Earth running a race with the Sun at one point and the same thing happened? I digress…)

Advanced Connection Builders

Here on Earth though, we have friction. The car will drift only until the turn ends.

When it does, the 'drifting' force diminishes, friction becomes relevant again, and the car returns to normal operation.

It is the same in our techniques. The 'drift', cannot last for a full loop around us, but we will not need it to.

If that makes no sense, there is no need to unravel the description, just rely on the drill and on what you felt when Uke went 'light'.

All we need know is that, by walking a circle, we can influence Uke to 'drift' and 'orbit' us.

In doing so, Uke is no longer moving directly toward us, but around us; running past us, and…

Uke running past us makes for big fun!

Consider your ability to disrupt Uke as they move around you.

Here is an idea using our old friend the 'elephant arms drill'…

Instead of using the elements of 'albatross and cheat' to make Uke 'roll into' a path across us, let us instead use their influence to make Uke 'roll away' or 'roll toward' us? (i.e., tilt our hands and move our hands outward or inward using the elements of the albatross and cheat, Uke rolling into a void.)

Our new 'toward and away, albatross and cheat' would encourage Uke's path around us to expand or contract; to spiral out away from us (like a hammer throw or a rocket breaking free of gravity) or be caught in a vortex that wraps itself ever closer to us, much like a string on a Maypole or being sucked into a whirlpool.

Would Irimi-Nage not become effortless if, as Uke passes by, we slow down Uke's upper body while at the same time allowing their legs to launch? (Should slowing Uke's upper body be performed with the Tomoe (wave) version of our 'catch the pinball' rolling drill in our arms? You bet!)

Could we wrap Uke around ourselves and 'roll' them down our torso?

Would any of this take much exertion on our part?

I have watched persons perform Irimi-Nage many, many times, and now with your new awareness, you too will recognize who is doing this effortlessly and why.

The magic is in the 'launch', the 'drift' you produced in Uke as you walked in a circle backward.

In the case of typical Irimi-Nage practice, Tori begins by walking the circle in the same direction as Uke, but then Tori quickly turns around to apply the throw. (We can consider that a type of Ba Gua 'Palm Change')

Although Irimi-Nage often starts with us facing in the same direction as Uke, and our drill starts the opposite way (with us facing Uke and walking backward), the same principle of Orbiting is generated by our moving around the 'Sun', not by which direction we face.

Yes, we could even do it sideways. Soon enough you will, once you build the ability to create the orbit in just a weight shift versus a few steps.

The 'Launch' within the 'walking the circle' orbit is itself very disconcerting to Uke. It will steal their ability to control their motion. In a sense, it is a loss of footing, thus, it is a method to de-stabilize Uke solely with our lower body! (technically, without using our hands! [Our hands/arms were limp])

In Aikido, 'Look Mom! No hands!' requires that Uke is intent upon holding onto us (e.g., grabs our Gi and will not let go). It is not outside the realm of reality that Uke has that kind of intent.

Many a wrestler or Judo practitioner has this kind of intent when setting up their attacks. We can take advantage of that intent by using nothing more than our 'Zhan Zhuang' and 'walking the circle' to generate the 'orbital launch' and legitimately throw Uke. (Real Aikido, not/no Uke-do)!

It typically means we will have to step across or in front of Uke, or we might even have to bend over at the waist to get the throw (odd/non-traditional Ba Gua 'Palm Changes'), but hey, no hands!

You might be thinking, 'Can we create orbits in our upper body as well?'

Sure! As below, so above! (reversed it!)

Your hands can create 'drift'. The action mimics a siege weapon called a trebuchet. It is a special type of catapult used in castle siege warfare.

Even if you have never seen a trebuchet before (go to the Internet!), it is easy to wrap your mind around how it functions, for, at its root, the trebuchet is just another teeter-totter.

We know all about those!

The part you can easily guess is that the person firing the trebuchet performs a sudden drop of a whole bunch of weight on one end of the trebuchet arm.

Well, when that happens, what is the other end of the lever doing?

Right! The other side is lifted upward in what direction/shape? A circle! (ok, maybe better described as an 'arc')

Just as you dragged Uke by stepping in a circle, the top of the trebuchet drags a sling (with an projectile in it), and again, just as Uke did, the 'pocket' of the sling (that they loaded the projectile into) starts to 'drift' and accelerate into an orbit around the end of the lever.

There is a point where the sling is released and the projectile is launched.

Cue the heavens opening, beams of light, choir of angels, let there be Rock!

We can replicate this miracle, by dropping (or raising!) our weight on one side of a transmission.

We learned how convert our forearm into a first-class lever by creating a transmission within. Dropping our weight onto our elbow will create an arcing motion in our hands that will begin to drag Uke's arms and torso as if a sling; their COB nestled in the bottom of that sling. Uke will be urged to orbit our hands and we can throw them in any direction we select.

The trouble would then be to decide whether to throw Uke into the ground, into our 'Sun', or just as wildly, spiral them outward into space!

Sure, there is still the need to make Uke unstable, rolling them correctly, and a few other nuance, but the beauty of this trebuchet analogy is that the real siege weapon creates this 'orbital pull' within a very short few degrees of turn (less than a half of a circle, less than 180 degrees; i.e., less than from 12 to 6 on the clock), and we can too!

In other words, we do not need a full cut in our hands or three steps to generate this force in our body.

Let us start by drilling how to create orbits from our lower body in as little distance as a weight shift.

Walking the circle

You might think that walking in a circle is simple and holds no secrets, but alas, there are secrets.

The first is that not everyone can do it without some practice.

You may be shocked to find that many cannot walk backward in a circle without some sort of guide to keep them aligned.

Remove the trash lid, bosu ball, erase the chalk from the floor, or whatever you used to represent the 'Sun' that guided you as you walked around in a circle backwards in the previous drill and watch the inability ensue.

The ability to walk in a circle both forwards and backwards without the help of a prop must be mastered before bothering to move forward to the next drills. I strongly suggest that you learn to walk in a particular direction around your circle, e.g., clock-wise, and learn to switch between walking forward and walking backward at random intervals.

When you can do it clock-wise, perform the same drill moving counter-clockwise. Again, you may find weakness in one direction or the other, especially when changing which direction you face (forward or backward).

When you get good at these basic steps, feel free to learn to change directions between CW and CCW, backward and forward, at will and still move in that circular pattern.

Captain Sarcastic says, "Thank you, Captain Obvious!", but what is not so obvious is that what is really being taught is nothing more than trying to get you to walk the circle like it is a round sphere, and not an 'edgy' stop sign.

Therein lies the confusion, for although you may be stepping with your toes pointed outward or inward, you may still be moving in straight lines.

There is nothing straight about circles/arcs.

Additionally, do not confuse walking the circle with 'Bamboo stepping'.

Bamboo stepping is just swinging your feet as you walk. Your hips, and most importantly your COB, move linearly in Bamboo stepping. Instead…

Advanced Connection Builders

Remember learning to drive? How driving around a bend started out as a number of jerking motions on the wheel, until you got the hang of it and it became a smooth, constant pressure that kept you in synch with the curve?

The smooth transition is how walking a circle should feel, at least at first. A smooth path around the 'Sun' (trash can lid, Bosu ball) in lieu of a bunch of short straight lines; (i.e., No edges, walk a circle, not a stop sign.)

Walking smoothly around the circle is where some stop, but the second secret is, that, to gain the full martial benefit, you cannot just 'step' smoothly, you have to generate the 'drift' in your hips.

Yes, what I am asking you to do is to feel yourself 'spiral out' a bit as you walk that circle. Have some fun with it!

The 'spiraling out', 'drifting', 'oversteering' has a significantly different effect on anyone trying to impede you or trying to move along with you.

The drift in your hips is drastically different than pirouetting.

In a pirouette, we stay posted, COB in a constant position horizontally, as we rotate it (we spin). Yes, it is very powerful, but it cannot generate much more than a 'Spooling' effect where Uke wraps around you (if you are lucky and do it 'just right'), and it is extremely difficult to get something that weighs much more than a small fraction of your weight to orbit you as you stand in place.

To get a pirouette started, you will likely have to swing your arms a bit first, and Uke is not going to let you do that to them.

Even hammer-throw practitioners move. They spin a great number of times per linear foot of movement, and they certainly are not pirouetting.

What they are doing is more akin to those 'gear on gear', 'move a circle inside a circle with a pen' toys used to draw complex roulette curves; (i.e., graph spirals). (oh, the fun you will have if you look up how trochoids are drawn!

View the trochoids as what defines the difference between the linear force of Hsing-I and the rotational force of Ba Gua.)

Mostly, I hope you begin to see how they can be used to generate Aiki!)

Besides, it is not martially sound to stand in one place. You will likely be run over before you get Uke into the orbit with a pirouette.

Orbits

My point is that pirouette's do not generate Orbits easily or quickly. You want your weight shift to do the work.

Drill creating orbits by getting a durable bag and put a small bit of weight in it. A couple pounds at most. Preferably, a soft, but weighty object (e.g., orange or something wrapped in a towel)

Begin walking the circle again, with a bag/weight combo in hand (or hands), and as you proceed around the 'sun'/circle, put some drift into your hips and feel for the weight in the bag to begin to spin out.

This practice is not much more than a variant on the original walking backwards drill where we draw Uke around the circle with us.

The replacement of Uke with a small bit of weight in a bag (mimicking the sling and projectile in a trebuchet!) allows us the freedom to do more elaborate 'spinning' to elicit the orbit. Much more freedom than using a person.

You should realize quickly that it is difficult to maintain the orbit for more than a few seconds, or even just an instant at a time. Do not be frustrated, this is good and how nature intended it.

Feel for those snippets of success and pay close to attention to how you had to move to generate the orbit; e.g., which direction around the 'Sun' you were moving? CW or CCW? Which way were you facing? Forward, backward, or shifting from one to the other? Which way did you 'spiral out'? Left hip or right hip moving/rotating your body CW or CCW?

The beauty of the 'short burst of orbit' is that it allows us access to the last secret of Orbits I have to offer: we can grow or shrink the 'Sun'/circle.

Consider this, "just how big does that 'Sun'/circle really need to be?"

We entered our first drill with a circle the size of a trash can lid, but can the circle be smaller? Yes!

Can the circle be larger? Yes!

Hmm... "How about so large that the edge of the circle is practically straight?"

Yes..! "Huh?!?! Then how does the circle matter at all?"

Advanced Connection Builders

It does not!

It is all in the hips! (When you understand what was just said, you will have had one of the epiphanies that evolved Ba Gua into Tai Chi)

Reveal this to yourself by practicing orbits within a weight shift.

Go get that bag and small amount of weight again.

Again, I strongly suggest a small amount of weight and something that is not hard/rigid, for we will do things wrong and you do not want to swing a hard object into yourself.

If it helps, I use a towel, put the object in the middle, fold/twist it in there so it will not fall out, and then grab the two ends. (You have seen this done in many a martial arts film, typically with a pool cue or other hard object. Skip the cue and use something a bit softer)

Stand with your right foot forward and move all of your weight onto it.

Hold your bag/object with two hands and let it hang in front you.

The plan is to do two weight shifts, one to the back and onto your left foot, then immediately into a shift back to return your weight to your right foot in front. (Down, over and up! See Figure 108)

The trick is to bring the 'drifting' sensation you had while practicing walking the circle drill with the weighted bag, and recreate that sensation as you 'bow out' your hips as you make your weight shifts. (Bow as in 'bow and arrow', not bow as in 'bow to your partner')

If you are not doing this correctly, you will find the bag and weight simply rocking away from you as you shift backward and then slamming into your knees/shins as you shift forward.

To avoid this, you should take a more curved path.

In this case, we want to have our hips bow to the left of the line between the arches of our feet as we shift weight backward to our left foot, and then bow our hips to the right of the line between our arches was we perform the weight shift that returns us to our starting position. (Figure 108 again)

Do not forget your 'down, over, and up'!

Done correctly, that weight should swing around us leftward, CCW from the above us perspective as we weight shift backward.

We truly succeed when the object in the bag makes its way completely behind us, if not all the way around us as we weight shift to the front. (the distance the weight travels around us depends highly upon how flexible and how low the bag/towel hangs. Obviously)

See Figure 108 to see the path of the COB in reference to the line between our feet.

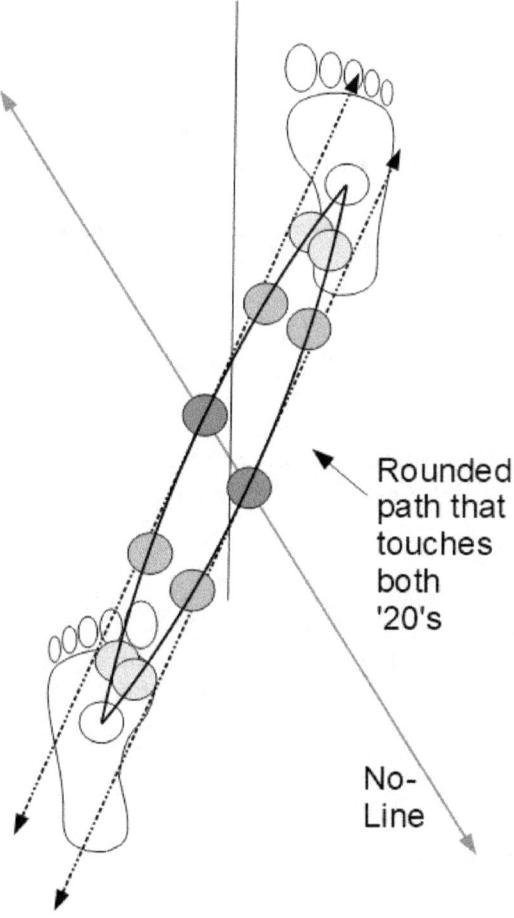

Figure 108: the ellipse you COB makes when you 'bow' outward as you shift weight from one foot to the other

The path of the COB is very much an ellipse; i.e., a very small, football-shaped 'circle' around a straight line.

Advanced Connection Builders

We do not always have to do a full circle.

You should have felt the orbit in both weight shifts. It was important to see the ellipse, the 'hairpin' that you can create to return to the original COB position on the front foot.

These actions equate to a 'step without stepping' and serve a purpose of learning to break free of having to root your COB in place to generate the 'spinning core' that we desire in Tai Chi.

We want both the mobility and the 'body that rotates as if posted'.

Are we still 'walking a circle'? Definitely!

We have the ellipse itself, but arguably, additionally, we are still walking around an extremely large circle around a 'Sun'. An extremely large circle looks like a straight line when you look at its edge. Our hips are 'lacing' arcs that return and leave the edge of the giant circle as we progress that circle.

Again, trochoids! Look them up.

For more martial fun, draw squares, or triangles on the floor as most Silat and Filipino Martial Arts practitioners might.

Each line drawn can be seen as short segment (tangent) of a large circle, and we can 'wrap' our COB's elliptical path around every one of those lengths of line.

That equates to a multitude of ellipses, or even half ellipses that our COB can follow in just a short weight shift and represents infinite opportunities to include Orbits in a combative way. Trust me, the best of the Silat and FMA practitioners already know this, they just do not call it Orbits, but they know what they are doing.

And with that thought…

    if you can believe it…

        we have reached the end.

Just as before, there is so much yet to share, but we should let these thoughts settle in before moving on to new ones (next book!).

Closing

So, there you have it.

Combine Orbits with the practice of moving 'down-over-and-up' while rolling, cutting, slicing, broom-ing, shoveling, shielding, albatross-ing, cheating, and hour-glassing Uke into the 'not-so' basic boxes and teeters to get them 'checked'. Be sure you have Presence as you 'put on your Zhan Zhuang shirt', 'grasp' Uke using your ligaments, apply a 'fork and knife' with your unbendable arm replete with a transmission, and watch Uke 'fall off the fence' created by your Hypotenuse, as Uke moves along their '20' and into an Aiki Joint Lock, that goes through their body from the CP on their wrist and all the way down to the ground-up locking, as you launch them toward a 'Dog spot' with a bit of orbit in your hands that sends them into the finest of aerial falls. Be sure to ask the observer in your study trio how it looked!

I am sure you made it look easy…

That sums up about a decade since we last talked about the Precepts.

There are many things I wanted to add, but again, at some point, you have to finish.

I hope this book helps you think through your performance plateaus, and maybe even takes you to a couple of places you might not have been to yet.

May it help you find common ground when teaching or exploring with others.

I also hope it helps inspire you to feedback constructively with your own concepts, most importantly, in words and pictures! The world does not need another video of 'do as I do'. If you do make videos, please convey with words what you are focusing on.

I know I said we have to stop at some point, but please feel free to check the appendix for some notes on what I might say about each section after we had reviewed the concept at least once. I could not help myself. It may be a while before the next book.

Yours in martial arts brotherhood,

Spiral out, keep going…

Appendix

There is still so much to share. What we did was just a first pass at these concepts. Yes, there is still great depth to explore on most of these topics.

Reading this appendix first will not hurt. It may set up the discussions in a better way, it all depends on the type of reader you are.

What follows is condensed 'author's commentary' on each of the concepts. Things we might share as we go past the introduction of the concept or even just a highlight to make sure we focused on the same aspects. Enjoy!

Basic Terms

<u>Willowed</u> – is becoming aware of what de-stabilizing Uke (not 'broke balance'! De-stabilized!) feels like. If it does not feel right, it likely is not. Oh! I also wanted to goad you into translating the word 'willow' into Japanese. Did I stumble onto something Yoshida Kotaro would agree with being significant? Was I hearing echoes of training passed down ultimately through Chris? or am I stretching? (Can we ever answer those questions? <wink>)

<u>Bounce it off the rock</u> – explored some basic applications of 'Precepts'. We emphasized the precepts lessons of 'Sagittal Plane', 'Direct' and 'Mostly Direct' touch and the usefulness of horizontal pressure on the cob. We started with isolating our hips to learn where power is generated from; then expanded the awareness with the weight shift inside a hip-wheel turn.

<u>Drop step, walking in the pool</u> – isolated how the legs 'swing our hips'. You will never generate your maximum power until you do this. This is the single most important part of the Tomiki Aikido drill called 'the walks', the foundation of all you do. Be sure to remember never to 'rock side to side' when you step! Get off the line of attack and stay off the line of attack. Never topple! (unless you are desperate for an excuse to get a nose job!)

<u>Silk reeling and the drop step</u> – the point of this one was to start adding to the drills we already have done; e.g., static hip-pop into a weight shift. We also started making the down, over, and up a bit more complex by adding our hands. This allowed us to point out a very common reflex flaw that allows our hands to lead the hips. We have to eliminate this tendency/comfort/reflex. In the second half, we again found not only the same 'hands lead hips' flaw, but lightly introduced 'mostly direct' touching as impetus for motion (Aiki-Age and Aiki-Sage). This is drill has a lot more depth if you once you are introduced to 'Rolling' Uke. Cutting rolls Uke!

Appendix

<u>Flying the plane</u> – we learned a way to shut off the biceps. It is most common in very strong persons, but even the many 'less colossal' people are in desperate need of this training. We also dropped a lot of hints in this one. Did you notice the remark on how the shift from straight back to forward C begins a path to projecting the hips through the hands? How about the brush/broom? Perhaps the 'At rest tension in your unbendable arm'? Flying the plane is a simple trick, but touches on so much theory.

<u>The timing of tick, tick, whoosh</u> – brought to light that no motion is wasted, from the very first instant of 'down' in 'down over and up', we are influencing Uke. Another insight on how to be like a wave (water)!

<u>Link your shoulders</u> – power stops at the shoulders if they are not joined, linking the shoulders lets us drive the power of our body into our arms so we can create the shield or blade we intend Uke to roll upon.

<u>Cheating the Circle (Arc)</u> – learned how to find 'the seam'; the motion best suited to spin Uke's center (and cause a throw), we also made note that practicing in three's is productive as it adds an insight (as observer, then moving in and out of the roles of Uke and Tori), and highlighted Uke's role is to guide, not give the technique. (What is Uke-do? It is the practice of Uke doing the throw for you. This nurtures false confidence.) The completely new aspect of this particular explanation was the 'Scratching' versus spinning. I wanted to set up 'the secret of the 20'.

<u>Peeling tape off of the wall</u> – we reinforced the Z/vertical dimension of our contact with Uke. In order to throw someone down, we have to move them downward!

<u>Rolling the sleeves</u> – small trick of focus that helps us maintain extension and compression connection. Technically, we should be thinking 'contract or extend Uke's Hypotenuse', but I could not have described it in that way so early in the book. We did not know what a Hypotenuse was at that point. The reference to 'the snakes' (from 'Six Precepts') at this point is a hint toward 'spinning Uke's Hypotenuse'.

<u>The Plank</u> – opened awareness of a little noticed weakness in our stance, what you might consider a kind of 'Joint Lock' on a joint we so rarely discus locking (the knee). Lots of old and new terms referred to in this one.

<u>Rolling the tack</u> – started investigation on how best to manipulate Uke into the desired 'pose', in this case, from Uke leaning forward (butt or belly outward) or when Uke is teetered backward, and we desire the other

condition. (Did you notice that, as we move from front to back teeter, we pass through the 'boxes'?)

The snow shovel – isolated the power of the vertical, what to do when you do not want to or cannot use Uke's no-line or 20.

Crush box, stretch box, the back teeter, with a twist, and checked – after long, drawn out explanation of why take the time to learn the poses, we learned a method to critique our performance, and add detail to the subtlety of what Sensei is doing. These basic terms actually open awareness to the basics of how to distort or stress Uke's QL's (although I could not call it that at first, we had not spoken of the 'chain' and the QL's [and the effect of torqueing the chain]. You also get a good workout on viewing how the COB moves around! This was the true introduction to how to study Aiki as a team of three, and give you the 'starter topic' of Uke's pose as the tinder that creates the fire for purposeful discussion as you practice. (Silence is deafening!)

Checked – We ended it with checked, and I took some liberty on writing about esoterical aspects of Aiki. I did this as I have frustration with many bloggers whose explanations of Aiki are effectively useless, either due to their lack of understanding (and pure posturing) or the inaccuracy of the architecture(s) they use to describe Aiki. You may not agree with me, but you should be able to follow my logic (even the didactic, esoteric digressions). It was also a good way to show just how deep the horse poo can get if you drift outside the physical aspects of Aiki and as we start mixing in incomplete aspects of Chinese wisdom. Aiki, physically, on the mat, on the field of battle, is expressed as the control of stability and balance. Nothing more, nothing less. It applies to mental stability/balance as well, as we discuss later in the Zhan Zhuang and Mushin section of 'Where We Are Strong'. Consider adopting 'control of stability and balance' as the pragmatic definition of Aiki. (nothing more, nothing less)

Where Uke is Weak

Secret of the '20' – I should have told you sooner.

Three-Legged Dog Spots – This section is the introduction to locking Uke's stance; i.e., the 'ground up' locking through Uke's leg/hip. Maeotoshi (#15) becomes a whole lot easier with this TLD focus. You will find Uke very close to these conditions in many techniques. Do you need to lock Uke's hips? No, but it is just another tool in the tool belt. Although, I suggest that many of the Judo leg sweeping actions get that much more effective when you pin Uke or lock their hip and tip them to the TLD's. Be sure to focus on 'Grabbing Uke's foot with the floor'. It can prove to be tricky at times.

Appendix

<u>Aiki Joint Locking</u> – To be completely transparent, I wrote this to help those suffering at the hands of Aikidoka, that due to lack of awareness, have caused more pain than those practicing dentistry without anesthesia! This aspect of this book should be shared very liberally. Many Aikidoka, even black belts of many years, do not have as firm a grasp on joint locks as they think they do. (Pun intended). I was challenged once in a Silat class by someone that felt they could hit/attack when in the Kotehineri hold. I revealed to that Uke that, when applied correctly, they were nowhere near as free to operate as they thought they were. This was not so much of a reveal of my performance that day, but the implied clinical evidence that the challenger had also met many persons that claimed ability, but were not applying the lock correctly.

May this explanation help you teach everyone you come in contact with and please teach joint locks as a method to willow Uke. Please do not skip teaching Pain Compliance, but learning Pain Compliance is a short lesson once you know how to do an Aiki Joint Lock. It helps greatly to know both of those joint locking intentions. We should be capable of holding back or relying upon pain compliance at will. Avoid Joint Destruction!

Remember 'The Chain':
- Pinky/Ulnar and Index/Radial pivots
- Hypotenuse
- SC joints
- Thoracic 1
- the QL's (the 'poses')

And do not forget to continue on through Uke's legs when you can. Thank you, Sean Schniederjan for helping me free myself from back pain and for unintendedly being the muse to revealing the key to controlling Uke's structure! (Look him up on the web! He has written some must-read books.) You will find that Uke's hypotenuse is very important. It is how strength is delivered within a push/punch. Take note of the position of Uke's hypotenuse when assessing "am I threatened by Uke's position?". This practice is especially useful when staying clear of Uke's strikes. Let me put it this way, stay off the 'end'/'tip' of Uke's hypotenuse and you will find yourself much safer than when Uke's hypotenuse is pointed at you.

Where We Are Strong

<u>Presence</u> – Please consider deeply the divide between 'Deadly' and 'Dangerous'. It may help to ask yourself, "How would I describe the perfect Guardian?" (i.e., Soldier, Officer, Parent, Friend). Be the kind of person that others can trust on the mat, as well as rely upon in the street.

<u>Tori's Hypotenuse</u> – one of a couple of dualities we touch upon in this book. What makes Uke weak, can be a great source of power for us. This advanced version of Daito Hands helps you generate a larger amount of power than just

using your muscles. Are you truly using a lever here? Yes, but it is a wheel/pulley. The drills of 'connecting via your hypotenuse' and then 'swinging your arm' (i.e., 'Rotating that hypotenuse' to move Uke) are the starting points for understanding 'the transmission' we create later in these lessons. Connection is applied through our pivots being directed at Uke's COB, the CP being elsewhere, gives us leverage to move our opponent.

Fork and Knife – the primary goal here was actually to extend the concept of 'Tick, Tick, Whoosh'. Once we became aware of the timing, it was important to know what to do with it; to recognize two elemental aspects of many techniques (the Fork and Knife). Once empowered with the awareness, use it to analyze and improve our coordination for performing techniques with an active intention. Stick around long enough and you will likely get tired of me pointing out that every action should be mindful (and by now, you should know what I mean by 'mindful').

Ligamentary Grasping – 'internal strength' is physically manifested by using your body's inherent desire to return to its 'rested' condition. Some describe this as 'soft style'. The 'softness' is the minimalistic use of exertion; i.e., using the minimal amount of 'muscle'. All martial arts exhibit some part of this 'soft, internal' aspect; e.g., we discussed how 'internal strength' is displayed within the hand of the expert Karateka. Although non-stressful for the Karateka, their hands become like hardened, shaped steel with which to apply force on their target. Therefore, do not conflate 'softness' with weakness or passivity, better to substitute 'efficiency' or 'precision'. A 'soft style' is a martial art that focuses upon using 'return to rest' as the primary motivator of action and are not actually that soft on Uke. Consider: what strike is the strongest of all, and at the same time, the least exertive? Answer: Hitting Uke with the ground! My hope is that you take to heart this introduction to internal martial arts, to recognize it as an invitation to dig deeper, go far beyond my awareness, and make your study in 'softness' a part of every physical action you perform. Internal strength is there for you until your last days, not just martially, but occupationally; (i.e., every day activities).

Zhan Zhuang – These were simple lessons, but I cannot stress just how deeply they engage your 'internal strength' (your focus upon using minimal muscular exertion, return to rest). Sounds strange, but the further from using your muscles you are, the more powerful you will become. In writing it, I have read this book more times than you will ever read it, and I have come to realize that I shot through this section quickly. I hope my intention of not boring you and keeping the book at a certain size did not downplay the grave importance of these Zhan Zhuang lessons. The order within which we flowed through the drills was/is important. It takes time to learn alignment, but the full benefit comes at the very end, when all drills are incorporated together. What is not so important in this context is Ki (Chi). Ki is important in many ways, but not here. The world itself is magical, but it does not support 'magic'. Mankind's desire for magic is our ego expressing its frustration with the lack of 'control'. (the perceived 'need' for instant gratification. On the positive side, it is great fodder for stories.) Ki is best applied to conversations surrounding coordination and health. Bringing that thought back around again, structuring your body should improve your overall wellness. Listening to your body will help you become aware of your physical condition so you can act appropriately. Listening to your body is the foundation of Mind/Body awareness; and the secret to Mind/Body awareness is understanding 'The Minds'. They do not tell you up front that they are talking about different 'Minds' do they? The word 'Mind' arguably holds the title for having the greatest lack of precision of any word in the English language, but yet we try to translate it between other languages. No wonder we are having difficulty. Ultimately, what I wanted to express is that Zhan Zhuang is an intense and wholistic focus on 'alignment'. We began outwardly, physically, aligning from our torso to our elbow and briefly touched upon integrating that alignment to our stepping. Inwardly, we distinguished the 'Minds' so we could align them in the first 'internal harmony' aka Mu-Shin, or in Chinese culture: Heart aligns with Intention. The 'listening to our bodies' develops the second harmony: the Mind/Body connection (Intention harmonizes with the Chi). Your intention (I) cannot be master of your body until it can sense it (we remove the 'numbness', 'ignorance' to our bodies feedback). Intention harmonizing with Chi is also 'learning new abilities' or better said 'perfecting' them, gaining coordination, programming your body/reflex. Did you really know how to efficiently, effectively deliver an Aiki Kotehineri before these drills were introduced, or not just how to cut, but that there are different types of cuts? It is hard to perform these skills at first. You may have had intention to, wanted to, directed your body to, but your awareness of what you were doing lacked precision to complete the technique. With Kung Fu (second harmony); i.e., correct, diligent, practice, we evoke the last internal harmony 'Chi harmonizes with movement', which essentially means you have trained your skills to the point of becoming useful and performed as reflex. The mind does less, the Mind/Body connection is such that when a punch is appropriate, the punch, that at first was 'more than a punch', is now autonomic and delivered through the minds notion that a punch should manifest. There is no 'mystery' to these three internal harmonies, they state that, through training, we become better at anything we put our heart and mind into.

<u>The Transmission</u> – no apologies on trying to teach you to be an engineer. It is not requisite to understand the details, but those that make the effort to

Appendix

perceive the science we are playing with will have an advantage. That said, maybe the numbers game did not get stuck in your head, but the simplicity of 'there are two different types of levers and your arm can be one or the other at your will' should have been relatable. Though the drill here seems basic, it is typically hard for many to coordinate. I used the wall at first and the friction/rigidity hurt, so that helped me learn to keep a light touch on the wall, and that helped keep my pivots in place when there was no wall. If you try this yourself with someone holding that jo staff, you may quickly find that your pivot is actually moving around a great deal, evidenced by shoving the jo staff around. You want to get this to the point that you can pick a point in space and have it become the pivot. It may help to imagine resting your arm on the top of a jo staff and using 'contact' with that imaginary jo staff as the pivot. (if you have one, actually using a short jo staff helps with the training as well) I kept this lesson brief because the magic is the connection builders.

Connection Builders

The Albatross – there are a lot of drills to leverage in this book (and in 'Six Precepts'), but the one with the greatest overall benefit is the elephant arms drill. Especially as we include the variants of 'albatross' and 'cheat'. (Heck, we added variants in 'Bounce It Off the Rock' and even the 'Prayer Wheel') Take to heart the 'Rolling Uke into a void' action. I will say it again, that action is the complete essence of Aiki. You have a few analogies offered to think through to help get you to awareness. Oh, and I hope you felt Uke 'launch' the first time you tried 'Albatross to Cheat'. It is crazy how the elephant arms drill looks so simple, but holds so much.

Broom and Shovel – Samurai lineage will obviously help you learn Aiki, but it is not necessary if you learned how to help clean up after yourself. The broom and shovel do not on their surface seem like martial training, but I had to hope that by the time you got to the broom and shovel section you realized that the detail presented always wraps back to complete another loop on the web of threads. Cutting, shielding, the hidden aspects of rolling; the root of Aiki is expressed in these tools. It was important that you knew this foundation before we tackled the rest.

Shield Theory – too many of us have very limited experience in using a shield. The good news is that it is a quick study. I threw a lot at you in this one, more than you might have noticed. Bringing the broom and shovel to life is enough, but tying it to your hip rotations is not necessarily easy to coordinate. That reverse x-block is much more pronounced outside of Aikido. My Silat friends use that kind of positioning plenty, especially when in-fighting. (close, 'corto' range, hand-to-hand, and bladed) It is important to learn the usefulness of it, and know that you cannot have your same-side hand and

Appendix

elbow on the same side of the centerline. If you insist that you can use one hand to defend yourself as in the middle picture in Figure 95 (i.e., when making 'half a triangle', where your hand does not cross the centerline) you should also admit that it often requires you to move your hand across the centerline or twist your hips which does what? Exactly! It moves your centerline and ultimately puts your hand across the centerline; i.e., your turn creates the 'half an X-block'. Why not simply start with your transmission on the center line, half an x-block pose, and learn to use your transmission. It was a brief topic, but the objective for most of your defensive stepping is to divorce/separate the line of attack from the centerline. The union between the centerline and line of attack should be temporal, only existing at the start of the attack, for if you move sideways before Uke attacks, you will likely find that Uke can adjust and track you accordingly. Again, you want that unity of lines, but only at the beginning, and only for a split second. Recognize that the hourglass action can be displayed in empty hands and weapons. You must have seen jo/bo-staff practitioners making those circles in the air. The same with fencers. Hopefully you see a bit deeper into what that action can become: broom and/or shoveling motivated by your hips without giving up the protection of the centerline. It is very important to understand the mechanic of the shield and becoming able to put it into your hands. We do a lot of blocking with our hands, but in order for it to really be effective, you have to put the transmission into your hand to keep the object you are blocking on the end of your hand, and not sliding around your block. Most already do a great job of this, but it helps to have an awareness of what you are doing (to the point of being able to talk about it and show others what is going wrong if they do not have the ability)

<u>Rolling</u> – the three aspects of controlling stability: Pinning, Structure, and Rolling (described as 'footing') were presented in the order in which they are typically uncovered by an Aikidoka. At least that is how it happened for me. 'Six Precepts' is all about Pinning and Spinning Uke's COB. It is a great revelation and does set you up for the structural attacks we explored in this book. Especially those QL's! Torqueing Uke's chain should be a focused/intentional event that leads easily to 'Pinning Uke against their stance'. Combing these first two stability controls is quite powerful, but neither compare to the effects of 'Rolling' Uke; i.e., destroying Uke's footing. Rolling is not much of a secret, you should have felt it in 'Six Precepts' when we first learned the 'elephant arms' drill. I know I did, but had no idea it was important (important does not capture it. Rolling is awesome!). The reason Rolling stays hidden/unnoticed/secret seems to be the subtlety of the action. It was five or so years after finding the elephant arms drill that Rolling became a 'thing' and actually had a name. The notice of Rolling happened shortly after the Albatross revealed itself. I guess I was distracted by that

'door opens unexpectedly from the other side' effect that lacks a continuous control of Uke's stability (creating Aiki!). Another reason that 'Rolling' goes unnoticed is that we most commonly using the 'almost rolling' condition. Uke 'wants to' roll, but is not actually rolling. The 'almost rolling' is my explanation for why so many say, 'you have to feel it to understand what Sensei is doing'. It also is a theorem for why we explain Tai Chi Chuan as a 'perceived power'. We cannot stop a moving truck with this skill, but we certainly can stop a human. I hope you came to enjoy the detail involved in chopping, slicing, and cutting. I admit, I love food and cooking (and eating!), but it really was the most relatable way to talk about how to use a blade. I actually love carrots, and yes, many of them were cut (and consumed) in the learning of this skill <smile>. I appreciate that it took a while to fully reveal the entirety of cutting, but understanding how wheel trundling is much like scoring an object, a ball bouncing is the same as a chop, a ball rolling down a slide is at least the start of the slice, and that it takes understanding of shield theory (the transmission), wheel trundling, 'Tomoe to the rear' and 'Tomoe to the front' to really own what a 'cut' truly is should have made the drudgery worth it. Aikido is a sword art and I will bet that you have spent a minor fraction of your time in class discussing cutting and blocking with the blade! (especially if you are in one of those 'silence is golden' dojos.). The Do cut really is the tattle tale move to reveal your understanding of cutting. I blitzed the suggestions on how to take this beyond the Do cut drills, but I had to save some space for other material. There is a lot to practice if you follow through on all of the suggestions. Do not forget to practice in the kitchen and at the dinner table!

<u>Wheels, Cams, and Tomoe</u> – by the end of this lesson, you should begin to free your mind of 'Circles, Squares, and Triangles', as well as 'Tomoes and Cams'. These shapes are not important, it is the application of force between the objects that matters; e.g., it is easy to understand the pinball flipper as a bat/stick/hammer, but can you use it, instead, as a way to 'sling' Uke? Can you use your concave arm as a wave? (i.e., a complex broom and shovel). The magic is in the touch. Aiki is a relationship, one that is expressed physically at the conjunction point (CP). If the relationship is correct, the form behind the CP means very little; i.e., all the advice, and I do mean ALL the advice, about 'move from your hips', 'step here, do this', means <u>nothing</u>. If the pressure at the CP is correct, you could be doing the technique in the most awkward pose you can imagine and it would not matter, you have accomplished creating a force that controls Uke's stability and balance. Sure, is it a risk, or could you be in a better condition to follow up on your success? Of course, but who cares if you are in a pose that looks exactly like in a kata if Uke is neutralized? Are we practicing Aiki or having a beauty/style contest? You can step as much as you like, but if you cannot put the force of those hips into your CP,

Appendix

which in most cases, is on your hands/arms, it does not matter at all. Power does come from your hips, but Aiki is generated at the point of contact; when done correctly, you do not need much force. Keep this in mind: learning to integrate your body requires paying attention to what is going on in your upper extremities just as much as your hips and footwork. To really beat this 'dead horse'… Aiki-Age (and Aiki-Sage) can be an expression of 'Catching the Pinball', but the secret is really about how we assert pressure/influence upon Uke at the CP. 'Catching the Pinball' is just an analogy, a guide, a 'finger pointing at the moon'. Do not get stuck analyzing the finger, look for the moon. The trick is understanding how these guides reveal the pressure at the CP, which is the expression of a relationship (the one between you and Uke; i.e., the moon). New topic. FWIW, we used to call the 'Drift' '60/40 pushing' or simply '60/40'. It equates to both sides of a shield moving forward but one side is a bit quicker. The solo drill on your own arm is there to illuminate how to push an object using a transmission (and thus, a shield) in your hand. It embodies 60/40 and all the concepts of how we cut (even though our palm cannot actually cut).

Extended Shield theory/Rolling, use no effort – I have seen many persons in many dojos exhibit tangibly rolling Uke along their body (via online video). There really is no secret to it, but it should be called out as many seem to forget about this option. It is just as important to understand the 'virtual line' that we can use to 'roll', or better yet, 'hairpin' Uke along. That 'rail twirler' type action expresses the inflection point, 'bottom of the U' feeling for Uke and really is a boon when you do not have a lot of room to move (e.g., when doing ground work or when your arms are already nearly fully extended).

Orbits – welcome to 'circular' Aiki. Orbits remind me of flying those model planes with two strings attached. Yes, long before we had remote controlled stepper motors, we used to fly motorized model planes in a circle around us and, just like a kite, you could manipulate the rise and fall of the toy by working the strings (and yes, the toys these days are SOOOO much more fun). The trick in Aikido is/was learning how to get into that condition. The Sun, Moon, and Tori is how. I do believe this is a rational explanation (theorem) for how the planets came to orbit the Sun (in the context of the 'Big Bang Theory'/'Expansion of the Universe'), and just as unauthoritatively, this is how/why Hsing-I differs to greatly from both Ba Gua and Tai Chi. (Linear [Hsing-I] versus the Circular (trochoids on a circle, Ba Gua), or Three Dimensional Circles [orbs/columns, Tai Chi]). It was suggested to me via a few channels that Ueshiba was introduced to Ba Gua while captive in China. Maybe it is all fictional, but the reality is, Orbits are there for our use. It all starts with walking the circle, and it is not as easy as it sounds. The focus is really your COB and how it is moving. Swinging your legs (bamboo stepping) and which way you point your toes is not the secret, for you very well may still be moving your COB linearly, e.g., directly from the arch of one foot to the arch of your other. The path of your COB is key. Think, 'COB in the arch of one foot, moving down and curving over to either of the 20's where they meet your No-Line; then up as you finish your curved path to the other foot's arch. It should feel as though you are 'drifting' from over-steering your body, just as a car might. How you form the CP (how you contact Uke) in that drift will be the study of how to apply that drift. Ultimately, it is a study of 'cutting', as the Front and Rear Cuts are both 'drifts'. On your upper body, you should be looking at how your forearm/shield drifts when applying pressure to Uke. Yes, this means revisit not just this book, but all of your techniques. Before you get too far with revisiting how your upper body should meet Uke's resistance, think about your lower body as well. Consider using the sides of the triangles and squares drawn on the floor and practiced upon by the FMA/Silat practitioners as 'direct lines between your arches'. Be sure that you create the ellipse (and half ellipses) with your COB as you shift weight between your feet and you be on your way to learning advanced Silat; i.e., arches on the ends, keep your COB off of those lines except at the arches (Figure 108). (I believe you will also be learning Tai Chi Chuan. Emphasis on the Chuan.)

<u>In Closing</u> – If there is any profundity within me, it is this: 'In all things, train awareness, appropriateness will follow'. Thus, it has been my passion to decode my Aiki experience. In seeking that 'Mind/Body Awareness', I have come to know this world as magical. I have a great sense of wonderment when contemplating and directly sensing bodies in motion. Aiki is pretty much the most fun you can have with your clothes on! I truly hope to meet you some day, and share our experience(s). We never will exhaust how deeply we can develop our understanding and awareness of Aiki. For myself, nearly 700 pages later, with at least another 250 more unwritten on related but different topics than those mentioned in 'Six Precepts' and here in 'Codex', I still feel like I have only just begun. I say that as, for every person I can help improve their own craft, there are at least as many that can still 'whoop my butt'. I think that could be why I keep going…

Expecting that you are one of the 'butt kickers', I humbly request that you also share, in a similar way as this book, how you got that far so I (and others) can try to catch up! Thank you in advance!

www.ingramcontent.com/pod-product-compliance
Lightning Source LLC
Chambersburg PA
CBHW050331230426
43663CB00010B/1813